T0155801

Soft Clay Engineering and Ground Improvement

Soft Clay Engineering and Ground Improvement

Edited by
Jay Ameratunga
Nagaratnam Sivakugan
Braja M. Das

CRC Press
Taylor & Francis Group
Boca Raton London New York

CRC Press is an imprint of the
Taylor & Francis Group, an **informa** business

First edition published 2021
by CRC Press
2 Park Square, Milton Park, Abingdon, Oxon, OX14 4RN

and by CRC Press
6000 Broken Sound Parkway NW, Suite 300, Boca Raton, FL 33487-2742

British Library Cataloguing-in-Publication Data
A catalogue record for this book is available from the British Library

ISBN: 978-1-138-31653-9 (hbk)
ISBN: 978-0-367-68198-2 (pbk)
ISBN: 978-0-429-45554-4 (ebk)

Typeset in Sabon
by SPi Global, India

Contents

Preface

This book covers the design and implementation of ground improvement techniques as applicable to soft clays in the eyes of experienced geotechnical personnel, from academics to practitioners. Soft soils and ground improvement have become major challenges to geotechnical engineers during the last three decades. Planners, architects, consultants, and contractors are now aware what soft soil is and the risks associated with development in areas where soft soil is encountered. They know that 'good' land is limited and, therefore, marginal lands need to be improved.

Ground improvement, in general, has been the theme for several international conferences, and there are dedicated journals and reference books in this area as well. However, very few books are available that are fully dedicated to the improvement of soft soils. Soft soil is very different compared to other weak geomaterials that require ground improvement. The objective of this book has been to cover this niche area in a very comprehensive manner, with numerical examples and case histories wherever appropriate.

As the title of the book emphasizes, soft clay engineering is intimately related to soft clay ground improvement. There are several ground improvement techniques available in the literature now that are applicable to soft clay. Some are unique; however, most are slight variations from one another. If one tries to cover all offshoots, the book will be extremely large and, more importantly, the reader will get distracted, since so much needs to be absorbed. Therefore, we have selected a few techniques that are most popular around the world and concentrated on those to provide design and construction methodologies. We have included some examples where it is considered important to relay the message.

This book will be of value to practitioners because of the past experiences of the contributing authors. All included have a wide experience in soft clay and ground engineering whose experience has not been limited to one country but across the world including Australia, Britain, Brunei, Malaysia, Hong Kong, Indonesia, Ireland, New Zealand, Singapore, and Sri Lanka. The three editors have a combined experience of over 120 years in academia, research, and in practice. The combined experience of all contributors would be close to 400 years. They have witnessed and participated in the rapid evolvement of ground improvement in soft soil in many countries in the world over the last few decades. They have used that experience to produce this book, which will be very valuable to all, because the industry practice is evident in all chapters.

Acknowledgements

The authors wish to acknowledge the following:

- Menard Oceania for provision of the cover photo of Controlled Modulus Column (CMC)™ Ground Improvement rig at the Brisbane International Cruise Ship terminal at Pinkenba, Queensland, Australia
- Zen Ng, Jarreau Alinur, Andy O'Sullivan, Australian Standards and Menard Vacuum™ who provided photos/figures
- Australian Geomechanics Society for allowing material to be taken for Chapter 17 from a paper by Bridges (2019)
- Neil McLennan from Port Lyttleton, Christchurch, New Zealand, for providing a case history and photos
- Tony Moore, Gabriella Williams, Frazer Merritt, and Prakash Jayaraman of Taylor and Francis for the support they provided

On a personal level, the authors wish to thank their wives, Kamali, Rohini, and Janice for their continuous emotional and practical support.

The first author wishes to acknowledge Professor Harry Poulos who has been a mentor for more than thirty years, and his former colleague Patrick Wong who has guided many engineers in dealing with ground improvement problems using the right tools.

Contributors

J. Ameratunga BSc(Eng)Hons, MEng, PhD (Monash), FIE Aust, FIE SL, CPEng, NER, RPEQ

C. Bridges BEng(Hons)(LondSB) MBA(UQ) MSc(Sur) MSc(Bolt) PhD(UQ) CEng FICE RPEQ

K. Chan BSc(Eng), MEngSc, MIEAust, CPEng, NER, APEC Engineer, IntPE(Aus), RPEQ

B.M. Das PhD (Wisconsin), PE, F ASCE

K. Dissanayake BSc (Eng) Hons, MPhil, DEng (Hiroshima), CPEng, CEng (SL), MIE Aust, RPEQ

S. Iyathurai BSc(Eng)Hons, PhD (Wollongong), MIEAust, CPEng, NER, APEC Engineer, IntPE(Aus), RPEQ

C. Lawson BE, ME, MASCE, MIE Aust, CPEng, MAID

T. Muttuvel BSc(Eng)Hons, PhD (Wollongong), MIEAust, CPEng, NER

Thayalan Nall BSc(Eng), MEng (AIT), MIE Aust, CPEng, NER, RPEQ

A. O'Sullivan BE(Hons), ME(Hons), CMEngNZ, IntPE(NZ)

B. Poon BEng(Hons), PhD (USyd), MIE Aust, CPEng

I. Shipway BSc, FIEAust, FGS, CPEng, RPEQ

N. Sivakugan BSc(Eng)Hons, MSCE (Purdue), PhD (Purdue), FASCE, FIE Aust, CPEng, RPEQ

Short biographies of contributing authors

Dr Chris Bridges: Chris Bridges has over 30 years of experience in geotechnical and civil engineering. After graduating and working in the UK, he has worked internationally in Australia, Hong Kong, and China. He is currently a Technical Principal with SMEC, Queensland Chair of the Australian Geomechanics Society and a member of the Standards Australia committee on earth-retaining structures. He has a PhD from the University of Queensland and has published on many aspects of geotechnical engineering.

Kim Chan: Kim Chan has over 40 years of experience and holds a BSc(Eng) degree from the University of Calgary and a MEngSc degree from the University of Sydney. He is the author of over 50 technical papers and book chapters and the invited speaker on various topics of geotechnical engineering. Kim is currently a Senior Technical Director in Geotechnics for GHD and previously held the position of Australasia and Asia Service Line Leader in Geotechnical Engineering for eight years.

Dr Kamal Dissanayake: Kamal Dissanayake obtained his first degree in 1994 and acquired a Doctor of Eng. Degree in 2002. He initially pursued his career as an academic and eventually joined the industry working as a fulltime consultant. Kamal has had involvements in a variety of large-scale projects in Japan, Sri Lanka, Australia, PNG, Hong Kong, and New Zealand. He has led many design teams and managed a number of linear infrastructure projects. He is currently a Principal Geotechnical Engineer at Queensland Department of Transport and Main Roads in Australia.

Dr Satha Iyathurai: Satha Iyathurai has over 20 years' experience and currently serves as Associated Director at AECOM. After completing his PhD from Wollongong University, he commenced his geotechnical journey in Australia in 2006. He worked for ten years at Coffey and undertook many major infrastructure projects across Australia and New Zealand. Satha has co-authored a number of technical papers on ground improvement and has been awarded the R.M. Quigley Honourable

Mention Award by CGS and Excellent Paper Award: Junior by IACMAG for some of his publications.

Chris Lawson: Chris Lawson has over 40 years' experience in geosynthetics, geotechnical, environmental, and coastal and marine engineering. Since receiving his engineering degrees at the University of New South Wales, Australia, he has worked in Australia, Asia, and Europe. He is a past council member of the International Geosynthetics Society and has published more than 50 technical papers. He has served as a keynote speaker at numerous international conferences and symposia.

Dr Theva Muttuvel: Theva Muttuvel has over 20 years' experience in geotechnical engineering. After graduating from University of Wollongong with a PhD in geotechnical engineering, he started working with Coffey Geotechnics in 2008. He is currently working as Principal Geotechnical Engineer in SMEC Australia. He has worked on several interesting projects in Australia and has gained extensive design and construction experience on major infrastructure, retaining walls, and building projects. He is currently a charted professional Engineer, Australia. He has authored and co-authored in publishing twenty technical articles in journals, conferences and books.

Thayalan Nall: Thayalan Nall has over 30 years' experience in geotechnical engineering. After obtaining his Masters from Asian Institute of Technology, Bangkok, he worked as a geotechnical engineer in southeast Asia for ten years. He commenced his career in Australia in 2006 and worked for 12 years at Coffey Geotechnics and Golder Associates. Now he serves as a Principal Geotechnical Engineer with SMEC. He has worked on projects in Australia, New Zealand, PNG, Singapore, Indonesia, Brunei, and Sri Lanka. He has published a number of technical papers.

Dr Bosco Poon: Bosco Poon is a specialist in geotechnical engineering with over 20 years' experience in major projects in all mainland Australian states and South East Asia. He holds a bachelor degree in civil engineering from the University of Adelaide and a PhD in geotechnical engineering at the University of Sydney. He is currently a Technical Director in geotechnical engineering with GHD. Bosco has had over 30 technical papers published in journals and conferences.

Andy O'Sullivan: Andy has 28 years' experience working on medium to large, heavy civil engineering projects with both contractors and consultants. Andy has worked on projects in Ireland, the UK, Germany, New Zealand, Australia, Fiji, and Singapore. Over the past 15 years, Andy has specialized in the design and construction of ground remediation projects including deep soil mixing, in-situ stabilization and slope remediation works. Throughout this time, Andy has published 20 technical papers and presented at a number of international conferences. In 2018, Andy established his own geotechnical engineering firm in New Zealand.

Ian Shipway: Ian Shipway is a Director of EDG Consulting in Brisbane, Australia. He has worked in geotechnical consulting for the last over 35 years, with the last 25 as a specialist engineering geologist for Golder Associates and Coffey Geotechnics. His main technical interests span engineering geological model development for major infrastructure, risk management, dams, and slope engineering. His technical leadership roles include contributions to expert review panels for water supply and tailings dams, work on geotechnical standards for Australia and ISO, and Project Director roles on large transport infrastructure projects in Australia and Asia.

Editors

Jay Ameratunga has over 40 years' experience in geotechnical engineering. He graduated from the University of Ceylon, Sri Lanka, and obtained his master's at AIT, Bangkok, and PhD from Monash University, Australia. Starting in 1989, he worked at Coffey Geotechnics for 25 years before moving to Golder Associates in 2015. Currently, he is an independent consultant. While based in Australia, he has worked on projects in Australia, Indonesia, Malaysia, New Zealand, PNG, Singapore and Sri Lanka.

Nagaratnam Sivakugan is an associate professor at James Cook University, Australia. He graduated from University of Peradeniya with First-Class Honours and obtained an MSCE and PhD from Purdue University. He is a chartered professional engineer and a fellow of the American Society Civil Engineers and Engineers Australia. His publications include nine books, 150 journal articles and 100 papers in refereed conference proceedings. He serves in the editorial boards of a few journals, including *Canadian Geotechnical Journal.*

Braja M. Das is Dean Emeritus at California State University, Sacramento. He received his Ph.D. from the University of Wisconsin, Madison. He is a Fellow and Life Member of the American Society of Civil Engineers and a Life Member of the American Society for Engineering Education. He is the author and co-author of several geotechnical engineering text and reference books and has more than 300 research papers published in journals and conference proceedings.

Chapter 1

Introduction

J. Ameratunga, N. Sivakugan and B.M. Das

1.1 GENERAL

This book covers the design and implementation of ground improvement techniques as applicable to soft clays. Soft clays are very different to other weak geomaterials such as loose sands that require ground improvement. Soft clays and ground improvement have become major geotechnical challenges in civil engineering during the last three decades. Not only civil engineers, but planners, architects, consultants, and contractors are now aware of the behavior of soft clays and the risks associated with developments in such areas. Soft clays are encountered in many large cities, including coastal regions and reclaimed lands. They occur naturally in the geological process of transportation and deposition or due to man-made activities such as reclamation.

Ground improvement in general has been the theme for several international conferences and there are dedicated journals and reference books in this area. The objective of this book is to cover this niche area in a very comprehensive manner, with numerical examples and case histories wherever appropriate.

There are two main concerns related to soft clays, viz., *strength* and *compressibility*. Such materials could lead to excessive settlements and instability under development loads if not properly designed. Soft clay is a term that could be misleading. In general, the term is associated with the consistency when describing a clay. The classification of clays is generally based on the undrained shear strength, c_u, and this is what is adopted by almost all Standards (e.g. Canadian Geotechnical Society 2006; AS 1726).

The title of the book and the above definition of soft clays do not limit the contents to very soft or soft clays, say, with c_u less than 25 kPa. The book covers all clays that are weak and/or compressible under the applied loads, where settlements or stability can be a concern. When the development load is heavy, even firm clays can undergo significant settlements and/or may experience stability issues. If the soil profile has deep strata and the loads are high, even stiff to hard clays can be classified as compressible clays. The reclamation work and ground improvement at Kansai airport in Japan is a classic example where the compressible upper Holocene layers were wick drained and surcharged, without any attempt to improve the underlying hard clay Pleistocene layer (see Mesri and Funk 2015). The reclamation load and thickness of the strata were large enough for the Pleistocene layer to undergo significant primary and secondary consolidation settlements. Hence, the theory and implementation discussed in this book are not limited to clays with a 'soft' consistency but includes clays that are compressible under the applied loads, with stability issues and/or excessive settlements (i.e., more than anticipated to meet the project requirements).

Ground improvement dates back to ancient times, where attempts were made to improve the ground conditions for building roads, dwellings and other infrastructure. The ancient history on ground improvement related to soft clay engineering can be found in Han (2015), Barends (2011), and Nicholson (2013). The recent history (last century) can be found in Brand and Brenner (1981) and Han (2015).

There has been substantial advancement in the past four decades, where previous techniques were refined and many new innovative methods were developed. These advancements were driven by sheer necessity, as we were forced to develop marginal and deleterious lands with better lands becoming scarce. These marginal lands are underlain by soft clays that are weak and highly compressible, and in most cases low lying, requiring extensive filling to raise the land above flood levels. In addition, there is increased tendency worldwide to increase the land footprint by reclamation, which gives rise to additional land for development, but this introduces soft, compressible soils with settlement and stability problems. The increased attention to ground improvement is clearly seen with international conferences dedicating sessions to this subject, journals and books solely dealing with the subject, and the introduction in the university curricula. For example, several decades ago, less than 10% of civil engineers would have heard about 'preloading' or 'surcharging' whereas it would indeed be a surprise these days if a civil engineer is unaware of such terms.

The soft clays are very different to other weak geomaterials requiring ground improvement, and certainly deserve special attention. This book has seen the input from academics and researchers as well as industry professionals who have extensive experience in ground improvement of soft clays. It is the intention of the book that it becomes a valuable tool for consultants, contractors, researchers and academics working on ground improvement of soft clays, and would be a good reference book for postgraduates in this area as well.

The book has been written in simple language with examples throughout the book with graphical illustrations and, most importantly, case histories that would provide an insight into innovative solutions and also lessons learnt where failures occurred. It will also be a book a graduate engineer or a civil engineer with limited knowledge of geotechnical engineering could easily read and digest and provide an insight into issues and solutions when dealing with soft clays.

Those who wish to enhance their knowledge acquired through this book are referred to the various references cited under each chapter. A web-based information and guidance system (SHRP2 2014), Geotechnical Solutions for Transportation Infrastructure (www.geotechtools.org/), presents information on geoconstruction technologies and provides a tool to assist in deciding which technologies are potentially applicable to site-specific conditions. It is understood that the website will be updated as more information becomes available.

1.2 WHAT IS GROUND IMPROVEMENT?

As the words imply, ground improvement is the improvement of the ground, which is needed when the in situ soils are not capable to either bear the development loads and/or will undergo deformations that would not be sustainable under the loads or likely to affect adjacent infrastructure. In soft clays, ground improvement is mainly carried out to achieve the following:

- Strengthen the soft clays to increase shear strength to improve stability and/or the bearing capacity and/or reduce expected settlements under the development loads;
- Control movements, especially where existing infrastructure could be affected;
- Reduce imposed load so that weak soils need to accommodate less load; and
- Accelerate settlement.

By nature, soft clays are weak and compressible and therefore these issues are relevant in deciding whether ground improvement is required and the type of improvement to be adopted. The developer or the client has the following options where soft clays are likely to be a problem (Mitchell and Jardine 2002).

1. Relocate the site or bypass it;
2. Transfer the development loads to a competent layer and avoid ground improvement;
3. Design the superstructure to accept movements expected;

4. Adopt an appropriate ground improvement; or
5. Abandon the project as it is not economical.

Whatever the outcome, ground improvement has become established as an efficient construction tool for marginal lands and routinely assessed as part of the value engineering associated with any major project.

1.3 GROUND IMPROVEMENT METHODS

It is not easy to list all ground improvement methods applicable to soft clay because of sheer numbers and because there are many variations. One document that has attempted to list and describe ground improvement applicable for all types of soils is www.geotechtools.org (SHRP2 2014). These methods are summarized in Table 1.1 excluding those methods that are not relevant to soft clay improvement.

It can be seen from Table 1.1, that there are many techniques to improve soft clays but some are similar in nature with slight differences in material used and/or method of construction.

Table 1.1 List of ground improvement methods applicable to soft clay (adapted from SHRP2 2014)

Aggregate Columns	Geocell Confinement in Pavement Systems	Mass Mixing Methods
Bio-Treatment for subgrade stabilization (Emerging technology)	Geosynthetic reinforced construction platforms	Mechanical stabilization of subgrades and bases
Chemical stabilization of subgrades and bases	Geosynthetic reinforced embankments	Micropiles
Column-supported embankments	Geosynthetic separation in pavement systems	Onsite use of recycled pavement materials
Combined soil stabilization with vertical columns	Geotextile encased columns	Preloading and prefabricated vertical drains (PVDs)
Compaction grouting	Hydraulic fill with Geocomposite and Vacuum consolidation	Sand compaction piles
Continuous flight auger (CFA) piles	Injected lightweight foam fill	Vacuum preloading with and without Prefabricated vertical drains
Deep mixing methods		
Electro-Osmosis	Jet grouting	Vibro-Concrete columns
Excavation and replacement	Lightweight fill	

1.4 SELECTION CRITERIA AND PERFORMANCE CRITERIA

When a ground improvement method is to be selected, there are many issues to consider. For an engineer, technical criteria could be the main interest, which is discussed in Section 1.5. However, there are other criteria that may influence the selection of a ground improvement method. Some of the criteria that are likely to be either imposed by the client or by project requirements can be summarized as:

- Cost – capital cost and maintenance cost;
- Time for ground improvement;
- Material availability;
- Effects on adjacent infrastructure;
- Environmental conditions; and
- Access and use of specific machinery.

Examples of the above six criteria are presented in Table 1.2.

Table 1.2 Examples of project criteria

Criterion	Examples
Cost – capital cost and maintenance cost	If a road embankment is designed allowing for settlement corrections every five years rather than for the full design life of 50 years, the capital cost may be low (e.g., lesser surcharge) but the maintenance cost is likely to be high because corrections (such as asphalt correction) may need to be carried out periodically.
Time for ground improvement	Say, ground improvement is required at a bridge abutment and surcharging is the cheapest option. However, bridge piles cannot be constructed until a substantial portion of the surcharging is completed because lateral movements can affect the bridge piles. If time is critical, the use of a more expensive technique such as a piled embankment would be the preferred solution rather than surcharging. If, however, the contractor could mobilize his workers and machinery in a different area of the project allowing time for surcharging, the economical option of surcharging would be possible.
Material availability	Say, the site is underlain by very weak soils and a high surcharge is necessary to improve the ground. From a stability point of view, large berms would be necessary to carry out the work. If the site is remote and borrow materials are scarce, high strength geotextiles can be used to reduce the earth volumes for berms.

(Continued)

Table 1.2 (Continued)

Criterion	Examples
Effects on adjacent infrastructure	Say, a service line adjacent to a road widening project is sensitive to lateral movement. If surcharging is adopted the assessed lateral movements are found to be too high. On the other hand, if a piled embankment is adopted, the differential settlement between the old and the new embankment is assessed to be a significant issue. To overcome such issues, an extra lightweight (e.g., polystyrene block) solution is likely to be the preferred solution.
Environmental conditions	Say, the proposed development is a warehouse structure on a site underlain by deep soft clays. Surcharging with wick drains is assessed to be the preferred solution because of the relative cost difference when compared to deep piles. However, additional soil investigations at the final design stage found the occurrence of contaminated soil within part of the footprint. These rule out the wick drain solution because of the environmental concerns. A piled structure or a ground improvement using semi-rigid inclusions appears to be the site specific solution.
Access and use of specific machinery	Say, a road alignment is adjacent to and almost parallel to a river. As the site is underlain by deep soft clays, significant ground improvement is needed. Surcharging with wick drains is the preferred solution based on economics, but stability is found to be an issue for the wick drain machine. A piled embankment is possible, but the piling machine still imparts high loads. However, stability issues could be overcome by the construction in a progressive manner (using the as completed piles as a working platform)

1.5 TECHNICAL CRITERIA

Technical criteria can be imposed by the principal, a design engineer who acts for the consultant, or based on government regulations. The main technical criteria can be summarised as follows:

- Stability; and
- Settlement.

1.5.1 Stability criteria

Stability criteria come into play when a road embankment is constructed or when a land development filling takes place. Typically, the stability criterion is defined by a factor of safety (FOS) that combines applied load and the soil resistance. Generally, two types of FOS values are considered:

- Short-term FOS (relevant to the construction phase) – Generally und-rained analysis in terms of total stress parameters (e.g., c_u and $\phi_u = 0$);
- Long-term FOS – Generally drained analysis in terms of effective stress parameters (e.g., c', ϕ').

where ϕ_u = undrained angle of friction, c' = drained cohesion, ϕ' = drained friction angle.

The consequences of not satisfying the criteria is simply excessive defor-mation leading to failure.

Typical FOS values presented by Hsi (2016) based on his experience are given in Table 1.3.

What FOS to be adopted should be based on several considerations, such as extent of site investigations, quality of data, conservative design or not, etc. Chapter 17 discusses risk management in geotechnical design and construction.

Table 1.3 Typical stability assessment criteria (after Hsi 2016)

Term	Condition	FOS	Parameters
Short term	Construction	1.2–1.3	Undrained
	Rapid drawdown	1.2– 1.3	Undrained
	Earthquake	1.0–1.1	Undrained
Long term	Post construction	1.4–1.5	Drained

1.5.2 Settlement criteria

There are, in general, two types of settlement criteria:

- Total settlement – For buildings, the consequences of exceeding the limit are only critical for service joints and steps formed adjacent to entrances/exits.
- Differential settlement/Change of grade/Deflection ratio/Angular defor-mation (see Figure 1.1 for definition) – These are critical for any type of construction, i.e. embankment or a building.

Usually, the settlement limits are imposed by structural engineers and civil engineers.

1.5.2.1 Settlement criteria – as applicable to structures

It is a known fact that settlement of a structure could lead to damage of the structure and therefore by controlling settlement you could avoid

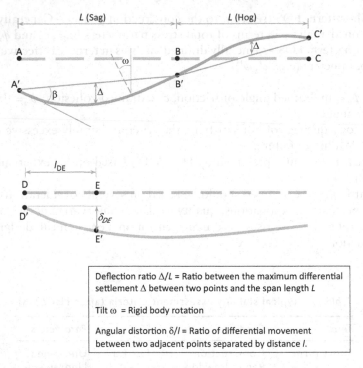

Deflection ratio Δ/L = Ratio between the maximum differential settlement Δ between two points and the span length L

Tilt ω = Rigid body rotation

Angular distortion δ/l = Ratio of differential movement between two adjacent points separated by distance l.

Figure 1.1 Schematic illustration of deflection ratio and angular distortion.

structural damage. Some cracking is unavoidable if the building construction is to be economical. Three broad categories of damage were cited by Burland (2012):

i. Visual appearance or aesthetics;
ii. Serviceability or function; and
iii. Stability.

Generally, differential settlement is more critical than total settlement as it gives rise to distortion of the structure leading to damage of the building. However, it is a common practice to limit total settlement, probably because it automatically limits differential settlement. Although total settlement limits are not critical to structural performance, excessive settlements could be aesthetically unpleasant, and could affect drainage and connections to services.

Where the structure footprint is underlain by soft clays, the thickness of the clay plays a major role in the expected settlement and therefore differential settlement. Although it is not possible to assess the clay thickness at each building column location, it may be necessary to draw thickness contours based on the available test data. This is why in areas where deep paleochannels are expected, additional testing such as cone penetration testing needs to be carried out.

Practical problem 1.1

A 60 m long warehouse is to be constructed in a low-lying area. It was expected that the area is underlain by soft soils and therefore a soil investigation was carried out to assess the thickness variation of the soft clay and to take samples and carry out laboratory testing. Only four boreholes were advanced, and their locations were about 50 m apart. As could be observed from Figure 1.2, the clay thickness at two of the test locations (points A and B) was of the same order, i.e. 5 m to 6 m. As the warehouse structure proposed is a light structure and the fill thickness was only 2 m or so, it was decided to go ahead with the development without any ground improvement. The engineers were aware of the soft clay but because all the boreholes showed similar clay thicknesses it was assumed the differential settlement was minimum and they could accommodate the uniform total settlement. Only after construction, when excessive settlements occurred on one side, were remedial measures discussed. Additional boreholes were advanced, which showed a paleochannel crossing the site with the thickness of the clay varying significantly across the site (see Figure 1.2). If the regional geology was studied and/or a more comprehensive investigation was carried out at the very start, this major issue could have been easily averted.

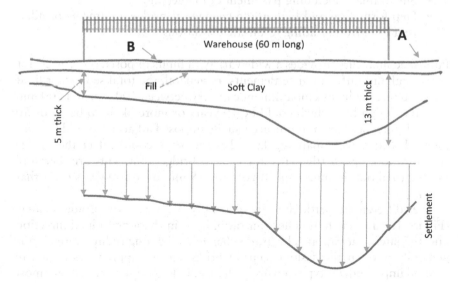

Figure 1.2 Example of a long warehouse underlain by clay of variable thickness.

In published literature, different damage criteria are provided for buildings and the reader is referred to the following:

- USACE (1994); and
- Canadian Geotechnical Society (2006).

More information on adopted practices are summarized by Kempfert and Gebreselassie (2006), Skempton and McDonald (1956), Bjerrum (1963), Burland et al. (1977), Smoltczyk (1996), Terzaghi (1948), Terzaghi and Peck (1961), Meyerhof (1953), and Polshin and Tokar (1957).

1.5.2.2 Settlement criteria – as applicable to highway embankments

In the case of a highway embankment, both total settlement and differential settlement can be important.

The total settlement may affect the following:

- Site drainage leading to water ponding and effects on pavements; and
- Reduced level of the site falling below flood level means the site could be under water if a flood event occurs.

Differential settlement or grade changes could affect the following:

- Overall rider comfort and safety on roads;
- Site drainage including pavement drainage; and
- Step formation and high grade changes at bridge abutments or adjacent to culverts leading to unsafe conditions.

Published literature provides a wide variety of limits imposed on projects for both total and differential settlements. In general, the total settlement limit imposed on a highway embankment can vary between 100 mm and 200 mm over the design life, which could be 10 years or more likely to be 40 or 50 years. Tighter settlement limits are usually imposed adjacent to bridge abutments because accompanying lateral movements could affect the bridge structure footings. Settlements adjacent to bridge abutments are therefore limited to about 50 mm over 50 years or 75 mm over 100 years or of that order.

The differential settlement is generally captured by grade change (Figure 1.3) rather than deflection ratio, both in the longitudinal direction and the lateral direction. The grade change in the longitudinal direction is generally not a major issue except at bridge/culvert approaches or where ground improvement types abruptly change. Lateral grade changes are most important when widenings are involved.

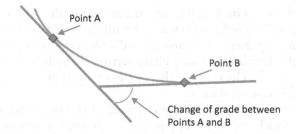

Figure 1.3 Change of grade.

1.5.2.3 Settlement criteria as applicable to land development projects

In land development projects, total settlement criteria are equally important because of issues related to services and drainage, and because by limiting total settlements one would automatically limit differential settlement and grade changes. However, in commercial property developments where large car parks and buildings are to be located, grade changes could be important as they could affect surface and pavement drainage leading to high consequences if significant maintenance issues develop.

1.6 CONTENTS OF THE BOOK

The main objective of the book is to provide professionals in the civil engineering industry, academics, postgraduates and undergraduates an introduction to ground improvement in soft clays, design procedures illustrated with simple examples, and lessons learnt from case histories. It is expected that the reader should be able to easily understand the steps involved in a design, carry out preliminary designs with the understanding of construction issues and major risks.

Chapter 1, as the title suggests, gives an introduction along with the historical development. A comprehensive list of ground improvement methods is provided applicable to soft clays. The reader is alerted to the importance of designing a ground improvement method to satisfy performance criteria, whether the development is a structure, service, or a highway.

Chapter 2 is reserved for geology. No geotechnical problem could be understood or solved if you don't get your geology sorted first. For example, the mention of alluvium could easily allow someone to believe the site is underlain by weak soils. However, the characteristics of alluvium would be dependent on so many factors including its age, process of deposition and the historical changes associated. A highly overconsolidated alluvial clay may not pose significant settlement issues depending on the development loads.

Chapter 3 discusses basic soil mechanics as an introduction to the general reader and provides a brief overview of the topics in soil mechanics that are relevant to soft clay engineering and ground improvement. It covers the definitions, classification and primary characteristics of soils including consolidation and strength. The chapter also covers main laboratory tests relevant to ground improvement design.

Chapter 4 gives a comprehensive coverage of in situ testing and laboratory testing, which are most important for ground improvement design work.

Chapter 5 is reserved for parameter derivation. The clay parameters and their derivation/selection are discussed. Empirical relationships are provided wherever possible.

Chapter 6 is a prelude to the next several chapters and discusses ground improvement methods for soft clays and strategies to avoid ground improvement.

Chapter 7 covers one of the simplest methods of ground improvement, i.e., excavation and replacement. Although it appears simple, the issues to consider and risks are clearly discussed here.

Chapters 8 and 9 discuss preloading as it is one of the most common ground improvement techniques used worldwide. The two chapters cover separately preloading without and with wick drains. Wick drains are discussed in detail and the design is explained in simple terms. Lessons learnt from other projects are discussed to educate the reader on the practical aspects.

Chapter 10 covers the design and construction of stone columns. A comprehensive coverage is presented on design approaches ranging from simple charts to numerical methods. Construction and environmental issues are also discussed. Advantages and disadvantages relative to other common ground improvement methods are summarized.

Chapter 11 is also fairly comprehensive and covers semi-rigid inclusions. Although several other names are given, this is essentially the introduction of stiff elements such as concrete columns to a soft clay deposit. Piled embankment design is also covered in this chapter.

Chapters 12 is reserved for lightweight fill including extra lightweight fill or polystyrene blocks.

Deep soil mixing is discussed in Chapter 13 with due reference to QA/QC procedures as they are most important in the implementation of this method.

Chapter 14 is reserved for high strength geotextiles. They assist in the implementation of some other ground improvement techniques and, in fact, are a critical element in pile embankments.

Chapter 15 covers mass stabilization as applicable to ground improvement projects. A detailed discussion is presented with special attention given to construction.

Chapter 16 is dedicated to the Observational Approach and Instrumentation, the most useful design and construction approach for ground improvement sites. The chapter discusses issues and remedies, and how instrumentation is used to deliver a safe, healthy product/project.

Geotechnical risk management is very important in construction in soft soils, and therefore an entire chapter, Chapter 17, is reserved to tackle this subject.

REFERENCES

Australian Standards (2017). *Geotechnical Site Investigations*, AS1726:2017.

Barends, F.B.J. (2011). *Introduction to Soft Soil Geotechnique: Content, Context and Application*. IOS Press, Amsterdam.

Brand, E.W. and Brenner, R.P. (1981). *Soft Clay Engineering*. Elsevier Scientific Publishing Company, Amsterdam.

Bjerrum, L. (1963). 'Allowable settlements of structures.' *Proceedings 3rd European Conference on Soil Mechanics and Foundation Engineering*, Wiesbaden, 83–118.

Burland, J.B. (2012). 'Building response to ground improvements.' *Chapter 26, ICE Manual of Geotechnical Engineering, Vol II Geotechnical Design, Construction and Verification*. Eds. Burland, J., Chapman, T., Skinner, H. and Brown, M. ICE Publishing, London, 281–296.

Burland, J.B., Broms, B.B., and De Mello, V.F.B. (1977). 'Behaviour of foundations and structures.' *Proceedings IX International Conference Soil Mechanics and Foundation Engineering*, Vol. 2, Tokyo, 495–546.

Canadian Geotechnical Society (2006). *Canadian Foundation Engineering Manual*, 4th edition.

Han, J. (2015). *Principles and Practices of Ground Improvement*. Wiley, Hoboken, NJ.

Hsi, J. (2016). 'Soft soil engineering in practice.' *Australian Geomechanics Society Sydney Chapter Symposium*, November 2016.

Kempfert, H.-G. and Gebreselassie, B. (2006). *Excavations and Foundations in Soft Soils*. Springer, Berlin.

Mesri, G. and Funk, J.R. (2015). 'Settlement of the Kansai International Airport islands.' *Journal of Geotechnical and Geoenvironmental Engineering*, 141(2), 1–16.

Meyerhof, G.G. (1953). 'Some recent foundation research and its application to design.' *The Structural Engineer*, 6, 151–167.

Mitchell, J.K. and Jardine, F.M. (2002). *A Guide to Ground Treatment*. CIRIA Report C573.

Nicholson, P. (2013). *Soil Improvement and Ground Modification Methods*. Elsevier, Woburn, USA.

Polshin, D.E. and Tokar, R.A. (1957). 'Maximum allowable non-uniform settlements of structures.' *Proceedings 4th International Conference on Soil Mechanics and Foundation Engineering*, London, 402–405.

SHRP2 (2014). *Geotech Tools: Geo-construction Information & Technology Selection Guidance for Geotechnical, Structural & Pavement Engineers*. http://geotechtools.org/

Skempton, A.W. and McDonald, D.H. (1956). 'The allowable settlements of buildings.' *Proceedings of the Institute of Civil Engineers* 5, 727–768.

Smoltczyk, U. (1996). *Grundbau-Taschenbuch*. Vol. I and II, 5th edition, Ernst & Sohn, Berlin.

Terzaghi, K. (1948). *Theoretical Soil Mechanics*. John Wiley and Sons, New York.

Terzaghi, K. and Peck, R.B. (1961). *Soil Mechanics in Engineering Practice*, 2nd edition, John Wiley and Sons, New York.

USACE – US Army Corps of Engineers (1994). *Engineer Manual 1110-1-1994. Engineering and Design – Settlement Analysis*. USACE, Washington, DC.

Chapter 2

Engineering geology of soft clay

I. Shipway

2.1 INTRODUCTION

An engineering geological model provides a representation of the ground conditions to the extent necessary for the ground behavior to be understood, and ultimately modelled and analyzed. A useful model will include all pertinent information and allow for consideration of uncertainties in stratigraphy, material characteristics and other aspects that will affect ground behavior. Soft, Holocene clay soils have high compressibility and low strength. The risks associated with design and construction over these deposits are significant, and consequently they are the materials most frequently subjected to ground improvement. Therefore, engineering geological models over areas of soft clay and other recently deposited materials must be detailed and robust if they are to provide the geotechnical engineer with an adequate understanding of the extents of the different material units and their behavior.

Soft clay deposits are often seen, and therefore considered, as a continuum with characteristics that can be defined by field or laboratory tests, and performance, in terms of strength and compressibility that can be defined by those tests. However, variability in strength and compressibility throughout any specific soft clay deposit is likely to be significant. This variability is a consequence of the geological setting, ongoing geological processes and stress history of the clay. An understanding of the geological origins of the deposit is essential to assess its physical extents and properties, and the potential variability of both.

Most soft clay soils are deposited in estuarine channels or certain parts of the near-shore environment. Relationships between different parts of the deposit are often complicated by changes in sea level leading to cross-cutting of alluvial channels, variation in flows in different parts of the channel or a complete change in depositional environment.

Understanding of the shape and variability of soft clay deposits can be achieved through the development of an engineering geological model that is developed throughout the design process and guides investigation and analysis. The model should also provide insight into sedimentary structures within the compressible deposits, which may have importance to drainage of pore pressures and stability under loading.

This chapter broadly describes the depositional environments and the morphology of soft clay deposits within those environments. It also provides guidance on the development of appropriately detailed engineering geological models for soft ground engineering and focusses on the practical steps required to develop a model that is adequate for engineering analysis and design.

2.2 GEOLOGICAL PROCESSES AND DEPOSITIONAL ENVIRONMENTS

Soft clay soils were generally deposited over the Holocene period, which covers the past 11,700 years of geological history since the end of the major glacial period of the Pleistocene Epoch. The Pleistocene marked an interval of global low sea levels and widespread glaciation within the northern hemisphere, and some parts of the southern hemisphere. Since the last glacial maximum, about 21,000 years before present (BP), the sea level has risen a total of about 130 meters, with numerous fluctuations. Figure 2.1 shows the approximate sea level curve over about the past 250 thousand years, illustrating the sharp rise until about 6000 to 7000 years BP when sea levels have varied only over a few meters. The curve shown is generalized, as there is evidence to suggest temporal sea level variability over several meters between different locations for many reasons, some of which are not well understood (Lambeck and Nakada 1990; Sloss et al. 2007).

When sea levels fell during the Pleistocene epoch, rivers cut down into the materials forming the coastline, incising channels and bays of varying shapes and sizes. As the climate rapidly warmed following the last glacial maximum, sea levels rose rapidly and sediments were deposited into the channels and bays cut into the coastal environment. Minor fluctuations in sea level

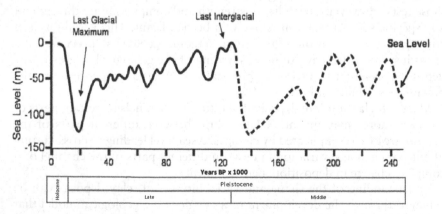

Figure 2.1 An interpretation of sea-level fluctuation from the middle Pleistocene to present day (Modified from Chappell et al. 1996).

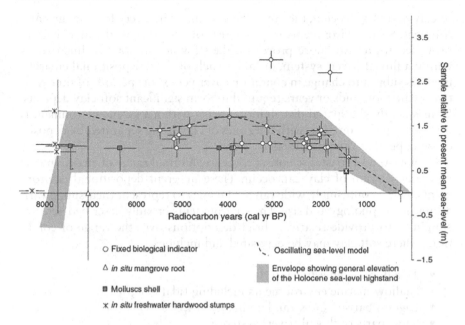

Figure 2.2 A relative sea-level versus time plot for the New South Wales south coast over the time period where deposition of soft clay would be expected based on radiocarbon dating of various biological elements (from Sloss et al. 2007).

allowed recently deposited material to be exposed, desiccated, re-eroded and redeposited.

It is often difficult to define past sea levels at any specific location, despite the many different methods that have been developed to provide correlations. Figure 2.2 shows a more detailed sea level curve for the south east coast of New South Wales, Australia, over the past 8000 years based on radiocarbon dating of various organisms and materials. This provides a good indication of the vertical complexity of depositional levels that is possible at just one location over a relatively short period of time.

The particle size of material that is deposited within a particular body of water depends on the energy associated with the currents and waves of the environment of deposition. When the various different depositional environments are considered in conjunction with a fluctuating sea level, it is evident that significant variability in the distribution of material types is likely.

Soft clay soils form through the sedimentation of very fine soil particles in water. The ability of moving water to transport a given particle depends on the velocity of the water, the magnitude of frictional forces between particles, and the particles and the bed of the water body. Clay particles are very small (less than 0.002 mm), remain in suspension for long periods, and therefore are readily transported by very slow-moving water. For deposition of clay particles to occur, they must be suspended in a very low energy environment. Such environments include areas of deeper coastal water that are protected from near shore processes, the slow-flowing or becalmed zones within a fluvial (river) system, or lakes. Each of these depositional environments is subject to change in conditions over very short periods of time relative to the thousands of years required to form significant soft clay deposits. Consequently, the spatial distribution of clay within a given environment is often extremely complex, with the zones of clay being separated by deposits of sand, peat, or other materials.

Geologists and coastal scientists recognize a broad range of environments where deposition of clay can occur. These different depositional environments have a confusingly wide range of names to represent their subtle variations in morphology and size. The environments broadly described here are simplified to provide relatively brief descriptions over the range of conditions where soft clay may be deposited and include:

- Tide dominated estuaries;
- Shallow marine environments including tidal flats;
- Lagoon barrier and strand plain systems;
- Some parts of fluvial (river) systems;
- Very sheltered bays; and
- Freshwater and saltwater lakes.

2.2.1 Shallow marine and estuarine environments

Figure 2.3 illustrates the near shore and estuarine depositional environments and the locations within each where soft clay (mud and possibly marsh in the Figure) is likely to be found. Others also illustrated include deltas and wave-dominated estuaries, both of which are generally higher energy environments dominated by sand or coarser materials, but which may include minor clay deposits.

The distribution of materials within a range of coastal environments is shown in Figure 2.3. In Figure 2.3, the main locations for the deposition are those indicated as 'Mud.' However, soft materials such as peat may also be present in the zones designated as 'Marsh.' Note that in some parts of the coastal area deposition of particles is dictated by the energy of the river system whereas in others tidal effects govern (modified from Siddiqui et al. 2017).

Figure 2.4 provides an example of a Holocene soil profile (stratigraphic column) through a shallow marine to estuarine environment. The profile is based on information provided in Boyd and Penland (1984) and Tucker (1982) and assumes a depositional environment where the sediment load is greater than the capacity of the coastal processes to remove it from the shoreline (i.e. a river-dominated, prograding shoreline). The main features of the stratigraphic column are the relatively thin clay deposit, which is intersected by beds of sand at intervals and a relatively thick layer of beach

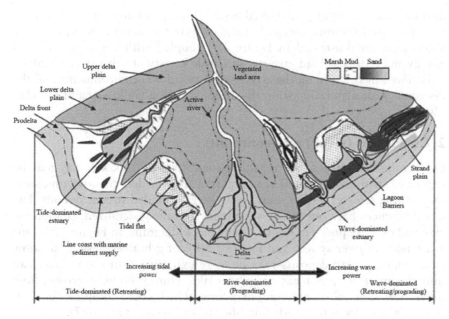

Figure 2.3 The distribution of materials within a range of coastal environments.

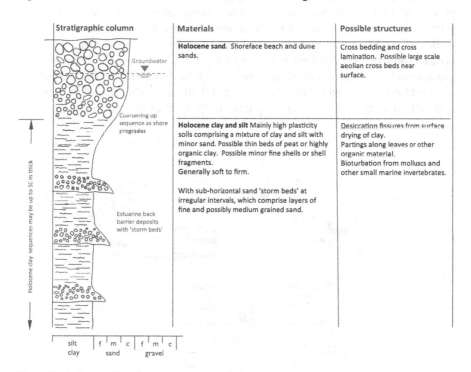

Stratigraphic column	Materials	Possible structures
Groundwater▼ Coarsening up sequence as shore progrades	**Holocene sand.** Shoreface beach and dune sands.	Cross bedding and cross lamination. Possible large scale aeolian cross beds near surface.
Estuarine back barrier deposits with 'storm beds'	**Holocene clay and silt** Mainly high plasticity soils comprising a mixture of clay and silt with minor sand. Possible thin beds of peat or highly organic clay. Possible minor fine shells or shell fragments. Generally soft to firm. With sub-horizontal sand 'storm beds' at irregular intervals, which comprise layers of fine and possibly medium grained sand.	Desiccation fissures from surface drying of clay. Partings along leaves or other organic material. Bioturbation from molluscs and other small marine invertebrates.

Holocene clay sequences may be up to 10 m thick

silt clay | f m c sand | f m c gravel

Figure 2.4 A simplified stratigraphic column through the Holocene deposits of a prograding barrier showing typical materials and potential structures for the different material units.

to dune sands overlying the clay. Although this type of deposit can be identified in many locations, the wide range of physical conditions in the near shore zone (as illustrated in Figure 2.3) coupled with variations in tidal energy and sediment load give rise to a wide variety of subsurface profiles.

Further information on these types of depositional environments and the distribution of materials within them is provided in Galloway and Hobday (2013).

2.2.2 Fluvial (river) environments

Further upstream from the immediate coastal areas, thick soft clay may be deposited in tide-dominated rivers, particularly in deeper, slower-moving, meandering streams, rather than higher-energy braided stream systems. The key differences between these two systems and the generalized location for clay and other deposits is shown on the block diagrams in Figure 2.5(a) for a meandering river system and Figure 2.5(b) for a braided system. In some circumstances, clay deposits may be more extensive than shown, and can underlie an entire point bar or oxbow lake. During flood, overbank flow deposits, or 'splays' (not shown) may form, which could comprise thin layers of clay or mixed materials (modified from Fookes et al. 2007).

A typical profile of materials through a deep clay estuarine deposit is shown on Figure 2.6. Such sequences are common within eastern Australia with Holocene soils over 30 m thick and in other locations in South East Asia such as Malaysia. Deposition of Pleistocene-aged soils commences in a scoured bedrock channel with a fining upwards sequence, which eventually grades to clay. Although there may be a break in deposition at the boundary of the Holocene marked by a coarser band of sand, the boundary can be difficult to detect in boreholes, particularly in deeper clay deposits. The

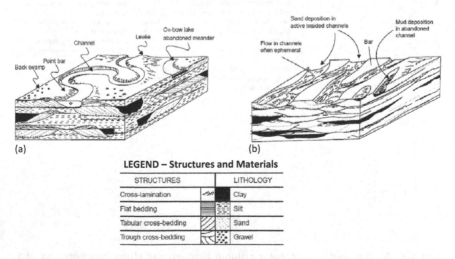

Figure 2.5 Distribution of different sedimentary structures and material sizes within a meandering river system (a) and a braided stream system (b).

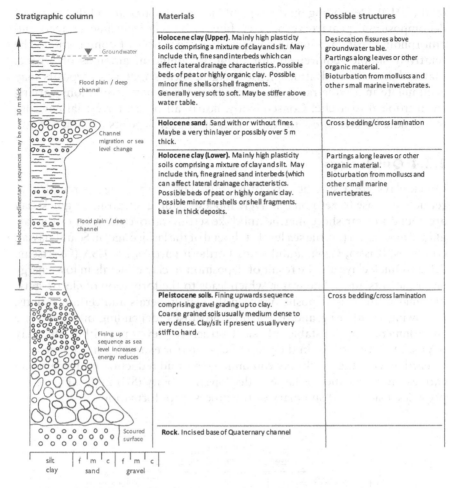

Stratigraphic column	Materials	Possible structures
Groundwater Flood plain / deep channel	**Holocene clay (Upper).** Mainly high plasticity soils comprising a mixture of clay and silt. May include thin, fine sand interbeds which can affect lateral drainage characteristics. Possible beds of peat or highly organic clay. Possible minor fine shells or shell fragments. Generally very soft to soft. May be stiffer above water table.	Desiccation fissures above groundwater table. Partings along leaves or other organic material. Bioturbation from molluscs and other small marine invertebrates.
Channel migration or sea level change	**Holocene sand.** Sand with or without fines. Maybe a very thin layer or possibly over 5 m thick.	Cross bedding/cross lamination
Flood plain / deep channel	**Holocene clay (Lower).** Mainly high plasticity soils comprising a mixture of clay and silt. May include thin, fine grained sand interbeds (which can affect lateral drainage characteristics. Possible beds of peat or highly organic clay. Possible minor fine shells or shell fragments. base in thick deposits.	Partings along leaves or other organic material. Bioturbation from molluscs and other small marine invertebrates.
Fining up sequence as sea level increases / energy reduces	**Pleistocene soils.** Fining upwards sequence comprising gravel grading up to clay. Coarse grained soils usually medium dense to very dense. Clay/silt if present usually very stiff to hard.	Cross bedding/cross lamination
Scoured surface	**Rock.** Incised base of Quaternary channel	

Holocene sedimentary sequences may be over 30 m thick

silt clay	f	m	c	f	m	c
	sand			gravel		

Figure 2.6 A simplified stratigraphic column through the Quaternary deposits associated with a meandering river system showing typical materials and potential structures for the different material units.

boundary is often relatively easy to detect within the plots of cone penetrometer tests as there is generally a significant increase in shear strength associated with the overconsolidated Pleistocene soils.

Further information on fluvial and near coastal depositional models can be found in Summerfield (1991) and Walker and James (1992).

2.2.3 Peat deposits

In geotechnical terms, peat is a soil with an organic content of over 25% dry mass (Standards Australia 2017). Peat deposits form in anaerobic environments where the accumulation of biological remains is greater than the capacity of the environment to destroy or recycle those components (Stracher et al., 2010). Depending on the type of biological material and the degree of decomposition, peats may be very fibrous or relatively fine grained and amorphous. Peats are highly compressible regardless of composition. The relatively simple requirement for input of significant amounts of organic material with limited atmospheric oxygen, means that peat deposits may accumulate in a wide range of locations under climatic conditions varying from tropical to arctic. Consequently, nearly all of the terrestrial environments noted above allow the development of peat deposits to some extent.

2.2.4 Quick clays

Quick clays are a specific type of sensitive clay with a shear strength that reduces to close to zero on relatively modest loading. Deposits of quick clay are found in near shore marine and lake settings above the level of the last glacial maximum (i.e. the sea level at the end of the last ice age) in Scandinavian countries, Russia, Canada and some northern parts of the USA (Geertsema 2013). Quick clays are the result of deposition of clay minerals in low-energy environments and saline water, which leads to the formation of electrostatic bindings in the clay deposits between the clay minerals and certain cations. Following weathering and/or leaching processes (depending on the precise environment), an unstable physical structure develops within the clay deposit. Where stresses in the ground change (for example, through erosion or earthquake), the quick clay can undergo a rapid reduction in strength and can lead to issues such as large-scale slope instability (SGI 2004). Figure 2.7 provides a simplified summary of the processes of formation.

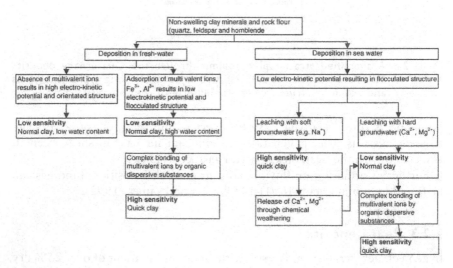

Figure 2.7 Outline of the principles of quick clay formation (reproduced from Brand and Brenner 1981)

Quick clay deposits are the subject of many detailed, specialized studies and may require knowledge specific to their location to conduct an adequate assessment of their properties and behavior. As such, their treatment is not covered in detail within this book.

2.3 THE ENGINEERING GEOLOGICAL MODEL

Identification of the type of depositional environment(s) at a specific site or location requires interpretation of the geographic and geomorphological features, in addition to developing knowledge and understanding of the near surface materials and their geological age. This should be conducted within the framework of developing an engineering geological model.

2.3.1 Model types and components

What is evident from previous illustrations of the various depositional environments is that the subsurface spatial distribution of soft clay deposits can be highly complex. Further, the variation in age of the individual zones of clay within a specific bay, fluvial system or other environment suggests the potential for the clay to behave differently in different parts of the deposit. Consequently, an engineering geological model is an essential starting point for ground engineering involving soft clay.

Parry et al. (2014) suggests that 'an engineering geological model is any approximation of the geological conditions, at varying scales, created for the purpose of solving an engineering problem.' This implies that engineering geological models have common characteristics with other forms of models such as numerical models in that they are not necessarily accurate, despite their potential usefulness, and may be produced to different levels of detail for different purposes.

As detailed information on the distribution and characteristics of materials is generally required for soft ground engineering, the models developed usually need to be detailed and based on direct observation and measurement of conditions.

As a broad classification, engineering geological models can be divided into Conceptual Models and Observational Models (Parry et al. 2014).

Conceptual Models can be developed without any subsurface information. A Conceptual Model for a given road corridor or linear structure may be as simple as an annotated sketch based on the topography and map conditions showing where the alignment may intersect one or more of the depositional environments outlined above, or other higher areas that are more likely to be underlain by rock. That Conceptual Model could then be tested through investigation of ground conditions and further refined to an

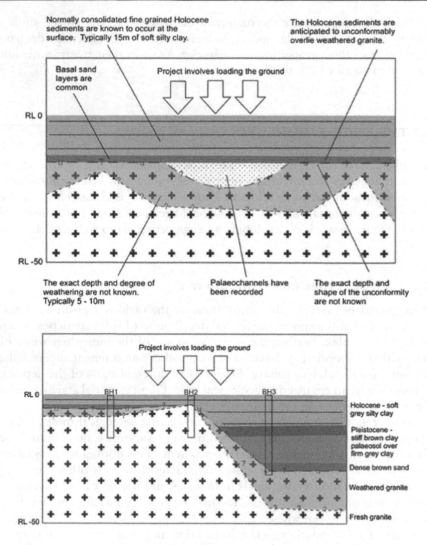

Normally consolidated fine grained Holocene sediments are known to occur at the surface. Typically 15m of soft silty clay.

The Holocene sediments are anticipated to unconformably overlie weathered granite.

Basal sand layers are common

Project involves loading the ground

RL 0

RL -50

The exact depth and degree of weathering are not known. Typically 5 - 10m

Palaeochannels have been recorded

The exact depth and shape of the unconformity are not known

Project involves loading the ground

RL 0

BH1   BH2   BH3

Holocene - soft grey silty clay

Pleistocene - stiff brown clay palaeosol over firm grey clay

Dense brown sand

Weathered granite

Fresh granite

RL -50

Figure 2.8 Comparison between a Conceptual and an Observational Model for an area of ground underlain by soft clay over granite (from Parry et al. 2014).

Observational Model with the requisite robustness. The two broad model types are illustrated by example in Figure 2.8:

a. A Conceptual engineering geological model for an area where soft clay is thought to overlie granite; and

b. An Observational engineering geological model for the same area based on mapping and boreholes.

Despite the Conceptual Model overpredicting the presence of soft clay on some parts of the site, it is still useful in that the potential for the clay to be present is understood. The boreholes merely better define the extent of soft clay.

The Conceptual Model may be used to define locations for investigation and testing. As additional data from further research and investigation is interpreted and incorporated into the model, it becomes primarily an Observational Model. As the model is intended to guide each stage of investigation and assist in the identification of geotechnical issues and risks, in broad terms it must address the following aspects:

- The distribution of materials including the relative age of the different soft clay deposits within the study area;
- The properties of the different soft clay units and other materials including their moisture content, consistency limits, shear strength, stiffness and compressibility characteristics;
- The presence of geological structures such as partings, fissures, sheared surfaces or sand layers within the soft clay deposits and other materials; and
- The levels of seismicity associated with the site.

Figure 2.9 shows the stages of model development and the components required to develop models from simple through to detailed.

2.3.2 Assembling the observational model

Figure 2.10 shows details of the sequence of activities that would generally be required to obtain adequate data and develop a detailed Observational Model, and how the different components are partitioned into the different stages of geotechnical investigation.

For the model to be useful, all steps of investigation must build on the preceding work. This requires close communication and understanding between those working on the various stages of model development, and those conducting the site investigation activities themselves. During the later stages of investigation, it is also essential that those who are to use the model to provide the engineering for the project, are closely involved in model development as they may have a better understanding of which model components will affect analysis.

Figures 2.9 and 2.10 illustrate that the process of model development commences with broad considerations over a wide geographic area, with the focus slowly narrowing to detail as the work proceeds. The importance of maintaining this sequence of development and not becoming trapped in excessive detail at an early stage of investigation cannot be over emphasized. The following key considerations and associated processes should be followed through model development:

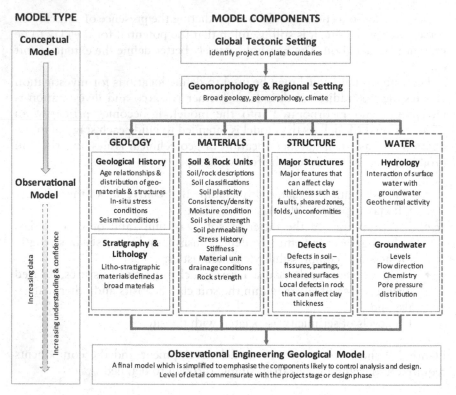

Figure 2.9 The components that are assembled to develop an engineering geological model for a soft ground engineering project from a conceptual
model to a detailed observational model (Modified from Bertuzzi 2019).

*Identify the type of depositional environment within which the study area
is situated.*

Through published data and field reconnaissance review the broad tectonic, geological and climatic setting. Consider how surface processes (rivers, tides, wind) have defined and altered the landscape. Create sketches of
block models that may aid understanding of the overall environment.

*Broadly define the physical extents of Quaternary materials, which will
be present within alluvial channels or low energy embayments.*

Consider information on the potential extents of historic sea level to gain
an indication of the possible base of the Quaternary/Holocene materials.
The boundary of the Quaternary materials (probably a geological unconformity) is generally defined by the extent of rivers/streams cutting down into
older soil/rock materials. Depending on the size of the area of interest, the
Quaternary deposits may not be laterally constrained by the site boundaries.
Where this is the case, it is important to extend the *model* boundaries to
allow the likely extent of deposition to be understood.

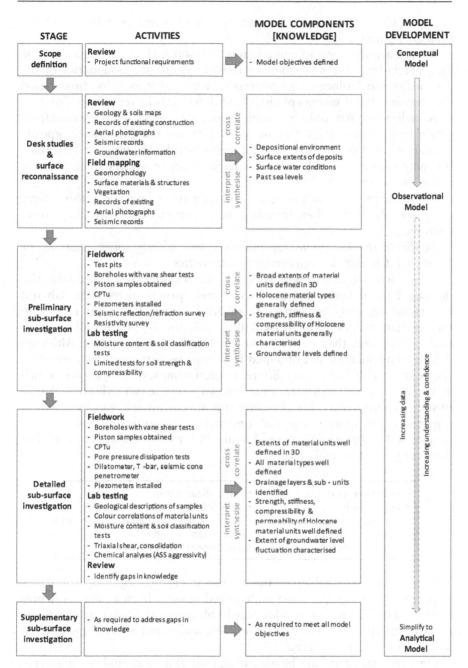

Figure 2.10 The staged sequence of activities required to develop a detailed observational engineering geological model suitable for soft ground engineering.

Define the extents of the potentially soft Holocene materials with the Quaternary deposits.

Use surface mapping in conjunction with sub-surface investigation (test pits, boreholes, cone penetrometer tests) to define the boundaries of the potentially soft Holocene age materials. Pick investigation locations based on observations of geomorphology and differences in vegetation. Some of these indicators may only be visible from above or in aerial photographs. Correlate data in a series of cross sections. Divide the Quaternary deposits into units based on age and material type. Consider the use of a 3D model, but only where a clear benefit in interpretation or understanding can be gained by its adoption.

For linear projects such as roads and railways, remember to think three dimensionally rather than focusing on a long section down the project corridor.

Subdivide the Holocene age materials into older/younger channels or zones and define their engineering characteristics.

These two aspects should generally be considered concurrently. Use detailed logs of boreholes, analysis of cone penetrometer data, lab test results and detailed observations to subdivide materials into sub-units with unique characteristics. This work will typically involve plotting all results in such a way that they can be compared and cross-correlated. Although detailed data such as that obtained from cone penetrometer tests are often the most useful in identifying different material units, simple properties such as variations in soil moisture content down a profile can indicate a change in soil behavior when other characteristics appear similar. Start by considering grain size to separate materials into units of fine-grained soil and coarse-grained soil. Properties that could be useful for further correlation against depth or each other include:

• moisture content	• vane shear strength
• liquid limit	• coefficient of consolidation
• plasticity index	• OCR or preconsolidation pressure
• organic content	• compression index
• liquidity index	• recompression index

In addition to considering the measured engineering properties, it is important to cut open and log undisturbed samples from a range of depths so that subtle differences in color may be observed, and observations can be made of structures such as thin sand layers, which may affect drainage characteristics of soft clay layers.

Once material units can be identified, their extents should be considered both vertically and laterally taking into account both the observed material properties and basic geological principles.

Material units can then be assigned characteristic properties using the data and plots previously prepared to allow unit identification. Usually, the model can be presented using a series of long sections/cross sections and a detailed geological plan. It is often advantageous to further development of the model using 3D software, which will allow sections to be cut at critical points for analysis. If this is undertaken it is important that the 3D model incorporates all of the data points including surface mapping information, rather than just sub-surface data, and that the understanding of model uncertainty is not lost. The detailed engineering geological model may then need to be simplified to some degree to allow analysis.

2.4 SUMMARY

When conducting design for projects over compressible or otherwise weak ground, an understanding of the extents of different material types and their potential behavior under loading is critical. The development of an engineering geological model places the geotechnical information that is available to the designer into a geological framework, which allows an informed prediction of material properties within space.

Development of the engineering geological model should commence with an understanding of the overall tectonic setting and the depositional environment for the Quaternary-aged deposits within that setting. Models should be developed in stages in conjunction with site investigation, with preceding stages progressively informing the scope of work for each subsequent stage. Material properties relevant to soft ground engineering and ground improvement should be incorporated into the model. This requires close collaboration between engineering geologists and geotechnical engineers throughout model development. Ideally, the cooperative approach to model development should continue from the very early stages of the project through to design, at which point the detailed engineering geological model may be simplified to allow practical analysis.

REFERENCES

Australian Standards (2017). *Geotechnical Site Investigations, AS1726:2017.*
Bertuzzi, R. (2019). *Tunnel Design Handbook.* Sydney: Pells Sullivan Meynink.
Boyd, R. and Penland, S. (1984). 'Shoreface translation and the Holocene stratigraphic record: Examples from Nova Scotia, the Mississippi Delta and eastern Australia.' *Marine Geology,* 60(1–4), 391–412.
Brand, E.W. and Brenner, R.P. (1981). *Soft Clay Engineering,* New York: Elsevier Science.
Chappell, J., Omura, A., Esat, T., McCulloch, M., Pandolfi, J., Ota, Y. and Pillans, B. (1996). 'Reconciliation of late Quaternary sea levels derived from coral terraces at Huon Peninsula with deep sea oxygen isotope records.' *Earth and Planetary Science Letters,* 141, 227–236.

Fookes, P.G., Griffiths, J.S. and Lee, E.M. (2007). *Engineering Geomorphology: Theory and Practice*. London: Whittles.

Galloway, W. and Hobday, D. (2013). *Terrigenous Clastic Depositional Systems*. Berlin: Springer Berlin.

Geertsema, M. (2013). 'Quick clay.' In: Bobrowsky, P.T. (eds.) *Encyclopedia of Natural Hazards. Encyclopedia of Earth Sciences Series*. Dordrecht: Springer.

Lambeck, K. and Nakada, M. (1990). 'Late Pleistocene and Holocene sea-level change along the Australian coast.' *Palaeogeography, Palaeoclimatology, Palaeoecology*, 89(1–2), 143–176.

Parry, S., Baynes, F., Culshaw, M., Eggers, M., Keaton, J., Lentfer, K., Novotný, J. and Paul, D. (2014). 'Engineering geological models: An introduction: IAEG commission 25.' *Bulletin of Engineering Geology and the Environment*, 73, 707–707. doi:10.1007/s10064-014-0614-8.

Siddiqui, N.A., and Rahman, A.H.A., Abdul H., Sum, C.W., Yusoff, W.I.W. and Ismail, M. (2017). 'Shallow-marine sandstone reservoirs, depositional environments, stratigraphic characteristics and facies model: A review.' *Journal of Applied Sciences*, 17, 212–237. doi:10.3923jas.2017.212.237.

Sloss, C., Murray-Wallace, C. and Jones, B. (2007). 'Holocene sea-level change on the southeast coast of Australia: A review.' *Holocene*, 17, 999–1014.

Stracher, G.B., Prakash, A. and Rein, G. (2010) *Coal and Peat Fires a Global Perspective*. New York: Elsevier Science.

Summerfield, M. (1991). *Global Geomorphology*. London: Routledge.

Swedish Geotechnical Institute (2004). *Report N065 – Quick Clay in Sweden*. Linköping: Swedish Geotechncial Institute.

Tucker, M. (1982). *The Field Description of Sedimentary Rocks*. London: Open University Press.

Walker, R.G. and James, N.P. (1992). *Facies Models: Response to Sea Level Change*. St John's, Newfoundland: Geological Association of Canada.

Chapter 3

Basic soil mechanics

N. Sivakugan and J. Ameratunga

3.1 INTRODUCTION

The objective of this chapter is to give a brief overview of the topics in soil mechanics that are relevant to soft clay engineering and ground improvement. Most readers will have some background in soil mechanics, hence this review is only the essence of the practical aspects, which one should understand. For a detailed overview of the principles of soil mechanics, readers may refer to textbooks in soil mechanics or geotechnical engineering (e.g. Das and Sivakugan 2017).

3.2 PHASE RELATIONSHIPS

Soil is a three phase medium, which may consist of soil grains (solids), water and air. The voids between the soil grains are filled with water and/or air. The relationships between the masses and volumes of the different phases define the parameters such as water content (w), void ratio (e), porosity (n), degree of saturation (S), density (ρ) and unit weight (γ), and soil mechanics text books (e.g. Das and Sivakugan 2017) cover such relationships in detail. While all such parameters as well as relationships are important in dealing with soft soils and ground improvement, saturation is an important parameter one has to consider. Saturation is governed by the degree of saturation, S, and it is related to basic soil parameters as shown in Eq. (3.1). In this book, only saturated soils are discussed and considered.

$$w = \frac{Se}{G_s} \tag{3.1}$$

where, G_s = specific gravity of the soil grains. Specific gravity says how heavy a material is compared to water, which has a specific gravity of 1.0. For soil grains, G_s varies between 2.5 and 2.9, with higher values for mine tailings rich in heavy metals and lower values for organic soils and materials such as fly ash. For clays, in the absence of any test results, a value of 2.7 could be adopted.

3.3 ATTERBERG LIMITS

When the water content of a fine-grained soil is increased, its state changes as shown in Figure 3.1, where the three borderline water contents separating two different states are the Atterberg limits. These limits play a prominent role determining the behavior of saturated clays.

Among the three Atterberg limits, it is the Liquid Limit (LL) and Plastic Limit (PL) that are commonly used in soil classifications and deriving other parameters from empirical correlations. The plasticity index (PI) is simply the difference between LL and PL. The liquidity index (LI) is a measure of how close the current water content (w) is to the liquid limit. It is defined as

$$LI = \frac{w - PL}{LL - PL} \tag{3.2}$$

The natural water content of a soft clay is generally very high and can exceed the liquid limit, with liquidity index being greater than unity. Such soils are more sensitive, and sensitivity plays an important role in selecting a ground improvement method. The variation of the natural water content with depth in a soil profile is sometimes shown in bore logs with the range of LL to PL, giving a clear indication of the liquidity index. More details on Atterberg limits can be found in Das and Sivakugan's (2017) work.

3.4 SOIL CLASSIFICATION

A soil classification system is a systematic way to group soils of similar behavior and describe soils without any ambiguity. The person who identifies the soil at the site or the one who does the tests in the laboratory is often different to the one who carries out the designs and analysis. Therefore, it is necessary that the description coming from one end conveys a clear message to the other end.

There are several soil classification systems used in different disciplines dealing with soils. The Unified Soil Classification System (USCS) is widely used in geotechnical engineering, while AASHTO (American Association of State Highway and Transportation Officials) is used mainly in roadwork.

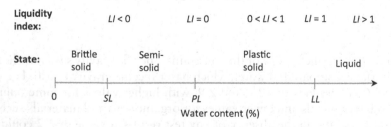

Figure 3.1 Water contents, Atterberg limits and the different states.

In any such system, the coarse-grained soils are classified mainly based on the grain size distribution and the fine-grained soils are classified mainly based on the Atterberg limits. The cut-off between *coarse-grained soils* and *fine-grained soils* is 0.075 mm (75 μm).

The cut-off to describe a soil as a coarse-grained or a fine-grained varies between countries. As previously described, in the Unified Soil Classification System, this value is 50%, i.e., soils where more than 50% of the grains are fine are classified as fine-grained soils, otherwise they are coarse-grained soils. A cut-off of 30% (i.e., >30% of fines is classified as a fine-grained material) or 35% is adopted by some other countries.

Fine-grained soils are identified easily by their dry strength, by crushing a dry lump of soil. Silts have low dry strength and clays have high dry strength. In addition, clays feel sticky between the fingers and silts feel gritty. Fines can also be identified as clays or silts based on their dilatancy. The test for dilatancy involves placing a moist pat of fines in the palm and shaking it vigorously to see how quickly the water rises to the surface of the soil pat. Dilatancy is quick in silts and very slow in clays.

3.5 STRESSES IN SOILS

In most soft clay and ground improvement projects, the current overburden stresses in the ground and the stresses imposed by the structures being placed in the ground play an important role in the design. The techniques to determine the vertical stress increases due to the different types of surface loads are discussed in this section.

3.5.1 Terzaghi's effective stress principle

In a saturated soil mass, the normal stress (σ) at a point is shared by the soil grains and the pore water. The normal stress carried by the soil grains is known as the *effective stress* or *inter-granular stress*, denoted by σ'. The normal stress carried by the water is known as the *pore water pressure* or *neutral stress*, denoted by u, which is hydrostatic (i.e., the same in all directions). This is known as Terzaghi's effective stress principle, and can be written as

$$\sigma = \sigma' + u \tag{3.3}$$

Eq. (3.3) is valid at all points within a soil mass, at all times and in all directions. During consolidation of clays, the total stress at a point within the clay remains the same, while the effective stress increases and the pore water pressure decreases, such that Eq. (3.3) is valid at all times.

In unsaturated soils, the voids also contain air and the equation becomes more complex, involving pore water pressure u_w and pore air pressure u_a. Unsaturated soils are not covered in this book.

3.5.2 Vertical stress increase due to surface loads

The initial overburden stress at a depth within the soil mass is the sum of the products of the unit weights and heights of the layers above the point. When a building or an embankment is placed at the ground level, the stresses within the soil mass change in horizontal and vertical directions. These changes are determined by elastic analysis, where the soil is treated as a homogeneous and isotropic continuum.

An assessment of vertical stress increase due to external loads is a key point in soft clay and ground improvement designs. The vertical stress at a point within the soil mass can be determined using the expressions proposed by Boussinesq (1885) or their extensions as described in many soil mechanics text books (e.g. Das and Sivakugan 2017) for the following cases:

- Point load;
- Line load;
- Strip load;
- Rectangular load; and
- Uniformly loaded circular load.

An approximate estimate of the vertical stress increase beneath a uniform rectangular load can be made by assuming that the load spreads with 2 (vertical):1 (horizontal) distribution on both sides as shown in Figure 3.2. When a uniform pressure q is applied over an area $B \times L$, the vertical stress increase at depth z is given by

$$\Delta\sigma_v = q\frac{BL}{(B+z)(L+z)} \tag{3.4a}$$

For two-dimensional plane strain loading (i.e., strip) where $L = \infty$, Eq. (3.4a) becomes

$$\Delta\sigma_v = q\frac{B}{(B+z)} \tag{3.4b}$$

Others have proposed the stress distribution with depth spreading at an angle of 30–45 degrees with vertical (Bowles 1997), the latter probably for sands and gravels. Bowles (1997) further states that the 2:1 method compares

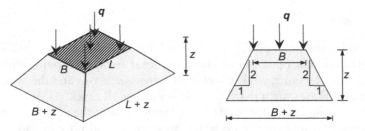

Figure 3.2 Approximate 2:1 stress distribution.

reasonably well with more theoretical methods for depths between B and $4B$ below foundations and cautions that it should never be used for the depth zone from 0 to B.

Sometimes the loaded area is not rectangular. When the shape is irregular, the most popular is the Newmark's chart (Newmark 1942; Sivakugan and Das 2010).

Embankment loads are most common on linear projects such as highways. The vertical normal stress at a depth below a long embankment along the centerline could be assessed using influence factors proposed by Osterberg (1957), which is discussed in several soil mechanics books (see Das and Sivakugan 2019).

3.6 CONSOLIDATION

When loads are placed at the ground level, or the water table is lowered, the underlying soil layers are subjected to stresses that are in addition to the current overburden stresses. In saturated clays, the loads produce excess pore water pressures, which take several months or years to fully dissipate in a process called consolidation. As excess pore water pressure dissipates, the effective stress increases by the same magnitude until the additional load is fully transferred to the soil grains.

Prior to consolidation, there is immediate settlement that occurs instantaneously upon loading under undrained conditions. It is generally computed using elastic analysis. The magnitude of immediate settlement in a saturated clay is relatively small compared to the consolidation settlement.

In geotechnical textbooks, consolidation is generally discussed in two parts:

a. computation of final consolidation settlement, and
b. time rate of settlements.

In the first part, no consideration is given to how fast the consolidation is occurring. It is considered mainly in the second part. The two parts are discussed separately in Sections 3.6.2 and 3.6.3.

Figure 3.3 shows a typical consolidation curve of void ratio (e) against vertical effective stress (σ'_v). The *maximum past pressure* (σ'_p), also known as the *preconsolidation pressure*, indicates when the soil moves from an *overconsolidated* state (line AB) to a *normally consolidated* state (steeper line BC known as the *virgin consolidation line*). It is the largest vertical stress the soil specimen has experienced in its geological history. The *compression index* (C_c) controls the magnitude of primary consolidation when subjected to external loading. The *Compression Ratio* (CR) is directly related to C_c by the following equation:

$$CR = \frac{C_c}{1 + e_0} \tag{3.5}$$

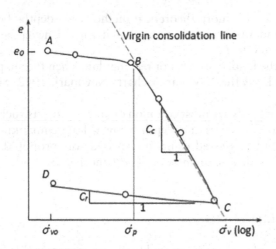

Figure 3.3 Typical void ratio versus effective stress plot due to incremental loading in an oedometer.

If the clay is overconsolidated, the stress state would be on the line AB in
Figure 3.3 and therefore the *recompression index* (C_r) or *recompression
ratio* (RR) is applicable instead of C_c or CR.

$$RR = \frac{C_r}{1 + e_0} \tag{3.6}$$

3.6.1 Effect of sample disturbance

Sample disturbance has a major effect on the derived soil parameters, and
this is particularly important in a consolidation test. Sample disturbance
occurs in soft clays during sampling in the field, in transportation and later
in laboratory handling.

In Figure 3.3, the consolidation test result shows two linear segments (AB
and BC) intersecting at the maximum past pressure, σ'_p (point B in the
Figure). In practice, because of sample disturbance, it is quite common to get
curved segments for AB and BC. As the magnitude of settlement in the NC
region is usually an order higher than in the OC region, erroneous line gra-
dients (AB and BC) could significantly affect settlement predictions.

Schmertmann (1955) presented a simple procedure to correct the laboratory consolidation curve as discussed by Sivakugan and Das (2010). A typical reconstruction of a laboratory consolidation curve using Schmertmann's
method is shown in Figure 3.4.

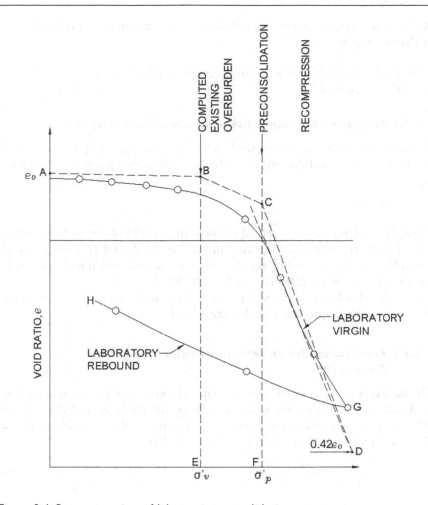

Figure 3.4 Reconstruction of laboratory consolidation curve.

3.6.2 Computation of final consolidation settlement (s_c)

The magnitude of the final consolidation settlement of a saturated clay layer subjected to a surcharge depends on the following factors: thickness of the layer (H), initial (natural) water content (w_n), overconsolidation ratio (OCR) and magnitude of the surcharge (q), C_c and C_r. In the case of thick clay layers, these parameters can change with depth and hence it is a good practice to subdivide the thick layer into few sublayers, compute the settlement for each sublayer, and add them up to arrive at the total consolidation settlement.

Within a sublayer, the final consolidation settlement can be estimated by one of the two ways:

a. Using the coefficient of volume compressibility m_v; or
b. Using CR and RR.

3.6.2.1 Final consolidation settlement determined using m_v

Finding the final consolidation settlement of a sublayer using the coefficient of volume compressibility m_v is fairly straightforward from the following equation.

$$s_c = m_v \Delta\sigma_v H \tag{3.7}$$

Here, $\Delta\sigma_v$ = change in the vertical stress at the middle of the sublayer determined using the methods discussed in Section 3.5.2, and H = thickness of the sublayer. The larger the m_v, the softer is the soil skeleton. Typically, it varies in the range of 0.01×10^{-3} to 2.0×10^{-3} kPa^{-1}.

One drawback is that m_v is stress dependent and therefore it is necessary to use the value appropriate to the stress level.

3.6.2.2 Final consolidation settlement determined using CR and RR

The final consolidation of a sublayer depends on several parameters such as H, OCR, CR, RR and $\Delta\sigma_v$ at the middle of the sublayer and the current effective vertical overburden stress (σ'_{vo}). For normally consolidated clays, the consolidation settlement can be determined as

$$s_c = H \times \frac{C_c}{1+e_0} \log \frac{\sigma'_{vo} + \Delta\sigma_v}{\sigma'_{vo}} \tag{3.8}$$

In overconsolidated clays, where the clay remains overconsolidated even after the application of the load (i.e., $\sigma'_{vo} + \Delta\sigma_v < \sigma'_p$), the final consolidation settlement can be determined as

$$s_c = H \times \frac{C_r}{1+e_0} \log \frac{\sigma'_{vo} + \Delta\sigma_v}{\sigma'_{vo}} \tag{3.9}$$

In overconsolidated clays, where the clay becomes normally consolidated after the application of the load (i.e., $\sigma'_{vo} + \Delta\sigma_v > \sigma'_p$), the final consolidation settlement can be determined as

$$s_c = H \times \frac{C_r}{1+e_0} \log \frac{\sigma'_p}{\sigma'_{vo}} + H \times \frac{C_c}{1+e_0} \log \frac{\sigma'_{vo} + \Delta\sigma_v}{\sigma'_p} \tag{3.10}$$

It is necessary to know which of the three equations (3.8), (3.9), or (3.10) to use in determining settlement.

3.6.3 Time rate of settlement

The interrelationship among the *excess* pore pressure u within the clay at a point lying at depth z within the clay layer and at time t is given by

$$\frac{\partial u}{\partial t} = c_v \frac{\partial^2 u}{\partial z^2} \qquad (3.11)$$

where, c_v is the *coefficient of consolidation*, defined as

$$c_v = \frac{k_v}{m_v \gamma_w} \qquad (3.12)$$

where k_v = vertical permeability, and γ_w = unit weight of water (= 9.81 kN/m³).

The preceding Section 3.6.2 focused on determining the final consolidation s_c. How fast or slow the settlement occurs depends on the coefficient of consolidation (c_v) and whether the layer is singly or doubly drained. The larger the c_v, the faster is the consolidation process. Generally, c_v of an over-consolidated clay is an order of magnitude larger than the c_v of the same clay if it is normally consolidated. Thus, overconsolidated clays consolidate faster than the normally consolidated clays.

3.6.3.1 Degree of consolidation

Solving Eq. (3.11) with appropriate boundary conditions, it can be shown that the degree of consolidation $U(z, t)$ at a certain depth z within the layer at time t can be written as

$$U(z,t) = 1 - \sum_{m=0}^{m=\infty} \frac{2}{M} \sin(MZ) e^{-M^2 T} \qquad (3.13)$$

where $M = (\pi/2)(2m+1)$. Z and T are the dimensionless *depth factor* and *time factor* defined as

$$T = \frac{c_v t}{H_{dr}^2} \qquad (3.14)$$

$$Z = \frac{z}{H_{dr}} \qquad (3.15)$$

where, H_{dr} is the average maximum length of drainage path for the consolidating clay layer, which is H for singly drained layers and $H/2$ for doubly drained layers, where H is the thickness of the clay layer. The U-Z-T relationship is shown graphically in Figure 3.5.

So far, we have discussed only consolidation due to dissipation of excess pore water pressure in the vertical direction. In some circumstances, radial

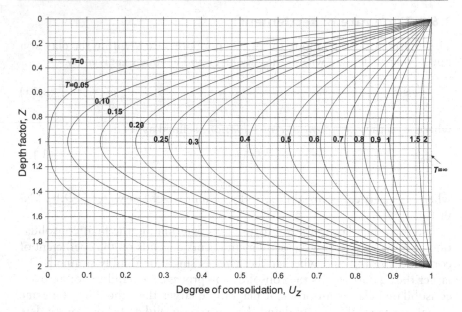

Figure 3.5 **U – Z – T** relationship.

consolidation does occur. Especially in the case of ground improvement using the preloading technique with wick drains, horizontal dissipation is significantly greater than in the vertical direction. The chapter on preloading discusses in detail the assessment of horizontal and the combined dissipation of excess pore pressure.

3.6.3.2 Average degree of consolidation

During consolidation, the degree of consolidation varies with depth. The overall average degree of consolidation U_{avg} for the clay layer at a specific time is computed as the ratio of the dissipated excess pore water pressure distribution with depth to the initial excess pore water pressure distribution with depth. This is the same as the ratio of the consolidation settlement at that time to the final consolidation settlement. U_{avg}-T relationship is shown in Figure 3.6.

It can be seen from Figure 3.6 that consolidation literally goes on forever. However, when the time factor reaches 1.0, consolidation is more or less complete with $U_{avg} = 93.1\%$.

The $U_{avg} - T$ relationship can be approximated as

$$T = \frac{\pi}{4}\left(\frac{U_{avg}}{100}\right)^2 \text{ for } U_{avg} \leq 60\% \tag{3.16a}$$

$$T = 1.781 - 0.933\log\left(100 - U_{avg}\right) \text{ for } U_{avg} \geq 60\% \tag{3.16b}$$

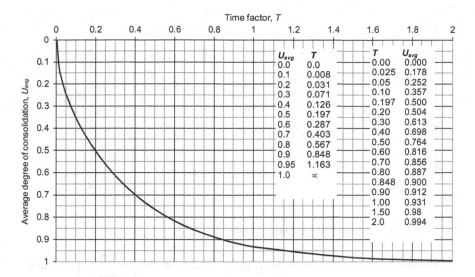

Figure 3.6 $U_{avg} - T$ relationship.

$$U_{avg} = \sqrt{\frac{4T}{\pi}} \times 100\% \text{ for } U_{avg} \leq 60\% \tag{3.17a}$$

$$U_{avg} = 1 - \frac{8}{\pi^2} \exp\left(-\frac{\pi^2}{4} T\right) \text{ for } U_{avg} \geq 60\% \tag{3.17b}$$

3.6.3.3 Coefficient of consolidation of multi-layered soils

When a thick deposit of clay consists of layers of clay with varying values of c_v, US Navy (1982) suggests a procedure to transform the clay layers into a single homogeneous layer with specific value (assigned by the user) of coefficient of consolidation c_v^*. Let there be n layers, where the thickness and the c_v of the ith layer are H_i and c_{vi}, respectively (Figure 3.7). Here, the ith layer is transformed into a layer with thickness of H_i^* and coefficient of consolidation c_v^*. Here, H_i^* is given by

$$H_i^* = H_i \sqrt{\frac{c_v^*}{c_{vi}}} \tag{3.18}$$

Once all the layers are transformed, the total thickness of the consolidating clay deposit is given by

$$H^* = \sum_{i=1}^{n} H_i^* \tag{3.19}$$

which has one value of c_v^* throughout.

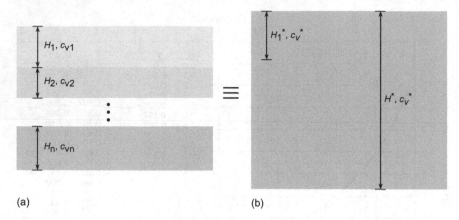

(a)   (b)

Figure 3.7 (a) n-layers, (b) equivalent layer.

Alternatively, you can keep the original thickness $(H)\left(=\sum_{i=1}^{n}H_i\right)$ the same and obtain an equivalent coefficient of consolidation $c_{v,\text{equivalent}}$ using Eq. (3.20).

$$c_{v,\text{equivalent}} = c_v^* \left(\frac{H}{H^*}\right)^2 \tag{3.20}$$

3.6.4 Secondary compression

Secondary compression, also known as *creep*, is a time-dependent process that produces settlements in addition to the consolidation settlements. Although it can occur during consolidation, for simplicity, it is assumed to start at the end of consolidation and to continue for a long period as shown in Figure 3.8. Secondary compression occurs due to realignment of the clay

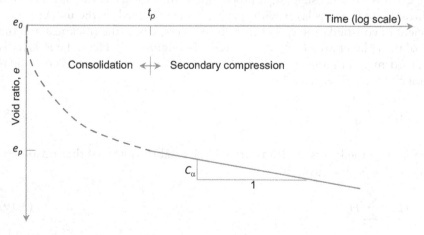

Figure 3.8 Secondary compression.

grains, which results in a change in the clay fabric and reduction in void ratio. There is no excess pore water pressure to dissipate, and the effective stress remains the same throughout the secondary compression process.

Prior to consolidation, the clay layer has a thickness of H_0, and void ratio of e_0. At the end of consolidation, at time t_p, they reduce to H_p and e_p, respectively. The slope of the void ratio versus log time plot (Figure 3.8) is the coefficient of secondary compression C_α, which remains the same during the secondary compression. It can be defined as the reduction in the void ratio per log cycle of time.

$$C_\alpha = \frac{\Delta e}{\Delta \log t} \tag{3.21}$$

It can be determined from the tail end of the void ratio versus log time plots used for determining c_v by Casagrande's (1936) method.

As discussed in Section 3.6.2, the void ratio in the vertical axis in Figure 3.3 can be replaced by vertical strain ε:

$$\varepsilon = \frac{\Delta H}{H} = \frac{\Delta e}{1+e} \tag{3.22}$$

where, H = reference (initial) height, e = reference (initial) void ratio, ΔH = reduction in height and Δe = reduction in void ratio. Here, the slope of the secondary compression line is called the *modified secondary compression index* $C_{\alpha\varepsilon}$ defined by

$$C_{\alpha\varepsilon} = \frac{\varepsilon}{\Delta \log t} = \frac{C_\alpha}{1+e} \tag{3.23}$$

The secondary compression settlement (s_s) during the time period from t_p to t can be written as (see Sivakugan and Das 2010)

$$s_s = C_\alpha \frac{H_0}{1+e_0} \log\left(\frac{t}{t_p}\right) \tag{3.24a}$$

or

$$s_s = C_{\alpha\varepsilon} H_0 \log\left(\frac{t}{t_p}\right) \tag{3.24b}$$

3.6.4.1 Hypotheses A and B

There are two schools of thought on the secondary compression associated with soft clays based on the assumption whether secondary compression

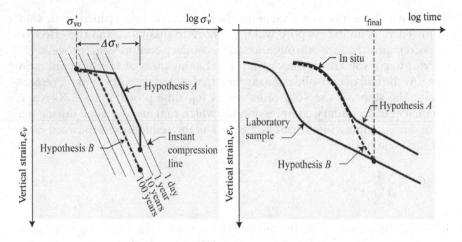

Figure 3.9 Hypothesis A versus Hypothesis B (after Ladd et al. 1977).

occurs at the completion of primary or simultaneous with primary. Ladd et al. (1977) named the two as Hypotheses A and Hypotheses B:

- Hypotheses A – Secondary compression occurs at the end of primary;
- Hypotheses B – Secondary compression occurs in tandem with primary.

This is illustrated in Figure 3.9.

If Hypotheses A is used for back analysis of an embankment, one would expect a larger value for CR compared to Hypotheses B. This may be on the conservative side if C_{α}/CR ratio is the parameter adopted by the designer to predict the future long-term settlements after primary consolidation is completed.

3.7 SHEAR STRENGTH

All soils follow the Mohr-Coulomb failure criterion, where the shear strength is derived from two separate components, namely cohesion and friction. The failure criterion is expressed as

$$\tau_f = c + \sigma \tan \phi \tag{3.25}$$

The frictional component of the shear strength is proportional to the normal stress. With appropriate values of the shear strength parameters, the failure criterion is valid in terms of effective stresses or total stresses.

Shear strength parameters are obtained either from laboratory tests such as triaxial tests and direct shear tests or in situ tests (e.g., vane shear test) and/or empirical relationships.

3.7.1 Undrained and drained loading

When a saturated clay is loaded, it will not let the water out immediately (i.e., it remains undrained) and that is when most of the embankment failures occur. In the short-term, the clay can be treated as an undrained homogeneous material without separating the grains and water. Here, we carry out the undrained analysis in terms of total stresses, using undrained shear strength c_u ($\phi = 0$).

In the long term (after some months or years), the clay will drain out some water until the excess pore water pressure is fully dissipated and the pore water pressure is in equilibrium with the in situ conditions. Now, it is prudent to carry out an effective stress analysis using c' and ϕ', where we separate the stresses acting on the pore water (pore water pressure) and the grains (effective stresses).

Undrained analysis is often much easier to carry out, inexpensive to get the design parameters, and is necessary to assess the short-term stability, which can be more critical than the long-term stability. For undrained loading, the failure envelope in terms of total stresses is horizontal and hence we only need one parameter, undrained cohesion c_u ($\phi = 0$). Undrained cohesion can be derived from an unconfined compression test, unconsolidated undrained triaxial test, vane shear test (laboratory or field), or simply using a pocket penetrometer. Drained analysis needs c' and ϕ', which are derived from more expensive consolidated drained or undrained triaxial tests or in situ tests (and estimated using correlations). They are necessary when working with effective stresses.

Granular soils drain very quickly, and hence they are always treated as drained and analyzed in terms of effective stresses using ϕ' ($c'=0$).

3.7.2 Drained parameters

Figure 3.10 shows typical stress-strain plots of two different types of soils: (a) normally consolidated clays or loose sands, and (b) overconsolidated clays or dense sands. The first one exhibits contractive (i.e., volume reduction) behavior throughout the shearing process, and the shear stress peaks at critical state, which occur at relatively large strains. At critical state, the soil deforms under constant volume or void ratio. The second type of soil exhibits dilative (i.e., volume increase) behavior during the shear after some initial contraction, and the peak value of the shear stress is reached at relatively small strain in the order of 1–2%, where the friction angle is known as the *peak friction angle* ϕ'_{peak}. The shear stress reduces on further straining, and reaches the critical state, which occurs at strains in the order of 10–20%. Further straining in clays can take them to residual state, which occurs at strains in the order of 100%.

Irrespective of the initial state (i.e., normally consolidated or overconsolidated clay; or loose or dense sand), the shear stress and the void ratio reach a constant value at critical state. The friction angle and the void ratio at the

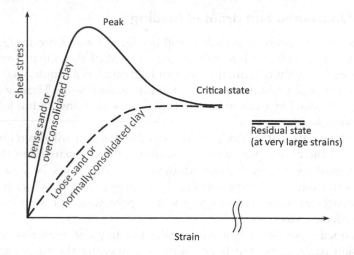

Figure 3.10 Stress-strain plots showing peak, critical and residual states.

critical state are known as the *critical state friction angle* ϕ'_{cv} and *critical state void ratio* e_{cv}, respectively, where 'cv' stands for constant volume or critical state void ratio. ϕ'_{cv} largely depends on the mineralogy of the soil grains.

At very large strains, clay soils would go from critical state to *residual state*, where the friction angle is known as the *residual friction angle* ϕ'_{res}. In clays, strains in the order of 100% or more are required to remold the clay and reach residual state (Kulhawy and Mayne 1990). In granular soils, the residual state is not far from the critical state, where it can be assumed that $\phi'_{res} \approx \phi'_{cv}$. For cohesive soils, ϕ'_{res} can be several degrees lower than ϕ'_{cv}. For all soils in general, $\phi'_{res} < \phi'_{cv} < \phi'_{peak}$. ϕ'_{res} and the residual shear strength depend largely on the mineralogy of the grains, grain sizes and shapes, clay fraction and plasticity.

3.7.3   Undrained parameters

Since the friction angle is zero for the undrained analysis, it is only required to determine the undrained cohesion (c_u), also known as the undrained shear strength. A crude and preliminary estimate of c_u can be made at the site or from an intact sample using a pocket penetrometer, torvane or a laboratory vane. Better estimates can be made through an unconfined compression test or an unconsolidated undrained triaxial test on good quality samples, or a field vane shear test. Alternatively, empirical correlations can be used for estimating c_u.

3.7.4   Stress points and stress paths

Shear strength parameters c' and ϕ' are determined from Mohr's circles or stress points. When a series of triaxial tests is carried out, it is possible to

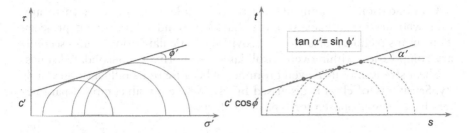

Figure 3.11 Mohr's circles and stress points at failure along with failure envelopes.

draw the line of best fit that is tangent to the Mohr's circles. When there is a significant scatter in the test data, using stress points may work better. The stress point is the top of the Mohr's circle, which is defined by the following coordinates.

$$s' = \frac{\sigma'_v + \sigma'_v}{2} \tag{3.26}$$

$$t = \frac{\sigma'_v - \sigma'_h}{2} = \frac{\sigma_v - \sigma_h}{2} \tag{3.27}$$

where, σ'_h and σ'_v are the horizontal and vertical effective stresses, respectively, on the soil specimen. At failure, it can be shown that

$$t_f = s'_f \sin\phi + c'\cos\phi' \tag{3.28}$$

where, the subscript f denotes the failure values. When t_f is plotted against s'_f, the slope and the intercept of the line of best fit are given by $\sin \phi'$ and $c' \cos \phi'$, respectively. What is discussed here is applicable in terms of total or effective stresses. Figure 3.11 shows the Mohr's circles and the stress points at failure in terms of effective stresses, along with the failure envelopes. It can be seen that the stress points are simply the top of the Mohr's circles.

3.8 SUMMARY

Chapter 3 provides a review of soil mechanics theories essential for ground improvement design. A brief attention has been given to physical characteristics and highlighted their importance to derive other parameters such as strength and consolidation.

Stresses imposed by external loads that transmit down to soil layers are not constant as they depend on the plan area of the load, subsurface profile and material parameters. As this is a subject well covered in soil mechanics and foundation engineering books, only a brief description has been given.

Consolidation is examined in detail with adequate information to any user who needs the background to tackle ground improvement projects. This means technical direction is provided to calculate primary and secondary compression within overconsolidated and normally consolidated states.

Shear strength is also given prominence because of its influence on stability. Sensitivity of clays is discussed briefly. Where possible, typical parameters have been proposed for typical weak clays.

REFERENCES

Bowles, J.E. (1997). *Foundation Analysis and Design*, International edition, McGraw-Hill, Singapore.

Boussinesq, J. (1885). *Application des Potentiels à l'étude de l'équilibre et du Mouvement des Solides E'lastiques*. Gauthier-Villars, Paris, 30.

Casagrande, A. (1936). 'The determination of pre-consolidation load and its practical significance.' Discussion D-34, *Proceedings of the 1st International Conference on Soil Mechanics and Foundation Engineering*, Cambridge, III, 60–64.

Das, B.M. and Sivakugan, N. (2017). *Fundamentals of Geotechnical Engineering*, 5th edition, Cengage, Boston.

Das, B.M. and Sivakugan, N. (2019). *Principles of Foundation Engineering*, 9th edition, Cengage, Boston.

Kulhawy, F.H. and Mayne, P.W. (1990). *Manual on Estimating Soil Properties for Foundation Design*, Final report EL-6800, Electric Power Research Institute, Palo Alto, CA.

Ladd, C.C., Foott, R., Ishihara, K., Schlosser, F., and Poulos, H.G. (1977). 'Stress-deformation and strength characteristics.' *Proceedings of the 9th International Conference on Soil Mechanics and Foundation Engineering*, Tokyo, 2, 421–494.

Newmark, N. M. (1942). *Influence Charts for Computation of Stresses in Elastic Soils*. University of Illinois Experiment Station Bulletin, Series No. 338, 61(92), Urbana, Illinois, reprinted 1964, 28 pp.

Osterberg, J.O. (1957). 'Influence values for vertical stresses in semi-infinite mass due to embankment loading.' *Proceedings of the 4th International Conference on Soil Mechanics and Foundation Engineering*, London, 1, 393–396.

Schmertmann, J.M. (1955). 'The undisturbed consolidation of clay.' *Transactions ASCE*. 120, 1201.

Sivakugan, N. and Das, B.M. (2010). *Geotechnical Engineering: A Practical Problem Solving Approach*. J.J. Ross Publishing, Florida, USA.

U.S. Navy (1982). *Soil Mechanics, Foundations and Earth Structures*, NAVFAC Design Manual DM-7, Department of Navy, Washington, DC.

Chapter 4

Geotechnical testing

J. Ameratunga and N. Sivakugan

4.1 INTRODUCTION

For any ground improvement project, it would be necessary to use several tools available, including geotechnical testing, to develop a site model, and define the soil profile and the necessary parameters for the different soil layers. The five geotechnical tools available to the designer to carry out ground improvement can be summarized as follows:

- In situ testing;
- Laboratory testing;
- Geotechnical trials;
- Empirical correlations; and
- Observational approach.

Depending on the project, stage of the design and available information, a combination of the above tools along with experience could lead to deriving the most appropriate parameters in an efficient manner. Each method has its advantages and disadvantages. It is the responsibility of the geotechnical engineer to strike the right balance between the various methods. It is always recommended to derive parameters by at least two different methods to provide a check on the parameters derived. Empirical correlations are mostly used in such situations, although conducting a large number of in situ and laboratory tests would certainly help but at a cost.

While all types of methods and tests are available to the engineer, in situ testing and laboratory testing are the most widely used methods in general. Therefore, they always compete against each other, although they should complement each other rather than at the expense of the other. There are advantages and disadvantages for both laboratory and in situ tests and are summarized in Table 4.1.

All methods discussed in this Chapter, either solely or in combination, may assist the derivation of geotechnical parameters for soft clays depending on the type and size of the project, access conditions to site, time available to carry out investigation and design work, subsurface conditions and cost.

Table 4.1 Advantages and disadvantages of laboratory and in situ tests

Laboratory tests on specimens	In situ tests
More rational interpretation	Empirical and semi-empirical interpretation
Smaller volumes tested	**Larger volumes tested in relatively shorter period. More representative of the soil mass.**
Well-defined boundary conditions	Poorly defined boundary conditions
Better control of the boundary conditions	Little control of the boundary conditions
Stress relief and mechanical disturbance in sampling	**Testing in situ; no stress relief and limited sample disturbance**
Very few points tested along the depth	**Continuous profiling possible**

Note: Advantages are shown in bold.

In addition, experience of the individual consultant and/or the organizational culture would also come into play in the selection of the type of investigation to be carried out.

Of the five tools available to the geotechnical designer discussed above, this chapter discusses the following:

- In situ testing;
- Laboratory testing; and
- Geotechnical trials.

Empirical correlations are introduced later in this chapter and discussed in detail in Chapter 5. The Observational Approach is discussed in detail in Chapter 16.

4.2 IN SITU TESTING

In situ testing has evolved over the years with new techniques becoming available along with the refinements to existing techniques. For example, the standard penetration test was available from the 1940s but the dilatometer was introduced to the world several decades later. In a typical in situ test, a probe (e.g., cone, T-bar, split-spoon, or vane) is inserted into the ground and the resistance is measured and interpreted into soil parameters. While the standard penetration test (SPT) is one of the oldest and most popular in situ tests, it is of limited value when it comes to soft soils. The cone penetration test, vane shear test and dilatometer test are the most popular in situ tests for soft soils.

Lunne et al. (1997) presented a summary of their assessment of the applicability and usefulness of in situ tests, which were used at the time of their work. Since then, i.e., over two decades, further developments have taken place, a classic example being the dilatometer.

Considering most useful parameters for ground improvement design of soft clays are related to strength (especially undrained shear strength) and compressibility, some of the more popular tests (e.g., standard penetration test) are not useful in soft clays. Therefore, only the following in situ tests are discussed in this chapter:

- Vane shear test;
- Cone penetrometer test (CPT);
- Self-boring pressuremeter test; and
- Dilatometer test.

A summary of advantages and limitations relevant to these tests are presented by Ameratunga et al. (2016) and the derivation of parameters from these tests are discussed in Chapter 5.

4.2.1 Vane shear test

The vane shear test is a popular in situ test used in the geotechnical engineering practice to assess the undrained shear strength of cohesive materials in a saturated state. It cannot be used in materials where undrained conditions cannot be maintained, i.e., in cohesionless materials such as sands or gravels, and is a concern in some types of peats. However, it is generally recognized that a field vane shear test has many advantages over other in situ tests as well as some laboratory tests.

Vane shear equipment consists of two thin rectangular blades perpendicular to each other (cruciform) connected to a solid pushing rod (see Figure 4.1). The test can be performed by pushing through the base of an auger hole or a bore hole, directly through the overburden, or by drilling through a vane housing. Once the vane is in place at the nominated depth, it is rotated at the nominated speed and the torque is recorded. At the end of the test, to obtain the residual/remolded strength, the vane is rotated very rapidly and the torque is measured. The maximum torque under slow rotation is used to obtain the peak undrained shear strength while the other is used to obtain the residual/remolded undrained shear strength. The test procedure is found in ASTM D 2573-08.

Unlike a piezocone, the vane shear test is not performed continuously with depth. Usually, tests are conducted at 0.5 m or 1 m depth intervals or at depths selected by the geotechnical designer. As the test does not provide any meaningful result if sandy lenses/layers are within the profile, sometimes it becomes necessary (especially in a layered profile) to first carry out a CPT before deciding on the depths to be tested by the vane.

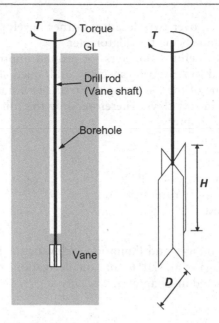

Figure 4.1 Typical geometry of a field vane.

One of the reasons the vane shear test is popular is because it has a sound theoretical basis that can be found in soil mechanics books (e.g. Das and Sivakugan 2017; Sivakugan and Das 2010).

According to Bjerrum (1972), the field vane shear strength measured may not necessarily be equal to c_u. He found this based on analyses of several embankment failures and back calculations suggesting that the factors of safety using field vane shear test results are significantly higher than 1.0, although the embankments have failed, implying that the field vane shear tests overestimate the undrained shear strength. His research concluded that soil characteristics have a role to play and that soil plasticity has direct influence on the field vane shear c_u values. He advised that c_u obtained from a field vane should be corrected to derive design parameters when the design involves embankment loading or excavation stability. Bjerrum thus suggested that values obtained should be multiplied by a correction factor (μ), popularly known as Bjerrum's correction factor, to be applied to the field vane shear c_u prior to using in embankment or excavation stability design. μ is a function of plasticity index, *PI*, and found in Bjerrum (1972), which is extensively repeated in many soil mechanics books (see Das and Sivakugan 2017; Sivakugan and Das 2010).

4.2.2 Cone penetration test (CPT)

The *cone penetration test* (CPT), originally known as the *Dutch cone penetration test*, is a versatile in situ test that can be carried out in most soil conditions, for determining the soil profile and various soil parameters.

The interpretation of the test is more rational than in the standard penetration test (SPT). SPT and CPT are used in more than 90% of the soil exploration programs worldwide.

The original Dutch cone developed in Holland in 1932 was a 60° *mechanical cone* (ASTM D 3441) that is labor intensive and limited in accuracy. It is currently used only in developing countries.

Fugro developed the first *electric cone* (ASTM D 5778) in 1965 as an improvement of the mechanical cone, which almost gives continuous measurements with depth. The modern cones also have the pore water pressure measurement facility and are known as *piezocones* (CPTu). Piezocones are the modern electric friction cones. Figure 4.2a shows a piezocone with a filter (for pore pressure measurement) located between the cone and the shaft. In piezocones, pore water pressure is measured at three possible locations, u_1 at the cone itself and, u_2 and u_3 as shown in Figure 4.2b. Of the three, u_2 (Figure 4.2b) is used the most in computations. When profiling stiff clays, $u_2 \approx 0$, and it is recommended to use u_1. With an addition of a geophone approximately 500 mm above the cone, *seismic piezocone*

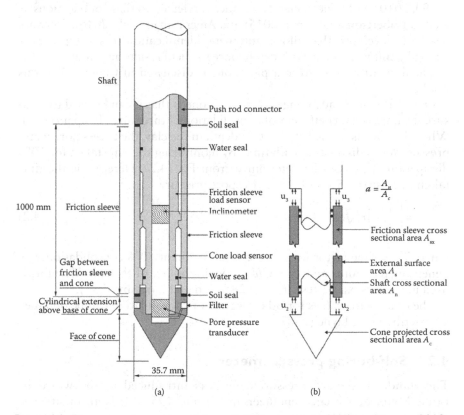

Figure 4.2 Piezocone (reproduced from *Soil Mechanics* by William Powrie, 2013, by permission of Taylor & Francis).

penetration tests (SCPTu) are nowadays possible, which can also measure shear wave velocity v_s and hence the shear modulus G.

Depending on the location of the pore pressure transducer, the cone resistance (q_c) and the sleeve friction (f_c) have to be corrected to become q_t and f_t, respectively, and details are presented in Lunne et al. 1997.

One of the most attractive features of the cone is in profiling the soil, where it enables identification of the soil type in real time based on q_t and f_t.

Friction ratio (R_f) is defined as

$$R_f (\%) = \frac{\text{Corrected sleeve friction}, f_t}{\text{Corrected cone resistance}, q_t} \times 100 \tag{4.1}$$

R_f varies in the range of 0–10%, with the lower end of the range for granular soils and the upper end for cohesive soils. Typically, the corrected cone resistance q_t is high in sands and low in clays. The friction ratio R_f is low in sands and high in clays. This distinction makes it possible to identify the soil type based on the two values of q_t and R_f. The most commonly used *soil behavior type* (SBT) chart for cone penetration test is the one suggested by Robertson et al. (1986), consisting of 12 zones. This was updated by Robertson (1990, 2010) with nine zones. The reader is referred to these publications as well as Robertson and Cabal (2015) and Ameratunga et al. (2016). Software has been developed that allows automatic identification of soil type, shear strength, unit weight etc. with depth, once the field results are input.

The derivation of c_u from a piezocone is discussed in Chapter 5 of this book.

One of the many advantages of a piezocone is that it can be used to measure the horizontal coefficient of consolidation c_h through a dissipation test. When the cone is paused at a certain depth in the clay, the excess pore water pressure starts dissipating with time. By monitoring the time taken for 50% dissipation t_{50}, c_h can be determined from Eq. (4.2). Here, t_{50} is the time taken to dissipate 50% of the excess pore pressure.

$$c_h = \left(\frac{T_{50}}{t_{50}} \right) r_0^2 \sqrt{I_R} \tag{4.2}$$

where T_{50} = time factor for radial consolidation at 50% consolidation, r_0 = penetrometer radius and $I_R = G/c_u$ = rigidity index of the soil. When u_2 is used in the dissipation test, $T_{50} = 0.245$ (Teh and Houlsby 1991).

The time factor values for other degrees of consolidation can be obtained as described by Mayne (2007).

4.2.3 Self-boring pressuremeter

The standard Menard pressuremeter was introduced to the world by Louis Menard, a French engineer, in the mid-1950s. It is an in situ test carried out using a cylindrical probe with an expandable flexible membrane and a control unit. The probe is cylindrical in nature, with a guard cell at each end to eliminate the end effects, and radially expanded in a borehole with measurement of pressure applied to inflate recorded along with the volume change due to cavity deformation. It has gradually developed not only in terms of mechanics involved with the equipment but, equally important, theoretical derivation of the pressuremeter behavior in the 1960s and 1970s.

Whatever the advancement had been, the process has several issues relevant to soft soils because of the disturbance as the test is conducted in a pre-drilled borehole. The disturbance occurs due to soil relaxation and the mechanical disturbance created by the drilling operation. The disturbance is greater for softer soils compared to stiff or hard clays. While the disturbance created by drilling methods could be reduced by using appropriate equipment and experienced drillers who are cognizant of the effects of drilling disturbance, it is simply not possible to remove the effect of stress relief in a Menard pressuremeter test. This led to further independent research by several parties, which ultimately led to the birth of the self-boring pressuremeter (Baguelin et al. 1972; Wroth and Hughes 1973). The self-boring pressuremeter creates its own hole by a rotating a cutting shoe at the bottom of the probe, which removes the soil thereby creating the hole for the pressuremeter. This mechanism was able to significantly reduce the disturbance created due to stress-relief in a borehole.

The self-boring pressuremeter, as the name implies, self-bores the hole to the depth of the test before carrying out the test thereby eliminating or reducing the elastic stress relief. It is quite useful for softer clays where stress relief could affect the test results significantly. Details of a self-boring pressuremeter are illustrated in Figure 4.3.

The development of the self-boring pressuremeter occurred due to research carried out in Europe by two independent groups. One of the groups is based in France and Baguelin et al. (1978) described their version of the self-boring pressuremeter known as French Pressiomètre Autoforcur (PAF) while the Camkometer developed in Cambridge UK is described by Wroth and Hughes (1973).

Schnaid (2009) states that the important differences between the two types of self-boring pressuremeters include the following:

- The Camkometer measures the radial strain using strain gauged feeler arms, whereas in PAF, the radial strains are calculated by measuring volume changes; and
- The Camkometer is driven by rods extending to the ground surface, whereas the PAF has cutters that are driven by downhole hydraulic motors.

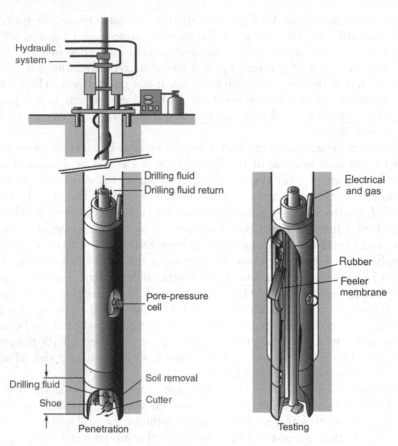

Figure 4.3 Details of a self-boring pressuremeter (reproduced from *In Situ Testing in Geomechanics* by Fernando Schnaid, 2009, by permission of Taylor & Francis).

The self-boring pressuremeter could be used to obtain the following geotechnical design parameters:

- In situ lateral stress (the lift-off pressure as measured by the pressuremeter);
- Modulus; and
- Strength parameters, c_u for cohesive soils and friction angle for cohesionless soils.

The derivation of modulus and c_u from a self-boring pressuremeter is discussed in Chapter 5.

The self-boring pressuremeter is also used to measure the horizontal coefficient of consolidation, c_h. The test is known as a holding test and as the name implies, by measuring pore pressure dissipation with time an assessment of c_h could be made as described by Bo et al. (2003). They also comment that for Singapore marine clay, c_h obtained from the self-boring pressuremeter test is much larger than that obtained from other in situ or laboratory tests.

4.2.4 Dilatometer

The flat dilatometer was developed by Dr. Sylvano Marchetti in 1975 in Italy. It consists of a 240 mm long, 95 mm wide and 15 mm thick stainless steel blade with a flat, thin and expandable 60 mm diameter and 0.20–0.25 mm thick circular steel membrane that is mounted flush with one face (see Figure 4.4). The blade has a cutting edge at the bottom end, tapered over 50 mm, with an apex angle of 24° to 32°.

The blade is generally pushed into the ground by the penetration test rig, at a rate of 20 mm/s, the same rate as a cone in CPT. A gas pressure unit at the surface is used to inflate the steel membrane when it is pushed into the ground.

When the dilatometer blade is pushed into the ground, three pressure readings are taken in a sequence (see Figure 4.4a):

a. The pressure required to bring the membrane in flush with the soil (i.e., to just move the membrane), known as lift-off pressure or 'A pressure';
b. The pressure required to expand the membrane against the soil laterally by 1.1 mm, known as 'B pressure'; and
c. The pressure when the membrane is deflated, known as the closing pressure or 'C pressure.' This is an optional reading.

The first two are the most used in the computations. C pressure is used for determining the pore water pressure in the ground. Pressures A, B and C are corrected for the membrane stiffness, determined through calibration. These three corrected values are denoted by p_0, p_1 and p_2, respectively. The test is generally carried out at 200 mm depth intervals.

In stiffer soils, p_0 and p_1 can be determined as (Marchetti 1980)

$$p_0 = A + \Delta A \tag{4.3}$$

$$p_1 = B - \Delta B \tag{4.4}$$

where, ΔA and ΔB are the calibration corrections applied on the A and B pressures. ΔA is the external pressure required on the membrane in free air to collapse it against its seating, in overcoming the membrane stiffness. It is determined through applying suction to the membrane. ΔB is the internal pressure that in free air would lift the center of the membrane by 1.1 mm

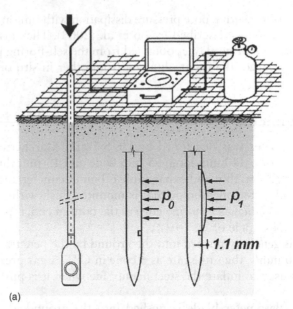

(a)

(b)

Figure 4.4 (a) Photograph of a dilatometer blade and the steel membrane; and (b) Schematic diagram of a dilatometer test (adapted from Marchetti et al. 2001).

from its seating, thus overcoming the membrane stiffness. It is obtained by pressurizing the membrane in the air.

In softer soils, the equations suggested by Schmertmann (1986) will give more realistic values. They are:

$$p_0 = 1.05(A + \Delta A - z_m) - 0.05(B - \Delta B - z_m) \tag{4.5}$$

$$p_1 = B - \Delta B - z_m \tag{4.6}$$

Here, z_m is the zero reading of the pressure gauge, which is generally zero for a new gauge. In free draining soils, the corrected equilibrium pressure, p_2, is given approximately by

$$p_2 = C + \Delta A - z_m \tag{4.7}$$

The seismic flat dilatometer was introduced in 2006, with two geophones located above the blade, 500 mm apart. When a plate is struck at the ground level, the signals are received at different times by the geophones. These data are used to compute the shear wave velocity v_s with 1–2% repeatability, and its variation with depth can be established. From the shear wave velocity, the small strain shear modulus G_0 can be computed as

$$G_0 = \rho V_s^2 \tag{4.8}$$

In addition, SDMT gives all other parameters derived from DMT.

Mayne et al. (2009) categorized all the in situ tests as 'old' and 'new' methods, with seismic cone penetration tests with pore pressure measurements (SCPTu) and seismic dilatometer tests (SDMT) falling under the new methods. A DMT can be used in very soft to very stiff clays or marls, with c_u = 2–1000 kPa.

4.2.4.1 Intermediate DMT parameters

The interpretation of dilatometer test data is essentially empirical, based on the two corrected pressures p_0 and p_1. Here p_0 is the corrected pressure A and p_1 is the corrected pressure B. The *three* intermediate parameters that are used for deriving the other soil parameters are the *material index I_D*, *horizontal stress index K_D* and *dilatometer modulus E_D*. They are computed empirically.

Material index I_D is defined as:

$$I_D = \frac{p_1 - p_0}{p_0 - u_0} \tag{4.9}$$

where, u_0 is the hydrostatic pore water pressure that was present prior to insertion of the blade. The material index, which is low for clays (<0.6), medium for silts (0.6–1.8) and high for sands (>1.8), is used to identify the soil. It is typically in a range of 0.1 to 10.

Horizontal stress index K_D is defined as:

$$K_D = \frac{p_0 - u_0}{\sigma'_{vo}} \qquad (4.10)$$

where, σ'_{vo} is the effective in situ overburden stress. K_D can be seen as the K_0 (coefficient of earth pressure at rest) amplified by the penetration (Marchetti et al. 1994). It is typically about 2 in normally consolidated clays, which is significantly larger than K_0. In overconsolidated soils, K_D is greater than 2. The horizontal stress index is used to determine horizontal stress and hence K_0, OCR and undrained shear strength (c_u) in clays and effective friction angle (ϕ') in sands. K_D is a measure of the soil's resistance to volume reduction, and is highly sensitive to aging and is sometimes called the stress history index.

The dilatometer modulus E_D is obtained from elastic analysis by relating the membrane displacement s_0 to the pressure difference $\Delta p = p_0 - p_1$. Assuming an elastic half space surrounding the membrane, using the theory of elasticity, the displacement s_0 can be expressed as (Marchetti 1980):

$$s_0 = \frac{2D(p_1 - p_0)}{\pi} \frac{1 - v^2}{E} \qquad (4.11)$$

where E = Young's modulus, D = membrane diameter and v = Poisson's ratio. For membrane diameter D = 60 mm and s_0 = 1.1 mm, Eq. (4.11) becomes

$$E_D = \frac{E}{1 - v^2} = 34.7(p_1 - p_0) \qquad (4.12)$$

The dilatometer modulus is used in determining the constrained modulus and hence modulus of elasticity. It must be noted that the soil is loaded laterally in determining the dilatometer modulus. In reality, the soil modulus is usually required for vertical loading. E_D computed from Eq. (4.12) is drained in sand, undrained in clays and is partially drained in sand-clay mixtures.

4.2.4.2 Soil type interpretation

One of the main uses of the dilatometer test is in identifying the soil type. Figure 4.5 shows the soil identification chart based on the material index I_D and the normalised dilatometer modulus E_D, where p_a is atmospheric pressure (101.3 kPa). The original chart was proposed by Marchetti and Crapps

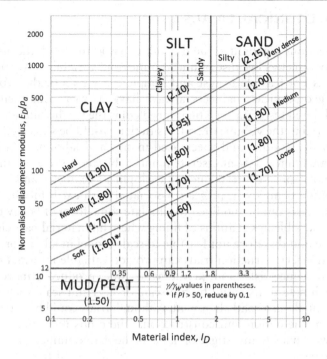

Figure 4.5 Soil identification chart (after Schmertmann 1986).

(1981), which was modified slightly by Schmertmann (1986). The chart also gives an estimate of the unit weights.

4.3 LABORATORY TESTING

Laboratory testing is carried out with disturbed or undisturbed samples collected from the field. Disturbed samples are generally used for index tests, moisture content, Atterberg limits and linear shrinkage, etc. However, undisturbed samples are required to obtain strength and stiffness parameters. Depending on the project, investigations carried out and samples collected, in general, a combination of in situ testing and laboratory testing could be adopted to assess soil parameters.

Chapter 3 provides a coverage of mostly used laboratory tests for soft clays, i.e., physical tests, consolidation and shear strength.

4.4 GEOTECHNICAL TRIALS

Previous sections highlighted the difficulties associated with the assessment of design parameters because of uncertainties and variability in ground conditions. It is not feasible to test and sample every inch of the ground and therefore geotechnical engineers always have to carry out their designs based on limited test data. The main problem is the small volume of soil we test compared to the volume of soil in situ that would be subjected to development loads. Routine and specialized in situ and laboratory testing can provide only limited data at discrete points subjected to stresses and strains that may not replicate effects due to development geometry and loads. This becomes more complex in projects associated with earth fill and rock fill. While some tests can be modified to simulate some form of behavior, generally it is impossible in situations of filling where fill material particles can be large in size and therefore not feasible to scale model in laboratory tests. In such situations, construction trials can provide valuable information to the designer to reduce risks and, at the same time, produce a more efficient design. Such field trials could be full scale or near full scale such as a trial embankment along a road alignment. Although trials could be carried out at any stage of the design and construction, where possible, trials should be included within a site investigation area as the data obtained are very valuable to the designer.

The question arises why such trials are not routinely conducted. This is because of:

- Time - Time for a trial and analysis of observations may be too late for the designer to complete the design before the construction commences. For example, if we have designed a road embankment surcharging for a period of six months, the trial will also need a similar order of time if good data is to be received.
- Cost - As the trial needs to be carried out prior to construction, the contractor and the machinery usually need another mobilization and demobilization, which could be very costly. Costs can be reduced if the trial becomes part of the main construction. A classic example is the location of trial embankment within the proposed road footprint so that it could be incorporated into the final design.

4.5 EMPIRICAL CORRELATIONS

Of the methods available to derive design parameters for soft clays, geotechnical trials are often not carried out because of the expense involved and time limitations. In situ testing and laboratory testing on samples recovered during a site investigation are considered the main weapons available to the practicing engineer to assess geotechnical design parameters for any project.

However, they have their limitations including the cost for site investigations, effect of sample disturbance and the resources and time required to carry out such investigations and testing. The scarcity of geotechnical information is more pronounced during the pre-feasibility, feasibility and preliminary design stages, although cannot be ruled out even during the detailed design stage. At the early stages, projects have very limited budget or no budget at all for site investigations except for a few index testing. Then the geotechnical engineer has the unenviable task of interpreting ground conditions and assessing parameters with very little information. In these instances, empirical correlations become very important and could be the only available tool to the geotechnical designer to set up his geotechnical models for the site. If you take a simple physical test result such as the natural moisture content of a soft clay, as is discussed in Chapter 5, it is remarkable that empirical equations are available to assess both strength and deformation parameters, i.e. consolidation parameters for magnitude and even for the rate of consolidation, and an overall idea of the undrained shear strength.

As Ameratunga et al. (2016) state, in geotechnical engineering, empiricism has a big role to play. In addition to giving preliminary estimates, the correlations can also be used to compare against the values determined from laboratory and in situ tests. There are many empirical equations and graphs available in the literature, which are regularly being used in the designs worldwide. These are derived based on laboratory or field data, past experience and good judgement.

Ameratunga et al. (2016) also provide useful literature easily available to engineers who are interested in empirical relationships:

- Design guidelines in the USA by the Army, Navy and Air Force who have many excellent design guides, which have been researched and tested (e.g. NAVFAC 7.1 – US Navy 1982);
- Canadian Foundation Engineering Manual (Canadian Geotechnical Society 2006), which is used worldwide; and
- Report by Kulhawy and Mayne (1990), which was comprehensive at the time and still popular, even long after it was published. In addition to collating valuable empirical relationships, they proposed their own.

Recent books on correlations include the following:

Ameratunga, J., Sivakugan, N. and Das, B.M. (2016). *Correlations of Soil and Rock Properties in Geotechnical Engineering*. Springer Publishing.

Carter, M. and Bentley, S.P. (2016). *Soil Properties and their Correlations* 2nd edition. John Wiley & Sons, Ltd.

Verbrugge, J.-C. and Schroeder, C. (2018). *Geotechnical Correlations for Soils and Rocks*. John Wiley and Sons.

4.6 SUMMARY

Chapter 4 covers geotechnical testing needed for ground improvement design and construction. It summarizes the available tools from in situ testing and laboratory testing to geotechnical trials, empirical correlations and the observational method and briefly discusses the advantages and disadvantages. Considering most useful parameters for ground improvement design of soft clays are related to strength and compressibility, only the vane shear test, CPT, self-boring pressuremeter and dilatometer are described in detail with a summary of advantages and limitations of each test method. Chapter 3 covers laboratory testing and Chapter 5 provides more insight into empirical correlations. The Observational Method is separately discussed in Chapter 16.

REFERENCES

Ameratunga, J., Sivakugan, N. and Das, B.M. (2016). *Correlations of Soil and Rock Properties in Geotechnical Engineering*. Springer Publishing, India.

ASTM D2573-08. Standard test method for field vane shear test in cohesive soil. Annual Book of ASTM Standards, West Conshohocken, PA.

ASTM D 3441-16. Standard test method for mechanical cone penetration testing of soils. Annual Book of ASTM Standards, West Conshohocken, PA.

ASTM D 5778-12. Standard test method for electronic friction cone and piezocone penetration testing of soils. Annual Book of ASTM Standards, West Conshohocken, PA.

Baguelin, F., Jézéquel, J.F., Le Mée, E., and Le Méhauté, A. (1972). 'Expansion of cylindrical probes in cohesive soils.' *Journal of Soil Mechanics and Foundation Engineering Division*, ASCE, 98(SM11), 1129–1142.

Baguelin, F., Jézequél, J.F., and Shields, D.H. (1978). *The Pressuremeter and Foundation Engineering*. Trans Tech Publications, Germany.

Bjerrum, L. (1972). 'Embankments on soft ground.' *Proceedings Performance of Earth and Earth-Supported Structures*, Vol 2. ASCE, Lafayette, Ind., 1–54.

Bo, M.W., Chu, J., Low, B.K., and Choa, V. (2003). *Soil Improvement – Prefabricated Vertical Drain Techniques*. Thomson, Singapore.

Canadian Geotechnical Society (2006). *Canadian Foundation Engineering Manual*, 4th edition, Canadian Geotechnical Society, 503 p.

Carter, M. and Bentley, S.P. (2016). *Soil Properties and their Correlations*, 2nd edition. John Wiley & Sons, Ltd, Chichester, UK.

Das, B.M. and Sivakugan, N. (2017). *Fundamentals of Geotechnical Engineering*, 5th edition. Cengage, Boston.

Kulhawy, F.H. and Mayne, P.W. (1990). *Manual on Estimating Soil Properties for Foundation Design*, Report EL-6800, Electric Power Research Institute, Palo Alto, CA.

Lunne, T., Robertson, P.K., and Powell, J.J.M. (1997). *Cone Penetration Testing in Geotechnical Practice*. Blackie Academic/Chapman-Hall Publishers, London, UK, 312p.

Marchetti, S. (1980). 'In situ tests by flat dilatometer.' *Journal of Geotechnical Engineering Division, ASCE,* 106(GT3), 299–321.

Marchetti, S. and Crapps, D.K. (1981). *Flat Dilatometer Manual, DMT Operating Manual.* GPE Inc., Gainesville, FL.

Marchetti, S., Monaco, P., Totani, G. and Calabrese, M. (1994). 'The flat dilatometer: Design applications.' *Proceedings of the 3rd Geotechnical Engineering Conference,* Cairo University, Keynote lecture, 26p.

Marchetti, S., Monaco, P., Totani, G. and Calabrese, M. (2001).'The flat dilatometer test (DMT) in soil investigations'. A report by the ISSMGE Committee TC16. In: Proc. In Situ 2001, Intl. Conf. on In Situ Measurement of Soil Properties, Bali, Indonesia.

Mayne, P.W. (2007). 'Cone penetration testing State-of-Practice.' *NCHRP Synthesis. Transportation Research Board Report Project* 20-05, 118p.

Mayne, P.W., Coop, M.R., Springman, S.M., Huang, A.B. and Zornberg, J.G. (2009). 'State-of-the-art paper SoA-1, Geomaterial behaviour and testing.' *Proceedings of 17th International Conference on Soil Mechanics and Geotechnical Engineering,* Alexandria, 42777–42872.

Powrie, W. (2013). *Soil Mechanics Concepts and Applications,* 3rd edition. CRC Press, London.

Robertson, P.K. (1990). 'Soil classification using the cone penetration test.' *Canadian Geotechnical Journal,* 27(1), 151–158.

Robertson, P.K. (2010). 'Soil behaviour type from the CPT: An update.' *2nd International Symposium on Cone Penetration Testing, CPT'10,* Huntington Beach, CA.

Robertson, P.K. and Cabal, K.L. (2015). *Guide to Cone Penetration Testing,* 6th edition. Gregg Drilling & Testing, Inc., California.

Robertson, P.K., Campanella, R.G., Gillespie, D. and Greig, J. (1986). 'Use of piezometer cone data.' *Proceedings ASCE Specialty Conference In Situ'86 Use of In Situ Testing in Geotechnical Engineering,* 1263–1280.

Schmertmann, J.H. (1986). 'Suggested method for performing flat dilatometer test.' *Geotechnical Testing Journal,* 9(2), 93–101.

Schnaid, F. (2009). *In Situ Testing in Geomechanics, The Main Tests.* Taylor & Francis Group, London and New York.

Sivakugan, N. and Das, B.M. (2010). *Geotechnical Engineering: A Practical Problem Solving Approach.* J. Ross Publishing, Florida.

Teh, C.I. and Houlsby, G.T. (1991). 'An analytical study of the cone penetration test in clay.' *Geotechnique,* 41(1), 17–34.

US Navy (1982). *Soil Mechanics – Design Manual 7.1,* Department of the Navy, Naval Facilities Engineering Command, US Government Printing Office, Washington, DC.

Verbrugge, J.-C. and Schroeder, C. (2018). *Geotechnical Correlations for Soils and Rocks.* Wiley-ISTE, London.

Wroth, C.P. and Hughes, J.M.O. (1973). 'An instrument for the in situ measurement of the properties of soft clays.' *Proceedings 8th International Conference on Soil Mechanics and Foundation Engineering, Moscow,* Vol. 1.2, 487–494.

Chapter 5

Parameter derivation

J. Ameratunga and N. Sivakugan

5.1 INTRODUCTION

Any ground improvement method needs a design, a concept, preliminary design and the final design before drawings for construction are issued. Depending on the type of ground improvement method, some of the above stages may not be required. For example, in the case of a simple shallow excavation and replacement, it may only need a concept that is likely to be unchanged to the final. However, if there is infrastructure nearby that could be affected, a lot more work would be necessary to advance it to the final design because of the effects (e.g., additional settlements of nearby buildings), which need to be assessed and remedial measures identified in addition to methodology for construction.

Soil parameters play a major role in the design development. They provide the designer the key tools in their assessment to initially assess whether the ground improvement would work and later, to design it to completion. It is the geotechnical designer's responsibility to analyze the data available, sometimes a large amount and sometimes highly variable, to come up with geotechnical models that could be used for the design. In some instances, based on the risks and consequences of failure, sensitivity of parameters also needs to be considered. At the end of the day, the design needs to be safe in the short term as well as in the long term and ensure that the performance criteria discussed in Chapter 1 are met.

Parameters generally required for soft clay engineering can be broadly classified as follows:

- Physical characteristics
- Strength
  - Drained strength (c' and ϕ')
  - Undrained strength (c_u and ϕ_u)
- Deformation
  - Drained modulus (E_d or E') and undrained modulus (E_u)

o   Consolidation characteristics related to magnitude (m_v, C_c, C_r, CR, RR, C_a, C_{ae}, OCR)
o   Consolidation characteristics related to rate of consolidation (c_v, c_h).

5.2 GEOTECHNICAL TOOLS AVAILABLE

Geotechnical tools available to the designer to carry out ground improvement can be summarized as follows:

- In situ testing– See Chapter 4
- Laboratory testing– See Chapter 4
- Geotechnical trials – See Chapter 4
- Empirical correlations – Current chapter
- Observational method – See Chapter 16.

5.3 STRENGTH PARAMETERS RELEVANT TO SOFT CLAYS

The strength parameters relevant to soft clay engineering could be summarized as follows.

- Drained shear strength (c' and ϕ'); and
- Undrained shear strength (c_u and ϕ_u).

In soft clay analysis, drained parameters are generally related to long-term behavior and to effective stress analysis. Opposite to that is the short-term behavior that is controlled by undrained strength parameters. However, in soft clay engineering, generally soft soils mostly occur below the groundwater table and are therefore fully saturated. Hence, the undrained friction angle, ϕ_u, will be zero. Throughout this book, soft soils are assumed to be saturated and therefore $\phi_u = 0$.

5.3.1 Drained shear strength parameters

Drained strength testing is traditionally carried out using direct shear or triaxial tests on undisturbed samples collected from the site. Drained shear strength is usually defined using the Mohr-Coulomb strength parameters c' and ϕ' with an assumed linear relationship between peak shear strength and normal stress. It should be noted that effective strength envelopes for soft soils are not necessarily linear and that a different strength will be obtained for peak strength, post-peak critical state and residual strength conditions as discussed in Chapter 3.

The drained shear strength parameters of soft clay can be obtained from several methods:

- Laboratory tests (direct shear, triaxial) – Section 5.3.1.1;
- Empirical relationships – Section 5.3.1.2; and
- In situ tests – Section 5.3.1.3.

5.3.1.1 Derivation of drained shear strength parameters from laboratory tests

Derivation of the drained shear strength from laboratory tests is discussed in Chapter 3.

5.3.1.2 Derivation of drained shear strength parameters from empirical correlations

5.3.1.2.1 Drained friction angle, ϕ'

The drained friction angle could have different definitions such as ϕ'_{cv} or ϕ'_{peak} as discussed in Chapter 3. Some of the empirical correlations for the different friction angles are summarized in Table 5.1 and definitions are provided in Chapter 3, Section 3.7.2.

Castellanos and Brandon (2013) showed from an extensive database of tests conducted on riverine and lacustrine alluvial intact specimens from New Orleans, USA, that the effective friction angle from the CU

Table 5.1 Empirical equations for friction angle ϕ' for clays

Equation	Remarks	Reference
$\phi'_{peak} = 39 - 11 \log PI$	NC clays; lower bound (conservative)	Sorensen and Okkels (2013)
$\phi'_{peak} = 43 - 10 \log PI$	NC clays; mean (best estimate)	
$\phi'_{peak} = 45 - 14 \log PI$	OC clays with $4 < PI < 50$; mean (best estimate)	
$\phi'_{peak} = 26 - 3 \log PI$	OC clays with $50 < PI < 150$; mean (best estimate)	
$\sin \phi'_{cv} = 0.8 - 0.094 \ln PI$	NC clays	Mitchell (1976); Kulhawy and Mayne (1990)
$\phi'_{IC(CU)} = 45 - \dfrac{PI}{0.5 + 0.04 PI}$	CU triaxial tests on undisturbed riverine and lacustrine alluvial deposits	Castellanos and Brandon (2013)
$\phi'_{ds} = 31 + 0.0017 PI^2 - 0.3642 PI$	Direct shear tests on undisturbed riverine and lacustrine alluvial deposits	

(NC – Normally Consolidated; OC – Overconsolidated; IC(CU) – Isotropically consolidated undrained)

(consolidated undrained) triaxial test is significantly greater than the ones from consolidated drained (CD) direct shear tests, and that they both decrease with increasing *PI*. The relationships are also given in Table 5.1.

5.3.1.2.2 Drained cohesion, c′

In terms of effective stresses, the failure envelope generally passes through the origin in the τ-σ' plane for most normally consolidated soils, suggesting $c' = 0$. Only in the case of cemented soils, partially saturated soils and heavily overconsolidated soils, can there be some effective cohesion. For uncemented soils including clays, the shear strength in terms of effective stresses is purely frictional. Based on the Danish code of practice for foundations, Sorensen and Okkels (2013) suggest that a conservative estimate of c' for overconsolidated clays can be obtained from

$$c' = 0.1 \ c_u \tag{5.1}$$

They also suggest that c' is poorly correlated to *PI*. With significant scatter, they suggested the following relationships for the lower bound value for c' in terms of *PI*.

$$c'(\text{kPa}) = 30 \quad \text{for } 7 < PI < 30 \tag{5.2a}$$

$$c'(\text{kPa}) = 48 - 0.6PI \quad \text{for } 30 < PI < 80 \tag{5.2b}$$

$$c'(\text{kPa}) = 0 \quad \text{for } PI > 80 \tag{5.2c}$$

Australian Standards for retaining walls (AS 4678) suggests the values for c' and ϕ' in Table 5.2.

Table 5.2 Typical values of c' and ϕ' (after AS 4678-2002)

Soil group	Typical soils in group	Soil parameters	
		c′ (kPa)	φ′ (deg.)
Poor	Soft and firm clay of medium to high plasticity; silty clays; loose variable clayey fills; loose sandy silts	0 to 5	17 to 25
Average	Stiff sandy clays; gravelly clays; compact clayey sands and sandy silts; compacted clay fills	0 to 10	26 to 32
Good	Gravelly sands, compacted sands, controlled crushed sandstone and gravel fills, dense well-graded sands	0 to 5	32 to 37
Very good	Weak weathered rock, controlled fills of road base, gravel and recycled concrete	0 to 25	36 to 43

For natural, intact, normally consolidated clays, ϕ' can vary from less than 20° to little more than 30°. For compacted clays, ϕ' is typically in the range of 25° to 30°, but can be slightly higher.

5.3.1.3 Derivation of drained shear strength parameters from in situ tests

The dilatometer can be used to obtain drained shear strength parameters for soft clay. Marchetti (1997) suggested an approximate equation for the effective friction angle, which he believes is a lower bound value that underestimates the in situ friction angle by 2–4°. The equation is

$$\phi'\left(\text{degrees}\right) = 28 + 14.6\log K_D - 2.1\left(\log K_D\right)^2 \tag{5.3}$$

where K_D is the horizontal stress index (defined in Chapter 4).

From tests conducted in ML(silt) and SP-SM (sand-silty sand) soils in Venice, Ricceri et al. (2002) suggested an upper bound as

$$\phi'\left(\text{degrees}\right) = 31 + \frac{K_D}{0.236 + 0.066K_D} \tag{5.4}$$

5.3.2 Undrained shear strength

Undrained shear strength (c_u) is the most important parameter required to assess short-term stability of embankments, and the design of several ground improvement methods. There are many methods that could be used to assess this important parameter, some are direct while some are indirect.

5.3.2.1 Direct methods to derive c_u

Direct methods include:

- Pocket penetrometer (not recommended for soft clays because it is a crude device), a laboratory vane or Torvane (only preliminary estimate possible);
- Laboratory direct shear and/or triaxial tests;
- Field vane shear test – very popular.

You could also use your fingers to make an assessment of consistency in soft clay, especially in the field, using the guidelines provided by Australian Standard 1726, Canadian Geotechnical Society (2006), etc. (see Table 5.3). All the above methods test a very small volume of soil and therefore are not considered accurate enough for soft clay design.

Table 5.3 Consistency terms for cohesive soils

Consistency*	c_u (kPa)	$(N_1)_{60}$	#q_c/p_a	##, **Field identification guide
Very soft	<12	0–2	<5	Exudes between fingers when squeezed in hand; easily penetrated several centimeters by fist
Soft	12–25	2–4		Can be molded by light finger pressure; easily penetrated several centimeters by thumb
Firm	25–50	4–8	5–15	Can be molded by strong finger pressure; can be penetrated several centimeters by thumb with moderate effort
Stiff	50–100	8–15	15–30	Cannot be molded by fingers; can be indented by thumb but penetrated only with great effort
Very stiff	100–200	15–30	30–60	Readily indented by thumb nail
Hard	>200	>30	>60	Can be indented by thumbnail with difficulty

* After Terzaghi and Peck (1967); **## AS 1726-2017, Canadian Geotechnical Society (2006); #McCarthy (2007)

Laboratory tests: Tests such as direct shear tests and triaxial tests are most useful to obtain shear strength of soft clays. These are discussed in detail in Chapter 3.

Field vane test: This is most popular to assess the shear strength of soft clays because it is an in situ test and therefore sample disturbance in collection, transportation and extrusion does not come into play. The method, its advantages and disadvantages, and field corrections are discussed in detail in Chapter 4.

5.3.2.2 Indirect methods to derive c_u

Indirect methods include some form of the use of empirical methods generally with in situ tests such as the CPT, self-boring pressuremeter and dilatometer discussed in this book.

5.3.2.2.1 Derivation of c_u from CPT

CPT provides a continuous resistance profile that is measured as a cone resistance (q_c), which is corrected to provide a corrected resistance (q_t) as discussed in Chapter 4. From the q_t profile, c_u could be obtained by the following equation where σ_{vo} is the total overburden pressure at the depth of testing and q_t is the corrected cone resistance at the same depth and the cone factor factor is N_{kt}:

$$c_u = \frac{q_t - \sigma_{vo}}{N_{kt}}$$

(5.5)

The factor N_{kt} is somewhat empirical but could be obtained by calibrating against laboratory tests such as triaxial or direct shear tests or in situ tests such as the vane shear test. Generally, the N_{kt} factor varies between 10 and 20 but literature has reported values outside this range. If a sample is collected adjacent to a CPT test location and tested in the laboratory, or a vane shear test is carried out next to the CPT, c_u obtained could be used with Eq. (5.5) to obtain N_{kt}. If there is no other data available to calibrate, a value of the order of 15 could be selected with sensitivity studies carried out over, say, ±20%.

To calibrate the vane shear test results and obtain the relevant N_{kt} values, the following process could be followed:

i. Carry out CPTs.
ii. Carry out a vane shear test adjacent to a CPT location to assess c_u at a specific depth. Generally, the Bjerrum correction factor for plasticity (Bjerrum, 1972) should be applied. According to Hight and Leroueil (2003: 257), 'the Bjerrum correction factor applied to vane strength was intended for use in the stability analysis of single stage embankment construction on non-stratified clays, assuming circular failure modes. It was not intended as a general correction factor to be applied, for example, to the strength of soft clays in the design of basically reinforced embankments'.
iii. Alternatively, carry out a laboratory test such as triaxial or direct shear test if an undisturbed sample is available.
iv. Use the c_u value thus obtained with Eq. (5.5) to obtain N_{kt} as the q_t at the depth of testing could be obtained from the CPT q_t profile.
v. Carry out further calibrations depending on the time and budget available.
vi. Use an appropriate value for N_{kt} based on calibration results to obtain c_u profiles from the CPT profiles.

If several calibrations were carried out it would be expected that the N_{kt} values to vary somewhat. The geotechnical engineer should select an average value or a reasonably conservative value for the design progress. If the N_{kt} values obtained for the same site show a wide variation, different N_{kt} factors could be applied to different areas of the site. It is also possible that N_{kt} values could be different for different layers of the soil profile.

5.3.2.2.2 Derivation of c_u from self-boring pressuremeter

The theory of cavity expansion (B-L-P theory) controlling pressuremeter behavior was developed in 1972 independently by Baguelin et al., Ladanyi and Palmer. The B-L-P theory has remained the same over 50 years and provides the tool to derive the undrained shear strength of saturated cohesive soils. The theory provides not just one value such as the peak strength in a vane shear test but the complete stress-strain curve. The theory, as is the case of most derivations for laboratory and in situ tests conducted in soils, is applicable to isotropic, homogeneous soils and assumes plane strain conditions. The latter assumption is reasonable considering the high length to diameter ratio of the probe, and hence the strains are confined to the horizontal planes. The theory behind the testing is provided by Baguelin et al. (1972), Ladanyi (1972) and Palmer (1972). The theory was developed with a simple assumption that the soil elements behave in the same manner and have a unique stress-strain curve for small strains and can be written as:

$$c_u = \frac{\sigma_r - \sigma_\theta}{2} = \varepsilon_o \frac{dp}{d\varepsilon_o} \tag{5.6a}$$

or

$$\frac{\sigma_r - \sigma_\theta}{2} = \left(\frac{\Delta V}{V}\right) \frac{dp}{d\left(\ln \frac{\Delta V}{V}\right)} \tag{5.6b}$$

where
c_u = undrained shear strength
$(\sigma_r - \sigma_\theta)/2$ = shear stress; σ_r = radial stress and σ_θ = circumferential stress
ε_o = cavity strain at borehole wall
p = applied pressure at borehole wall
$\Delta V/V$ = volumetric strain

The above equations provide the full stress strain curve as ε_0 or $\Delta V/V$ is varied (see Figure 5.1).

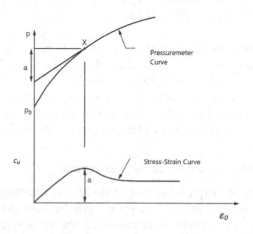

Figure 5.1 Undrained pressuremeter curve and the derived stress-strain curve (Adapted from Ameratunga et al., 2016).

If the designer is only interested in the peak shear strength value rather than the full stress-strain curve, Gibson and Anderson's (1961) equation given below could be used where it assumes the peak shear strength occurs at infinite expansion ($\Delta V/V = 1$) at the limit pressure.

$$p_L - p_o = c_u \left[1 + \ln\left(\frac{G}{c_u}\right) \right] \tag{5.7}$$

If

$$N_c = 1 + \ln\left(\frac{G}{c_u}\right) \tag{5.8}$$

Eq. (5.7) could be simplified to the following:

$$p_L - p_o = N_c c_u \tag{5.9}$$

As infinite expansion is not practical, it is standard practice to modify the above equation to obtain the following where a new limit pressure is defined as p_l, which is the pressure when the initial volume of the probe is doubled i.e., $\Delta V/V = 0.5$.

$$p_l - p_o = N_c^* c_u \tag{5.10}$$

Clarke (1995) provides empirical relationships to assess c_u based on p_l and are presented in Table 5.4.

Schnaid (2009) states that a word of caution is necessary when deriving c_u from self-boring pressuremeters. He states that there is sufficient evidence suggesting c_u obtained from pressuremeter to be higher than those assessed from other in situ and laboratory tests.

5.3.2.2.3 Derivation of c_u from dilatometer

According to Marchetti (1980), for uncemented clays with $I_D < 1.2$

$$OCR = (0.5 K_D)^{1.56} \tag{5.11}$$

Table 5.4 Empirical relationships between c_u and p_l (adapted from Clarke, 1995)

c_u	Clay type	Reference
$(p_l - p_o)/5.5$	Soft to firm clays	Cassan (1972); Amar and Jézéquel (1972)
$(p_l - p_o)/8$	Firm to stiff clays	
$(p_l - p_o)/5.1$	All clays	Lucas and LeClerc de Bussy (1976)
$(p_l - p_o)/5.5$	$c_u < 55$ kPa	Amar and Jézéquel (1972)

This was verified experimentally by Kamei and Iwasaki (1995) and theoretically by Finno (1993). Kulhawy and Mayne (1990) showed test data suggesting that the coefficient 0.5 in Eq. (5.11) could vary over a range of 0.27–0.75, depending on the degree of fissuring, sensitivity and geologic origin.

Noting that

$$\left(\frac{c_u}{\sigma'_{v0}}\right)_{OC} = \left(\frac{c_u}{\sigma'_{v0}}\right)_{NC} OCR^{0.8} \tag{5.12}$$

and assuming $\left(\frac{c_u}{\sigma'_{v0}}\right)_{NC} = 0.22$ \hfill (5.13)

as suggested by Mesri (1975), Marchetti (1980) proposed that Eq. (5.12) and Eq. (5.13) be re-written in the following form:

$$\left(\frac{c_u}{\sigma'_{v0}}\right)_{OC} = \left(\frac{c_u}{\sigma'_{v0}}\right)_{NC} (0.5K_D)^{1.25} \sim 0.22(0.5K_D)^{1.25} \tag{5.14}$$

Eq. (5.14) could be used to derive the c_u of the clay. Kulhawy and Mayne (1990) proposed that the application of this equation should be limited to $I_D \leq 0.6$. Similar derivations have been suggested by others including the following by Iwasaki and Kamei (1994):

$$c_u = 0.118E_D \tag{5.15}$$

$$\left(\frac{c_u}{\sigma'_{v0}}\right)_{OC} \sim 0.35(0.47K_D)^{1.14} \tag{5.16}$$

5.3.2.2.4 Empirical relationships

c_u/σ'_{v0} of normally consolidated clays in situ generally varies in the range of 0.2 to 0.3. Skempton (1957) suggested that for normally consolidated clays, based on vane shear test data,

$$\left(\frac{c_u}{\sigma'_{v0}}\right) = 0.0037PI + 0.11 \tag{5.17}$$

For overconsolidated clays, this ratio is larger, and it increases with the overconsolidation ratio. Ladd et al. (1977) showed that

$$\left(\frac{c_u}{\sigma'_{v0}}\right)_{OC} = \left(\frac{c_u}{\sigma'_{v0}}\right)_{NC} OCR^{0.8} \tag{5.18}$$

Jamiolkowski et al. (1985) suggested that for clays of low to moderate plasticity index

$$\left(\frac{c_u}{\sigma'_{v0}}\right)_{OC} = (0.23 \pm 0.04)OCR^{0.8} \tag{5.19}$$

For overconsolidated clays of low to moderate plasticity, the above equation can also be approximated as (Jamiolkowski et al. 1985)

$$\left(\frac{c_u}{\sigma'_p}\right)_{OC} = (0.23 \pm 0.04) \tag{5.20}$$

where, σ'_p is the preconsolidation pressure. Eq. (5.20) is in good agreement with Mesri's (1989) suggestion that $c_u/\sigma'_p = 0.22$.

5.3.3 Sensitivity and effects of disturbance

Soil sensitivity is an important consideration in soft clays subjected to ground improvement. It is the ratio of the undisturbed to disturbed shear strength and can be approximated as the ratio of the peak shear strength to the residual shear strength. The disturbance created by a ground improvement technique could result in worse conditions than natural, which needs to be considered in the ground improvement design. For example, when a wick drain mandrel is driven through soft clay, if the clay is sensitive it will create a disturbed zone having significantly less dissipation properties.

The level of sensitivity observed in the clays varies geographically. Significantly greater values of sensitivity have been reported from Scandinavian countries, compared to those from Canada or USA. As a result, there are slightly different classification scales, which are shown in Table 5.5.

High sensitivity is generally associated with high liquidity index. Scandinavian clays have a liquidity index significantly larger than 1.0.

Table 5.5 Sensitivity classification

Description	Sensitivity, S_t		
	US	Canada*	Sweden
Low sensitive	2–4	1–2	<10
Medium sensitive	4–8	2–4	10–30
Highly sensitive	8–16	4–8	30–50
Extra sensitive	16	8–16	50–100
Quick	–	>16	>100

* Canadian Geotechnical Society (2006)

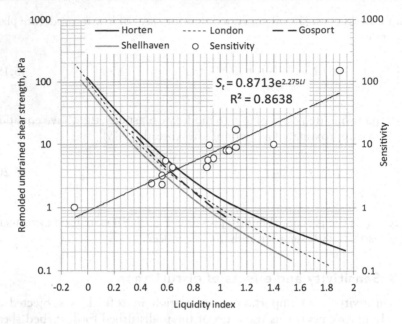

Figure 5.2 Remolded undrained shear strength, liquidity index and sensitivity relationship (adapted from Skempton and Northey 1952).

Skempton and Northey (1952) summarized some sensitivity – liquidity index (*LI*) data for some clays of moderate sensitivity, which are shown in Figure 5.2. The trend of sensitivity increasing with liquidity index is clear and can be related by

$$S_t = 0.87\exp(2.28LI) \tag{5.21}$$

It should also be noted that soft clays could be sensitive to mechanical disturbance either from sampling and handling, or from construction processes (e.g., stone column insertion) or due to stress relief. For highly sensitive clays, more attention should be given to the effects of disturbance.

5.4 SOIL STIFFNESS

In computing deformations under undrained loading using numerical models or otherwise, it is necessary to know the undrained Young's modulus, E_u, which can be obtained directly from laboratory tests.

E_u is also generally estimated from an appropriate value of the modulus ratio E_u/c_u, which is generally in the range of 100–1000. It can be derived from Figure 10.4 (in Chapter 10) proposed by Duncan and Buchignani (1976) and the US Army (1994).

Table 5.6 Typical values of E_u for clays (after US Army 1994)

Clay	E_u (MPa)
Very soft clay	0.5–5
Soft clay	5–20
Medium clay	20–50
Stiff clay, silty clay	50–100
Sandy clay	25–200

Typical values of E_u for different clay types, as recommended by the US Army (1994) are given in Table 5.6.

The drained Young's modulus (E') and undrained Young's modulus (E_u) are related (theoretically) by the following equations:

$$Shear\ Modulus = G = \frac{E_u}{2(1+v_u)} \tag{5.22}$$

$$Shear\ Modulus = G = \frac{E'}{2(1+v')} \tag{5.23}$$

By equating the expressions on the right hand sides of Eqs. (5.22) and (5.23), it is possible to relate undrained and drained moduli in terms of the Poisson's ratios. Undrained Poisson's ratio v_u can be taken as 0.5.

Drained Young's modulus (E') could also be derived from m_v obtained from the consolidation test using the equation below and discussed in the following sub-section (where $= m_v$ is the coefficient of volume compressibility and $v' =$ drained Poisson's ratio):

$$E' = \frac{(1+v')(1-2v')}{(1-v')m_v} \tag{5.24}$$

According to Look (2014), v' could vary between 0.25 and 0.4 for clays.

As mentioned in Section 3.6.2.1, it should be noted that m_v is stress dependent. This is further discussed in Section 5.5.1.1.

5.5 CONSOLIDATION PARAMETERS

As previously discussed, consolidation parameters needed for engineering design could either be related to magnitude or rate of settlement.

The magnitude related parameters are:

- m_v, in the NC state or CR $\{= C_c/(1+e_0)\}$
- m_v, in the OC state or RR $\{= C_r/(1+e_0)\}$

- $C_{\alpha\varepsilon} \{= C_\alpha/(1+e_0)\}$
- OCR

The parameter related to the rate of settlement is:

- c_v, c_h coefficient of consolidation (vertical, horizontal).

Generally, back analysis of trial embankments or monitoring during actual construction using settlement and excess pore pressure data should provide the most appropriate parameters related to the specific site. However, during the design phase, such information is not available for most projects. Therefore, designers have to rely on mostly laboratory and in situ test results as well as empirical relationships.

5.5.1 Consolidation parameters – magnitude

5.5.1.1 Primary consolidation

In the assessment of the primary consolidation, the following parameters are needed:

- m_v or
- C_c and e_0 {or $CR= C_c/(1+ e_0)$}

Table 5.7 shows a simple classification of clays based on m_v. Some value ranges for m_v of different soils, suggested by Domenico and Mifflin (1965), are shown in Table 5.8.

Table 5.7 Classification of clays based on m_v (after Bell 2000)

Type of soil	m_v (MPa^{-1})	Compressibility
Heavily overconsolidated clays	<0.05	Very low
Very stiff or hard clays, tills	0.05–0.10	Low
Varved and laminated clays, firm to stiff clays	0.10–0.30	Medium
Normally consolidated alluvial clays	0.3–1.5	High
Organic alluvial clays and peats	>1.5	Very high

Table 5.8 Ranges of m_v values for different soils

Soil type	m_v (MPa^{-1})
Plastic clay	0.26 to 2.1
Stiff clay	0.13 to 0.26
Medium hard clay	0.069 to 0.13
Water	0.44×10^{-3}

Source: After Domenico and Mifflin 1965.

There are many regional and global empirical correlations reported in the literature, which were summarized by Ameratunga et al. (2016). For normally consolidated clays, it can be shown theoretically that C_c and m_v are related by

$$m_v = \frac{0.434 C_c}{(1 + e_0) \sigma'_{average}}$$
(5.25)

where $\sigma'_{average}$ is the average value of the vertical normal stress during consolidation. The same equation can be used for overconsolidated clays, by replacing C_c with C_r.

It is generally accepted that the consolidation test is the best available tool to obtain consolidation parameters related to magnitude, i.e., m_v, C_c, C_r, CR or RR. However, the use of empirical relationships is widespread either because of limited investigations carried out, accuracy of the results and/or to check reasonableness of the parameters obtained.

Compression index, C_c, is strongly correlated with the natural water content, initial void ratio and liquid limit of the clay. It can be estimated as 1% of the natural water content (Koppula, 1981), i.e.,

$$C_c = 0.01 w_n$$
(5.26)

where, w_n is the water content in %. This relationship has been found to be applicable for many other clays including for Chicago clays (Holtz and Kovacs, 1981).

For saturated soils, assuming the specific gravity (G_s) of clay to be 2.7, Eq. (3.1) in Chapter 3 would allow the following relationship between C_c and e_0:

$$C_c = 0.37 e_0$$
(5.27)

According to Kulhawy and Mayne (1990), although hundreds of relationships are found in literature, correlations based on natural moisture content (w_n) are better than correlations based on LL or e_0. Balasubramaniam et al. (2010) cites four such relationships in Table 5.9.

A comprehensive list of empirical correlations for C_c and CR is found in Ameratunga et al. (2016).

Table 5.9 Compression ratio (CR) from natural water content (w_n)

Author	Formula	Range of w_n
Simons and Menzies (1975)	$CR = 0.006 w_n - 0.03$	$20 \leq w_n \leq 140$
Simons (1957)	$CR = 0.006 w_n^{1.68}$	$28 \leq w_n \leq 57$
Wilkes (1974)	$CR = 0.26 \ln(w_n) - 0.83$	$30 \leq w_n \leq 90$
Lambe and Whitman (1969)	$CR = 0.12 \ln(w_n) - 0.28$	$10 \leq w_n \leq 100$

Practicing engineers should always check reasonableness of parameters derived from in situ or laboratory tests, or empirical relationships. Eq. (5.28) proposed by Wroth and Wood (1978) could be easily used as a sanity check.

$$C_c = G_s \times \frac{PI}{200} \qquad (5.28)$$

where G_s is the specific gravity and PI is the plasticity index of the soil. Assuming G_s is 2.7, then

$$C_c = 1.35 \times \frac{PI}{100} \qquad (5.29)$$

While C_c typically varies in the range of 0.2–1.0, it can exceed 1.0 for soft or sensitive clays. C_r is generally in the range of 1/5-1/10 of C_c. In critical state soil mechanics using models such as Cam Clay, the natural logarithmic (ln) axis is used for the effective stress instead of the traditional base 10 logarithmic (log) axis. This makes the slopes of the virgin compression line and swelling line λ and κ, respectively. They are related to C_c and C_r by

$$C_c = \lambda \ln 10 = 2.3026 \lambda \qquad (5.30)$$

$$C_r = \kappa \ln 10 = 2.3026 \kappa \qquad (5.31)$$

The *compression ratio* (CR) or *modified compression index* (C_{ce}) is also widely used in computing consolidation settlements. It is similar to C_c, and is the slope of the virgin compression line when the *vertical normal strain* (instead of void ratio) is plotted against the logarithm of effective normal stress. It is defined as $C_c/(1+e_0)$ where e_0 is the initial void ratio. For most clays subjected to consolidation tests, it varies in the range of 0.2–0.4. Similarly, a *recompression ratio* (RR) or a *modified recompression index* is defined as $C_r/(1+e_0)$.

Based on the compression ratio or recompression ratio, the compressibility of a clay can be classified as shown in Table 5.10.

Table 5.10 Classification based on soil compressibility CR (C_{ce}) or RR (C_{re})

Description	$\dfrac{C_c}{1+e_0}$ For NC Clays or $\dfrac{C_r}{1+e_0}$ For OC Clays
Very slightly compressible	<0.05
Slightly compressible	0.05–0.10
Moderately compressible	0.10–0.20
Highly compressible	0.20–0.35
Very highly compressible	>0.35

5.5.1.2 Secondary compression

The consolidation test is the primary tool used worldwide to obtain C_a and $C_{a\epsilon}$. However, similar to other parameters such as C_c, empirical relationships are most sought after.

Modified secondary compression index increases with the natural water content. As a first approximation, the $C_{a\epsilon}$ of normally consolidated clays can be estimated as (US Navy 1982)

$$C_{a\epsilon} = 0.0001 w_n \, for \, 10 < w_n (\%) < 3000 \qquad (5.32)$$

where, w_n is the natural water content in percentage.

For most overconsolidated clays, $C_{a\epsilon}$ lies within the range of 0.0005–0.001 (Kulhawy and Mayne 1990). Figure 5.3 shows some typical values of $C_{a\epsilon}$ for normally consolidated clays plotted against their natural water content.

On the basis of the modified coefficient of secondary compression, Terzaghi et al. (1996) classified the clays as shown in Table 5.11.

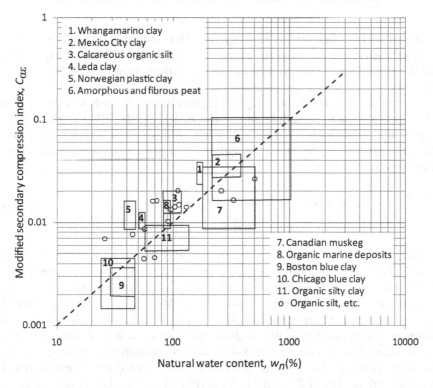

Figure 5.3 Modified secondary compression index versus natural water content for NC clays (Adapted from Holtz and Kovacs 1981; Data from Mesri 1973).

Table 5.11   Classification based on $C_{\alpha\varepsilon}$

Description	$C_{\alpha\varepsilon}$
Very low	<0.002
Low	0.004
Medium	0.008
High	0.016
Very high	0.032
Extremely high	>0.064

Table 5.12   Some typical C_a/C_c values (after Mesri et al. 1994)

Material	C_a/C_c or $C_{\alpha\varepsilon}/CR$
Granular soils including rockfill	0.02 ± 0.01
Shale and mudstone	0.03 ± 0.01
Inorganic clays and silts	0.04 ± 0.01
Organic clays and silts	0.05 ± 0.01
Peats and muskeg	0.06 ± 0.01

Mesri and Godlewski (1977) suggested that the ratio of C_a/C_c for cohesive soils generally varies in the range of 0.025–0.10, with an average value of about 0.05 and Table 5.12 gives the value ranges proposed by Mesri et al. in 1994.

For overconsolidated clays, C_α and $C_{\alpha\varepsilon}$ can be significantly less than when they are normally consolidated, in the order of 30–50% of the values reported for normally consolidated clays. Alonso et al. (2000) suggested that the ratio of C_α for overconsolidated and normally consolidated clays can be written as

$$\frac{C_{\alpha\varepsilon,OC}}{C_{\alpha\varepsilon,NC}} = \left(1 - m\right)e^{-(OCR-1)n} + m \tag{5.33}$$

where, m and n are constants. The constant m is the minimum possible value for the ratio, which applies for very large OCR, and is similar to the C_r/C_c ratio. The magnitude of n controls the rate of decay in the ratio with OCR, with larger values of n giving a faster decay. Alonso et al. (2000) suggested $m = 0.1$ and $n = 12$ from limited data. In practice, smaller values for n are being used conservatively. Wong (2006) suggested $n = 6$ in organic clays for preliminary assessments. Figure 5.4 shows values from projects in the vicinity of the Brisbane River in Queensland and the adopted model and reported by Ng et al. (2016) at Port of Brisbane (PoB) for PoB clay (Ameratunga et al., 2010). If samples are available, consolidation tests could be conducted for varying OCR values to establish a relationship site specific.

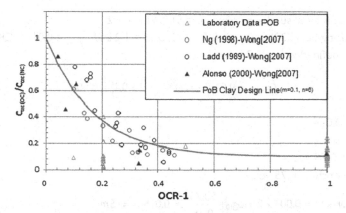

Figure 5.4 Variation of creep ratio with OCR (Adapted from Ng et al., 2016)

EXAMPLE 5.1

A 2 m thick very soft clay layer, assumed to be normally consolidated, is found at a depth of 2 m from the ground surface below 2 m thick fill layer. The water table is at the top of the soft clay. The site was subjected to a surcharge of 0.5 m over a period of 9 months when the primary consolidation was estimated to be almost over (based on Asaoka construction – see Chapter 8). Laboratory tests conducted on NC and OC samples provided the following relationship:

$$C_{\alpha\varepsilon,NC} = 0.01$$

Assuming a unit weight of 16 kN/m³ for the clay and 18 kN/m³ for the existing and surcharge fills, calculate the secondary consolidation after 50 years if the surcharge fill of 0.5 m was taken away.

Solution:

- At the center of the clay deposit, $\sigma_{vo}' = 18 \times 2 + (16-9.81) \times 1 = 42.2$ kPa
- Surcharge load $= 18 \times 0.5 = 9$ kPa
- Vertical stress at the middle of clay layer after surcharging $= 42.2 + 9 = 51.2$ kPa
- Vertical stress at the middle of clay layer after surcharge removal $= 42.2$ (i.e., as prior to surcharging)

- OCR after surcharge removal = 51.2/42.2 = 1.2
  Using Eq. (5.33) with $m = 0.1$, $n = 6$, OCR = 1.2 and $C_{\alpha eNC} = 0.01$ (OCR is defined in Eq. (5.34)

$$C_{\alpha\varepsilon,OC} = 0.01\left\{(1-0.1)e^{-(1.2-1)6} + 0.1\right\} = 0.004$$

$C_{\alpha\varepsilon,OC} = 0.004$

$H_0 = 2$ m

$t = 50 + 0.75$ years

$t_p = 9$ months = 0.75 years

Therefore, $S_s = 0.004 \times 2 \times \log_{10}\left(\dfrac{50.75}{0.75}\right) = 0.015m = 15m$

5.5.1.3 Overconsolidation ratio, OCR

Over consolidation ratio, OCR, has a significant impact on some types of ground improvement such as preloading. It is defined as the ratio of σ'_p (maximum past pressure the soil had been subjected to) and σ'_{vo} is the current vertical effective stress.

$$OCR = \frac{\sigma'_p}{\sigma'_{vo}} \tag{5.34}$$

When a soft clay sample is in situ and if the clay is in a normally consolidated (NC) state, σ'_{vo} will lie on the NC or virgin consolidation line on an $e - log$ σ'_v graph. When a sample of the clay is brought to the surface, the effective stress will be zero and will lie on the y-axis where the void ratio is e_o. When the sample is subjected to a consolidation test, the $e - log\ \sigma'_v$ plot provides both the OC (overconsolidated) and NC lines and where they meet is σ'_p.

EXAMPLE 5.2

The equation of the virgin consolidation line for a clay deposit is given by

$$e = -0.64\log\sigma'_v + 1.96$$

An undisturbed saturated specimen was recovered from the same clay, where the effective vertical overburden pressure (σ'_{vo}) is 50.0 kN/m^2 and the water content (w) is 27.0%. Assuming $G_s = 2.72$ and recompression index (C_r) to be 20% of the compression index (C_c), estimate the preconsolidation pressure and the OCR of the specimen.

Solution:

The in situ void ratio is given by

$$e = w\,G_s = 0.27 \times 2.72 = 0.734$$

Since the point P in the Figure 5.5 lies on the virgin consolidation line,

$$e_p = -0.64 \log \sigma'_p + 1.96 \qquad (5.35)$$

C_c = gradient of the $e - \log \sigma'_v$ graph = 0.64

Therefore, $C_r = 0.64 \times 20/100 = 0.128$, which is the gradient of the OC region of the $e \log \sigma'_v$ graph

Therefore, equating the gradient (see Figure 5.5),

$$\frac{0.734 - e_p}{\log \sigma'_p - \log 50} = 0.128 \qquad (5.36)$$

Solving eqs. (5.35) and (5.36), $e_p = 0.699$ and $\sigma'_p = 93.3 \, \text{kN}/\text{m}^2$

$$OCR = \frac{\sigma'_p}{\sigma'_v} = \frac{93.3}{50} = 1.87$$

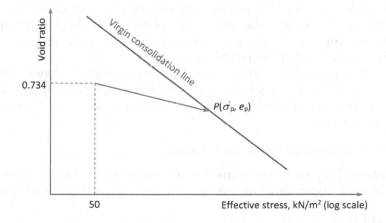

Figure 5.5 void ratio (e) vs log σ' curve for Example 5.3

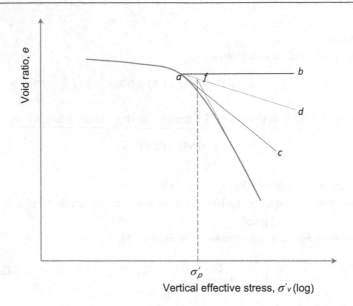

Figure 5.6 Casagrande construction

Generally, σ'_p is obtained from consolidation tests. However, if the samples used are not of good quality, the results may not represent the actual conditions to the accuracy required in a soft soil design. There are several methods of obtaining σ'_p from a laboratory test, the most widely used being the Casagrande's method (Casagrande, 1936), which is shown in Figure 5.6. However, some researchers have identified that Casagrande's method is unreliable (Umar and Sandrekarimi 2017), mostly because it is subjective and dependent on users. For instance, Clementino (2005) identifies that the x-y scale used to construct the e-log σ'_v plot could distort the plot and influence the determination of σ'_p. Tjie-Liong (2017) arrives at the same conclusion after a survey of 30 undergraduate students.

Two other promising techniques have been proposed by Oikawa (1987) and Pacheco Silva (1970). Readers are strongly encouraged to peruse these papers as they are likely to impact on some of the projects you will encounter over your career.

5.5.1.3.1 Estimating OCR using in situ test results

In situ tests also provide information to obtain the maximum past pressure or OCR using empirical relationships. Only a few methods are illustrated.

Method 1 (Mayne, 2007)

- *Estimate the total overburden pressure σ_v*
- *Calculate maximum past pressure, σ'_p, using Eq. (5.37)*

$$\sigma'_p = k\left(q_t - \sigma_v\right) \tag{5.37}$$

where a value of 0.30 to 0.33 could be initially adopted for the k factor, although a wider range is expected. As stated by Mayne (2007) k could vary between 0.2 and 0.5.

- *Assess σ'_v and calculate OCR = σ'_p/σ'_v*

Method 2

- *Estimate σ'_v*
- *Estimate c_u from a vane shear test or a laboratory test or assessed from a CPT*
- *Calculate c_u/σ'_v (which is $(c_u/\sigma'_v)_{OC}$)*
- *Adopt $(c_u/\sigma'_p)_{NC}$ (Mesri (1975) proposed that $(c_u/\sigma'_v)_{NC}$ for NC and slightly OC soils, is independent of PI and is a constant of 0.22 (±0.03))*
- *Calculate OCR using the following equation.*

$$OCR = \left[\left(\frac{c_u}{\sigma'_{v0}}\right)_{OC} \middle/ \left(\frac{c_u}{\sigma'_{v0}}\right)_{NC}\right]^{1/\lambda} \tag{5.38}$$

where :
$\lambda = 1 - C_r/C_c$ which is assumed to be 0.80 to 0.85

Method 3

- *Carry out a dilatometer test and obtain ID and KD (See Chapter 4)*
- *Calculate OCR using Eq. (5.11) proposed by Marchetti (1980) for uncemented clays with ID<1.2*

$$OCR = \left(0.5 K_D\right)^{1.56} \tag{5.11}$$

5.5.2 Consolidation parameters – rate

The magnitude of the final consolidation settlement (s_c) is influenced by several factors including the clay layer thickness, the pressure applied at the ground level, preconsolidation pressure, compression index, recompression index and the effective in situ overburden stress. Figure 5.7 shows five different scenarios, where the final consolidation is the same, but the consolidation is occurring at different rates – E being the fastest and A the slowest. The above parameters that influence the magnitude of the final consolidation settlement have little to do with the rate of settlement. The rate of

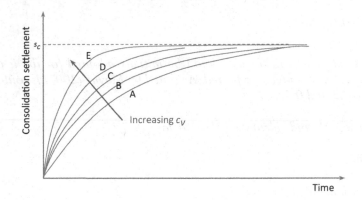

Figure 5.7 Effect of increasing c_v on consolidation settlement.

consolidation is controlled by the coefficient of consolidation c_v. The larger the c_v, the faster is the consolidation.

The parameter c_v in the field is significantly larger than that measured in the laboratory, due to the presence of sand seams and fissures that make the drainage path shorter in the field. Several authors have cited that laboratory c_v underestimates the field c_v value (Bergado et al. (1990), Bergado (1992), Hsi (2003) and Hsi and Martin (2005).

Intuitively, one can see that the consolidation settlements is greater in soft clays than in stiff clays. What is not obvious is that the soft clays settle slower than the stiff clays as shown in Figure 5.8. The coefficient of consolidation c_v comes into play here – it is generally significantly greater for stiff clays (i.e., OC clays) than for soft clays (i.e., NC clays), making the consolidation process a lot faster in stiff clays. The rate of consolidation has nothing to do with the magnitude of the consolidation settlement. Note that c_v depends on the compressibility of the clay skeleton and the permeability of the pore fluid, which is water. The parameter that controls time of consolidation, c_v, in the OC range is five to ten times greater than that in the NC range. The parameter that controls the magnitude of settlement, CR, is five to ten times greater in the NC range compared to that (RR) in the OC range.

While consolidation parameters related to magnitude could generally be assessed from laboratory tests and/or empirical relationships, very few options are available to assess c_v. It is also accepted that trial embankments and/or monitoring and back calculation using construction monitoring are the best methods available, the latter too late for the design. Therefore, the field c_v values could be significantly greater than laboratory consolidation test c_v values.

In situ tests such as piezocones are used to carry out dissipation tests, which could be closer to actual values because the tested soil volume is

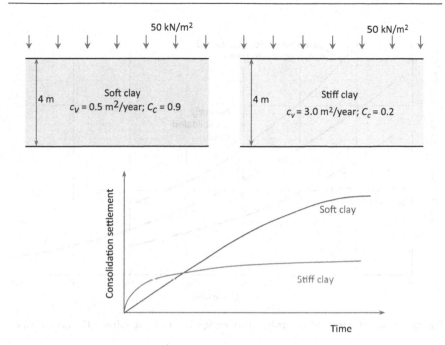

Figure 5.8 Magnitude and rate of consolidation settlement.

much larger and more representative. However, such tests have other issues such as the disturbance the cone creates during pushing. Some question the assumption that a piezocone dissipation test provides a horizontal coefficient of consolidation, which has to be increased by a factor to obtain the vertical c_v. This is generally accepted because of the natural anisotropic conditions in soft clay deposits. Jamiolkowski et al. (1985) provided data in Table 5.13 that indicates k_h/k_v ratio (which is similar to c_h/c_v) to lie generally between 1 and 4. Here, it is implied that m_v remains the same in all directions.

When there are no values at all for c_v, empirical relationships are the last resort. Figure 5.9 shows the relationship between c_v and LL, as suggested

Table 5.13 Range of possible field values of the ratio k_h/k_v for soft clays

Nature of clay	k_h/k_v
No microfabric, or only slightly developed macrofabric, essentially homogeneous deposits	1 to 1.5
From fairly well to well-developed microfabric, e.g., sedimentary clays with discontinuous lenses and layers of more permeable material	2 to 4
Varved clays and other deposits containing embedded and more or less continuous permeable layers	3 to 15

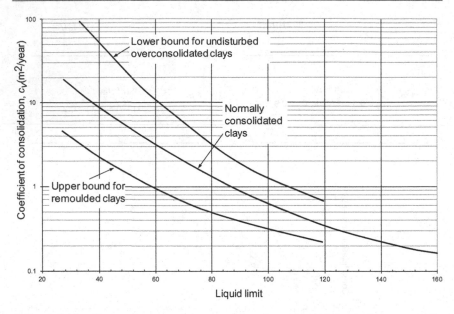

Figure 5.9 Coefficient of consolidation versus liquid limit (after US Navy 1982).

by the US Navy (1982). It is clear that c_v decreases with increasing LL. The figure also suggests that the undisturbed overconsolidated clays have significantly larger c_v values than the normally consolidated clays. Remolded clays have very low values for c_v, which implies that they will take longer to consolidate.

5.6 PHYSICAL CHARACTERISTICS – UNIT WEIGHT

One of the physical parameters often used in geotechnical calculations is the unit weight. In soft clay engineering, most often the saturated unit weight is an input for most ground improvement analyses. The unit weights of saturated soils vary between 15 and 22 kN/m³. Table 5.14 provides representative unit weights for common soils after AS4672-2002.

The unit weight of different materials could also be found from either laboratory or in situ testing. Empirical correlations are also used to convert in situ test results to unit weights. Ameratunga et al. (2016) provide several correlations of unit weight with SPT N value and empirical relationship using CPT test results.

Table 5.14 Typical values for bulk and saturated unit weights (after AS 4678-2002)

		Bulk unit weight (kN/m³)		Saturated unit weight (kN/m³)	
		Loose	Dense	Loose	Dense
Granular soils	Gravel	16.0	18.0	20.0	21.0
	Well-graded sand and gravel	19.0	21.0	21.5	23.0
	Coarse or medium sand	16.5	18.5	20.0	21.5
	Well-graded sand	18.0	21.0	20.5	22.5
	Fine or silty sand	17.0	19.0	20.0	21.5
	Rock fill	15.0	17.5	19.5	21.0
	Brick hardcore	13.0	17.5	16.5	19.0
	Slag fill	12.0	15.0	18.0	20.0
	Ash fill	6.5	10.0	13.0	15.0
Cohesive soils	Peat (high variability)	12.0		12.0	
	Organic clay	15.0		15.0	
	Soft clay	17.0		17.0	
	Firm clay	18.0		18.0	
	Stiff clay	19.0		19.0	
	Hard clay	20.0		20.0	
	Stiff or hard glacial clay	21.0		21.0	

5.7 SUMMARY

The importance of geotechnical parameters and their derivation plus typical values are presented in this Chapter. The main parameters needed are classified broadly as physical characteristics, strength parameters and deformation parameters.

Strength parameters relevant to soft clay including undrained and drained parameters are discussed. Their derivation from in situ and laboratory tests as well as empirical and semi-empirical means are discussed in detail. For some parameters typical values and/or typical ranges have been provided.

Consolidation parameters required to assess the magnitude of primary and secondary consolidation are discussed and their derivation explored. While laboratory tests are most useful and therefore discussed, empirical relationships are also important at least for a sanity check and discussed with appropriate examples.

REFERENCES

Alonso, E.E., Gens, A. and Lloret, A. (2000). 'Precompression design for secondary settlement reduction.' *Geotechnique*, 50(6), 645–656.

Amar, S. and Jézéquel, F.J. (1972). 'Essais en place et en Laboratoire sur sols coherents comparison des resultats.' *Bulletin de Liaison de LCPC*, Paris, 58, 97–108.

Ameratunga, J.J.P. (1986). *A Numerical Assessment of Pressuremeter Testing in Soft Rock*. PhD Thesis, Dept. of Civil Engineering, Monash University, Australia.

Ameratunga, J., Boyle, P., De Bok, C. and Bamunawita, C. (2010). 'Port of Brisbane (PoB) clay characteristics and use of wick drains to improve deep soft clay deposits.' *Proceedings of 17th Southeast Asian Geotechnical Conference*, Taipei, Vol. I, 116–119.

Ameratunga, J., Sivakugan, N. and Das, B.M. (2016). *Correlations of Soil and Rock Properties in Geotechnical Engineering*. Springer, India.

AS1726 (2017). Geotechnical site investigations. Australian Standard.

AS4678 (2002). Earth-retaining structures.

Baguelin, F., Jézéquel, J.F., Le Mée, E., and Le Méhauté, A. (1972). 'Expansion of cylindrical probes in cohesive soils.' *Journal of Soil Mechanics and Foundation Engineering Division*, 98(SM11), 1129–1142.

Balasubramaniam, A.S., Cai, H., Zhu, D., Surarak, C. and Oh, E.Y.N. (2010). 'Settlements of embankments in soft soils.' *Geotechnical Engineering Journal of the SEAGS & AGSSEA*, 41(2), 19 p.

Bell, F.G. (2000). *Engineering Properties of Soils and Rocks*. 4th edition, Blackwell Science, Oxford, UK.

Bergado, D.T. (1992). *Investigations of Effectiveness of Flodrain Vertical Drains on Undisturbed Soft Bangkok Clay using Laboratory Model Tests*. General Engineering Ltd and Geosynthetics Department, Bangkok, Thailand.

Bergado, D.T., Ahmed, S., Sampaco, C.L. and Balasubramaniam, A.S. (1990). 'Settlements of Bangna-Bangpakong Highway on soft Bangkok clay.' *Journal of Geotechnical Engineering Division*, ASCE, 116(1), 136–155.

Bjerrum, L. (1972). *Embankments on Soft Ground, Proceedings Performance of Earth and Earth-supported Structures*, vol 2, ASCE, Lafayette, Ind., pp. 1–54.

Boone, S.J. (2010). 'A critical appraisal of "preconsolidation pressure" interpretations using the Oedometer test.' *Canadian Geotechnical Journal*, 47(3), 281–296.

Casagrande, A. (1936). 'Determination of the preconsolidation load and its practical significance.' *Proceedings First International Conference on Soil Mechanics and Foundation Engineering*, Cambridge, MA, Vol. 3, 60–64.

Canadian Geotechnical Society (2006). *Canadian Foundation Engineering Manual*. 3rd edition, Canadian Geotechnical Society, 503 p.

Cassan, M. (1972). Corrélation entre essais in situ en méchanique des sols. Internal report, Fondsol, Avignon.

Castellanos, B.A. and Brandon, T.L. (2013). 'A comparison between the shear strength measured with direct shear and triaxial devices on undisturbed and remolded soils.' *Proceedings of the 18th International Conference on Soil Mechanics and Geotechnical Engineering*, Paris, 1, 317–320.

Clarke, B.G. (1995). *Pressuremeters in Geotechnical Design*. Blackie Academic & Professional, Glasgow.

Clementino, R.V. (2005). 'Discussion of "An oedometer test study on the preconsolidation stress of glaciomarine clays" by Grozic et al.' *Canadian Geotechnical Journal*, 42(3), 972–974.

Domenico, P.A. and Mifflin, M.D. (1965). 'Water from low-permeability sediments and land subsidence.' *Water Resources Research, American Geophysical Union*, 1(4), 563–576.

Duncan, J.M. and Buchignani, A.L. (1976). *An Engineering Manual for Settlement Studies*. Geotechnical Engineering Report, Dept. of Civil Engineering, University of California, Berkeley, USA, 94 p.

Finno, R.J. (1993). 'Analytical interpretation of flat dilatometer penetration through saturated cohesive soils,' *Geotechnique*, 43(2), 241–254.

Gibson, R.E. and Anderson, W.F. (1961). 'In situ measurement of soil properties with the pressuremeter.' *Civil Eng and Public Works Review*, 56(658), 615–618.

Hight, D.W. and Leroueil, S. (2003). 'Characterisation of soils for engineering purposes.' *Characterisation and Engineering Properties of Natural Soils*. Vol 1. Eds. Tan, T.S., Phoon, K.K., Hight, D. and Leroueil, S. Balkema Publishers, London.

Holtz, R.D. and Kovacs, W.D. (1981). *An Introduction to Geotechnical Engineering*. Prentice Hall, New Jersey.

Hsi, J. (2003). '*Risk management of highway embankments constructed over soft ground*.' *21st ARRB and 11th REAAA Conference*.

Hsi, J. and Martin, J. (2005). 'Soft ground treatment and performance, Yelgun to Chinderah Freeway, NSW, Australia.' *Ground Improvement – Case Histories*, Ed. Indraratna, B. and Chu, J., Elsevier Geo-Engineering Book Series, Oxford, Volume 3.

Iwasaki, K. and Kamei, T. (1994). 'Evaluation of in situ strength and deformation characteristics of soils using flat dilatometer.' *Journal of Geotechnical Engineering, JSCE No.* 499, 167–176.

Jamiolkowski, M., Ladd, C.C., Germaine, J.T., and Lancellotta, R. (1985). 'New developments in field and laboratory testing of soils.' *Proceedings of the 11th International Conference on Soil Mechanics and Foundation Engineering*, San Francisco, 1, 57–154.

Kamei, T. and Iwasaki, K. (1995). 'Evaluation of undrained shear strength of cohesive soils using a flat dilatometer.' *Soils and Foundations*, 35(2), 111–116.

Koppula, S. (1981). 'Statistical estimation of compression index.' *Geotechnical Testing Journal*, 4(2), 68–73.

Kulhawy, F.H., and Mayne, P.W. (1990). *Manual on Estimating Soil Properties for Foundation Design*, Final report EL-6800, Electric Power Research Institute, Palo Alto, CA, USA.

Ladanyi, B. (1972). 'In situ determination of undrained stress-strain behavior of sensitive clays with pressuremeter.' *Canadian Geotechnical Journal*, 9(3), 313–319.

Ladd, C.C., Foott, R., Ishihara, K., Schlosser, F. and Poulos, H.G. (1977). 'Stress-deformation and strength characteristics.' *Proceedings of the 9th International Conference on Soil Mechanics and Foundation Engineering*, Tokyo, 2, 421–494.

Lambe, T.W. and Whitman, R.V. (1969). *Soil Mechanics*. John Wiley & Sons Inc., New York, 553p.

Look, B.G. (2014). *Handbook of Geotechnical Investigation and Design Tables*. 2nd edition. CRC Press.

Lukas, G.L. and LeClerc de Bussy, B. (1976). 'Pressuremeter and laboratory test correlations for clays.' *Journal of the Geotechnical Engineering Division*, ASCE, 102(GT9), 954–963.

Marchetti S (1997) 'The flat dilatometer design applications.' Keynote Lecture, *3rd Geotechnical Engineering Conference*, Cairo University, Egypt, 421–448.

Marchetti, S. (1980). 'In situ tests by flat dilatometer.' *Journal of Geotechnical Engineering Division*, 106(GT3), 299–321.

Mayne, P.W. (2007). 'Cone penetration testing State-of-Practice.' *NCHRP Synthesis. Transportation Research Board Report Project* 20-05, 118 pages.

McCarthy, D. F. (2007). *Essentials of Soil Mechanics and Foundations*, 7th edition, Pearson Prentice Hall, New Jersey.

Mesri, G. (1973). 'Coefficient of secondary compression.' *Journal of the Soil Mechanics and Foundations Division*, 99(SM1), 123–137.

Mesri, G. (1975). 'Discussion of 'New design procedure for stability of soft clays'.' *Journal of Geotechnical Engineering Division*, 101, 409–412.

Mesri, G. (1989). 'A reevaluation of $s_{u(mob)} \approx 0.22\ \sigma'_p$ using laboratory shear tests.' *Canadian Geotechnical Journal*, 26(1), 162–164.

Mesri, G. and Godlewski, P.M. (1977). 'Time and stress compressibility interrelationship.' *Journal of Geotechnical Engineering Division*, 103(GT5), 417–430.

Mesri, G., Lo, D.O.K. and Feng, T.W. (1994). 'Settlement of embankments on soft clays.' *Proceedings of the Conference. on Vertical and Horizontal Deformations of Foundations and Embankments: Settlement '94*, Geotech. Special Publication No. 40, ASCE, 1, 8–56.

Mitchell, J.K. (1976). *Fundamentals of Soil Behavior.* John Wiley & Sons, New York.

Ng, Z., Dissanayake, K., Ameratunga, J. and Honeyfield, N. (2016). 'Port of Brisbane clay characteristics and correlations with back analysis results.' *19th Southeast Asian Geotechnical Conference & 2nd AGSSEA Conference*, Kuala Lumpur, Malaysia.

Oikawa, H. (1987). 'Compression curves of soft soils.' *Soils and Foundations*, 27(3), 99–104.

Pacheco Silva, F. (1970). 'A new graphical construction of the preconsolidation stress of a soil sample.' *Proceedings of the 4th Brazilian Conference on Soil Mechanics and Foundation Engineering*, Rio de Janeiro, Brazil, 2(1), 225–232.

Palmer, A.C. (1972). 'Undrained plane-strain expansion of a cylindrical cavity in clay: A simple interpretation of the pressuremeter test.' *Geotechnique*, 22(3), 451–457.

Ricceri, G., Simonini, P. and Cola, S. (2002). 'Availability of piezocone and dilatometer to characterize the soils of the Venice lagoon.' *Geotechnical and Geological Engineering*, 20(2), 89–121.

Robertson, P.K. and Cabal, K.L. (2010). 'Estimating soil unit weight from CPT.' *2nd International Symposium on Cone Penetration Testing, CPT'10*, Hungtington Beach, California.

Schnaid, F. (2009). *In Situ Testing in Geomechanics, The Main Tests.* Taylor and Francis, London and New York.

Simons, N.E. (1957). 'Settlement studies on two structures in Norway.' *Proceedings of the 4th International Conference on Soil Mechanics and Foundation engineering*, London, 1, 431–436.

Simons, N.E. and Menzies, B.K. (1975). 'Chapter 5 Settlement Analysis.' *A Short Course in Foundation Engineering.* GDS Instrument Ltd, England, 85–103.

Skempton, A. W. (1957). 'Discussion on 'The planning and design of the new Hong Kong airport'.' *Proceedings of the Institution of Civil Engineers*, London, 7, 305–330.

Skempton, A.W. and Northey, R.D. (1952). 'The sensitivity of clays.' *Geotechnique*, 3, 30–53.

Sorensen, K.K. and Okkels, N. (2013). 'Correlation between drained shear strength and plasticity index of undisturbed overconsolidated clays.' *Proceedings of the 18th International Conference on Soil Mechanics and Geotechnical Engineering*, Paris, Presses des Ponts, 1, 423–428.

Terzaghi, K. and Peck, R.B., (1967). *Soil Mechanics in Engineering Practice*, 2nd edition, John Wiley & Sons, New York.

Terzaghi, K., Peck, R.B., and Mesri, G. (1996). *Soil Mechanics in Engineering Practice*, 3rd edition. John Wiley & Sons, New York.

Tjie-Liong G. (2017). 'Consolidation parameters – Alternative to Casagrande and Taylor methods.' *Proceedings of the 9th International Conference on Soil Mechanics and Geotechnical Engineering*, Seoul.

Umar, M. and Sandrekarimi, A. (2017). 'Accuracy of determining pre-consolidation pressure from laboratory tests.' *Canadian Geotechnical Journal*, 54(3), 441–450.

US Navy (1982). *Soil Mechanics – Design Manual 7.1*. Department of the Navy, Naval Facilities Engineering Command, US Government Printing Office, Washington, DC.

US Army (1994). *Settlement Analysis*. Technical Engineering and Design Guides, ASCE, New York.

Wilkes, P.F. (1974). *A Geotechnical Study of a Trial Embankment at King's Lynn*. PhD Thesis, University of Surrey.

Wong, P.K. (2006). 'Preload design, Part 1: Review of soil compressibility behavior in relation to the design of preloads.' *Australian Geomechanics Society Sydney Chapter, 2006 symposium*, 27–32.

Wroth, C.P. and Wood, D.M. (1978). 'The correlation of index properties with some basic engineering properties of soils.' *Canadian Geotechnical Journal*, 15(2), 137–145.

Chapter 6

Ground improvement methods for soft clays

J. Ameratunga and N. Sivakugan

6.1 INTRODUCTION

Although ground improvement is not limited to any particular soil whether it is a clay or a sand, this book is limited to ground improvement techniques as related to soft clays. It becomes a critical design that must be confronted by the geotechnical designer as soon as he/she becomes aware of the issue. Ground improvement on a project could be implemented at different stages as summarized in Table 6.1.

In Chapter 1, a fairly comprehensive list of ground improvement methods by SHRP2 (2014) applicable to soft soils is presented. Some of the techniques have similar intentions but may use different materials and/or a different method of installation. Some techniques have dual functions, e.g., gravel columns that not only strengthen the subsoils but enhance the drainage properties of the soil block. Therefore, it is not easy to group them, although several have tried as given below (adapted from Han 2015):

- Hausmann (1990)   Process
- Mitchell and Jardine (2002)   Construction/function
- Ye et al. (1994)   Function
- Chu et al. (2009)   Soil type and inclusion
- SHRP2 (2014)   Application
- Han (2015)   Function
- FHWA (2017)   Function

Classifying the technique based on the function appears to be the most common. Technologies classified in terms of the function relevant to soft soils are presented in Table 6.2 (adapted from FHWA 2017).

The choice of a ground improvement technique may not be a simple and quick decision since it depends on the project size and several constraints. There are many factors to be considered at different levels of the decision-making process.

Table 6.1 The different stages for implementing ground improvement

Stage	Description
Design stage or early works	At the design stage, a ground improvement method could be trialed either on the footprint of the project or nearby to assess its feasibility. A good example would be trialing preloading where the subsoils remain slightly overconsolidated under the expected loads and therefore expected settlements are low. The trial would provide confidence in the design prior to roll out.
During construction	Ground improvement can be carried out during construction when unexpected soft soils are encountered, especially at shallow depth (e.g., excavation and replacement) or when the monitored behavior is different to predicted and therefore requiring remedial measures.
Post construction	Mainly remedial measures are required when the post construction behavior is different to the design. An example would be when actual settlements monitored indicate the post construction settlement is likely to exceed the performance criteria. This could be of serious concern, especially at the abutment of a bridge, as settlement would induce lateral movement leading to increased loads on the bridge footings.

6.2 SOFT CLAY IMPROVEMENT STRATEGIES INCLUDED IN THIS BOOK

The most popular ground improvement methods that are currently being used have been selected for the purpose of this book. The selected ground improvement methods and the relevant chapters are presented in Table 6.3.

In strictest terms, preloading and surcharging are different.

- Preloading means the application of a fill load equivalent to the development load and allowing it to settle and once sufficient consolidation occurs the preload is removed and replaced with the actual embankment. There is confusion as to whether compensating fill (the soil required to top up) for settlement is be included or not in estimating the settlements. The authors' opinion is that it should be included.
- Surcharging is similar to preloading but the fill load in this instance is greater than the development load. Once sufficient consolidation is achieved the surcharge is removed and replaced with the actual embankment. As the soft clay has been consolidated under a higher load than the development load, the clay is in an overconsolidated state and therefore future settlements are reduced.

Because of different definitions and assumptions, in this book, no differentiation is made between preloading or surcharging and only the term

Table 6.2 Technologies classified by function as applicable to soft clay (adapted from FHWA 2017)

Primary function	Technologies	Comment
Increase shear strength and bearing resistance	• Mixing methods • PVDs • Stone columns • Rammed aggregate piers • Chemical stabilization • Jet grouting	Some do dual functions. For example, stone columns increase drainage in addition to strengthening.
Increase density	• PVDs	
Decrease permeability	• Chemical grouting • Jet grouting • Deep mixing methods	Type of grouting dependent upon soils, depths, geology and design requirements.
Control deformations (settlement, heave, distortions)	• Column-supported embankments • Reinforced load transfer platforms • Non-compressible columns • Mixing methods • Stone columns • Rammed aggregate piers • Chemical stabilization • Encapsulation • Jet grouting	Technologies generally used to bypass or isolate soft ground or to modify and improve the soft ground.
Increase drainage	• PVDs • Aggregate columns • Geotextile encased columns • Electro-osmosis • Geosynthetics in pavement drainage	Generally, increase drainage by inserting a drainage path within the soil to be drained.
Accelerate consolidation	• PVDs • Aggregate columns • Geotextile encased columns	Accelerate pore pressure dissipation due to decreased length of flow path.
Decrease Imposed loads	• Granular fills of having lower density (wood fiber; blast furnace slag; fly ash; boiler slag; expanded shale, clay and slate; tire shreds) • Geofoam, foamed concrete	Lightweight fill use is subject to local availability. Density may vary. Lightest obviously is geofoam and foamed concrete having very low densities, at least one order less than common types of fill.
Transfer embankment loads to more competent layers	• Column-supported embankments • Reinforced soil load transfer platforms • Non-compressible columns • Compressible columns	A column-supported embankment may be constructed on a load transfer platform that is supported on columns.

Table 6.3 Soft clay improvement strategies included in the current book

Method	Book chapter	Brief explanation	To overcome stability or settlement issues
Excavation and replacement	7	Excavate soft clay, fully or partially, and replace with competent material, usually sand or gravel (Figure 6.1a).	Stability and settlement
Preloading/ surcharging	8	Preloading – apply a (fill) load equivalent to the final development load to remove all or some of the expected settlement under the proposed development, remove excess and apply the development load.	Generally, for settlement
Preloading/ surcharging with wick drains	9	Use of wick drains to accelerate settlement in preload projects.	Settlement and stability
Stone columns/ dynamic replacement	10	Aggregate columns inserted in a soft clay layer to increase overall strength/stiffness and to accelerate settlement through better drainage (Figure 6.1c).	Stability and settlement
Semi-Rigid Inclusions (SRI)	11	Columns made of concrete or similar inserted into the ground to take up the bulk of the development load and make the ground stiffer overall (Figure 6.1d).	Stability and settlement
Lightweight fill	12	Lightweight (e.g., fly ash) or extra lightweight material (e.g., polystyrene) used to reduce the load and therefore the expected settlement. (Figure 6.1b shows a road widening where lightweight fill is used to reduce stress increase in the footprint of the existing road embankment.)	Stability and settlement
Deep soil mixing	13	Similar to SRIs but columns made up of in situ mixing of cement or lime with the soft clay.	Stability and settlement
Basal high-strength geotextiles to assist ground improvement	14	Basal high strength geotextiles are extensively used to improve stability in soft clay sites, with or without ground improvement (Figure 6.1f).	Stability
Mass stabilization	15	Stabilize soft clays at shallow depth by using an additive such as cement or lime (Figure 6.1e After FHWA 2017).	Stability and settlement

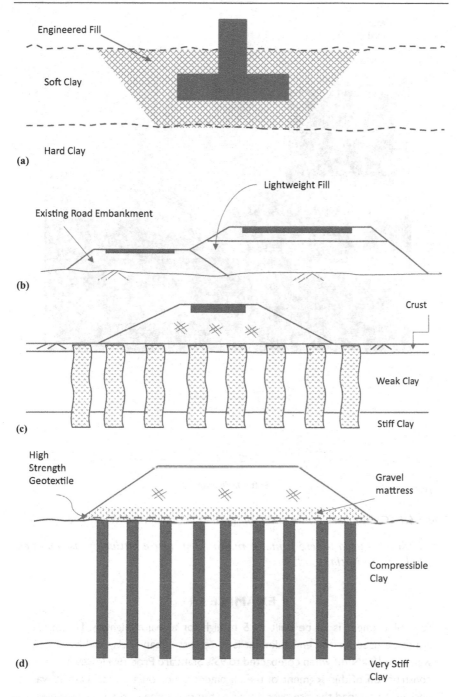

Figure 6.1 Some common ground improvement techniques

(*Continued*)

Mass
mixing
tool

(e)

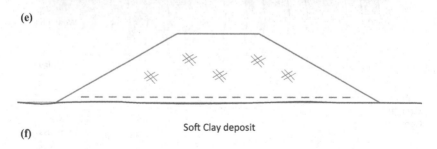

(f)   Soft Clay deposit

Figure 6.1 (Continued) Some common ground improvement techniques

'preloading' is used where some or majority of future settlement is removed by applying an initial load.

EXAMPLE 6.1

An embankment is to be built to 5 m high for a major highway. The material to be used for the embankment is to consist of sandy clay and has a unit weight of 19 kN/m³ when compacted to 95% Standard Proctor. However, as the construction of this segment of the alignment is not on a critical path, it was decided to preload the site over a year to remove some of the post-construction settlement. Excess material from another preload site was to be used. As no compaction machinery is on site, it was decided to place the preload but only light compaction was provided using site dozers. The unit weight of the material after light compaction was found to be 17 kN/m³, which is less than the final material unit weight of 19 kN/m³. What should the preload height be if the stress imposed by the preload needs to be equal to the stress from the final embankment? In this instance, ignore the settlement of the preload.

Load expected from the final embankment = 19 × 5 = 95 kPa

Height of preload required = 95/17 = 5.6 m

EXAMPLE 6.2

In the road embankment discussed in Example 6.1, if the upper 1 m consists of a pavement 1 m thick, what should the new preload height be? The average unit weight of pavement materials is 22 kN/m³.

Load expected from the final embankment = 19 × 4 + 22 × 1 = 98 kPa

Height of preload required = 98/17 = 5.8 m

6.3 PROCEDURE FOR GROUND IMPROVEMENT SELECTION

SHRP2 (2014) provides a step-by-step procedure to be used as a guideline to select and design ground improvement procedures for transportation projects covering varying types of soils, which has also been adapted by Caltrans (2014). FHWA (2017) also uses a slightly modified SHRP2 procedure but provides more explanation for each step. While the readers are advised to browse the SHRP2 website and FHWA manual, in this book, the authors have simplified the guidance offered to provide a soft clay specific investigation and design process to identify and select, and carry out preliminary designs for ground improvement. While this may appear to be biased towards embankment type construction, i.e., transportation projects, the guidelines are equally applicable for the construction of buildings and other infrastructure.

The steps proposed for the preliminary design of a ground improvement could best be summarized as shown in the flow chart presented in Figure 6.2 and discussed below.

Step 1:   **Identification of the presence and extent of soft clays**

The initial alert to the presence of soft clays should be made during the desk study of available information. As a start, regional geological maps may be used for locating the alluvial clays. Just because the soil is alluvial, it does

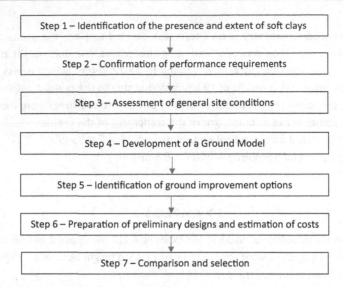

Step 1 – Identification of the presence and extent of soft clays

Step 2 – Confirmation of performance requirements

Step 3 – Assessment of general site conditions

Step 4 – Development of a Ground Model

Step 5 – Identification of ground improvement options

Step 6 – Preparation of preliminary designs and estimation of costs

Step 7 – Comparison and selection

Figure 6.2 Procedure for ground improvement selection.

not mean the soil is soft/compressible. The consistency can only be confirmed with test results. Aged alluvial clays can be competent and have a consistency of stiff to hard.

Aerial photographs of the recent past may provide an indication of old valleys, streams and waterways that may also indicate a weak subsoil profile that has been later filled.

Step 2: Confirmation of performance requirements

Each project may have unique project criteria related to total settlement, differential settlement, factors of safety, etc. as well as time for construction, as discussed in Chapter 1. Any ground modification proposed, to be successful, needs to satisfy the specified criteria. Some of the performance criteria relevant to buildings and highway embankments are discussed in Chapter 1.

If you don't know the performance criteria, it is not possible even to assess whether ground improvement is necessary. For instance, an internal road embankment within a mine site may not be imposed a future settlement limit or a maximum grade change but only a minimum Reduced Level to ensure the pavement will always be at a higher elevation than a 100-year flood level. In this case, the designer may opt to increase the height of the road embankment allowing for expected settlement without resorting to ground improvement.

An important criterion sometimes easily forgotten is the time available for construction. This could have a significant effect on the choice of a ground improvement method. As an example, if the building site needs to be handed over within a short time period, preloading is unlikely to be an option and more rigid solutions, including pile foundations, may need to be pursued.

CASE HISTORY 6.1 PORT OF BRISBANE (AMERATUNGA ET AL. 2010)

The Port of Brisbane (PoB) is located at the mouth of the Brisbane River at Fisherman Islands, Queensland, Australia. In 1999, PoB decided to expand into the sea using reclamation. To add 235 ha of reclaimed land, it was necessary to construct a seawall, 4.6 km long, to encompass the Future Port Expansion (FPE) area (see Figure 6.3).

The seawall project had significant challenges because of a weak subsurface profile greater than 30 m deep, and the water depth varying from 1 m closer to the land at low tide level to the other extreme of 6 m towards the Moreton Bay at high tide level. The embankment height could be as high as 8 m at the deeper end. When settlement assessments were made it was found that the total settlement is likely to exceed 3 m with 50% happening during construction unless some form of ground improvement method is used. The environment surrounding the proposed seawall is pristine and therefore environmental protection was paramount. This meant any ground improvement method involving the addition of additives will not be an option. Wick drains was a possibility, but installation would have required two different machine types to cater for the shallow end as well as the deeper end. After liaising between the client, designers and the contractor, the best solution was to do nothing, i.e., allow the bund to settle but manage topping up later on considering the bund is purely to contain the reclamation and the traffic on the bund is limited to earthworks machinery only. The cross-sectional geometry was modified to cater for settlement and once the work was completed the reduced level of the top of embankment was kept 0.5 m above the necessary height to allow for future settlements. A maintenance program was scheduled, and it is understood that the first topping up was completed ten years after completion of construction.

Step 3: Assessment of general site conditions

While geotechnical conditions could be the main criteria relevant to ground modification, other site conditions may also have a significant bearing on the selection of a ground improvement technique. These conditions include:

- Access and space – whether it is possible for the machinery to access the site, space for machines to work and for trucks bringing in materials or disposing, space for stockpiling;

Figure 6.3 Case History 6.1 Port of Brisbane FPE area (after Ameratunga et al. 2010).

- Services and other infrastructure – whether they could be affected by certain types of techniques. For example, wick drain machine mast height in an area of overhead services;
- Noise and vibration of machinery on residents in the locality or on buildings, services etc.;
- Environment constraints including fauna/flora especially in greenfield sites;
- Presence of acid sulfate soils;
- Presence of contaminated material;
- Need and the available space for disposal of excavated material, additives such as cement, bentonite, etc.;
- Ground water;
- Traffic constraints especially in the case of road widenings.

Step 4:   Development of a ground model

Before embarking on analysis or assessment, it is necessary to set up a ground model covering the site. The ground model needs to, as a start, incorporate the geology based on regional to local knowledge. This means the origin of the deposit is also important as well as recent fillings the site has undergone. As a minimum, the engineer should be able to draw up a cross section or cross sections as well as longitudinal sections of the project site imposing the proposed development details, which could include filling and/or cutting. The presence of the groundwater table and aquifer conditions should also be included in this presentation. If software is available and the project is a major one, 3D representation is most useful as any section could be derived very easily once the model is set up.

Once the geological conditions and variation are established, it is necessary to assess the material characteristics and assign material parameters. The parameters generally needed for ground improvement projects include physical parameters such as unit weight, strength parameters (undrained and drained), compressibility parameters and consolidation rate/permeability parameters. Groundwater chemistry may also be important for some type of ground improvement methods.

Methods to estimate geotechnical parameters are described in Chapter 5 of this book.

Step 5: Identification of ground improvement options

There could be several options of ground improvement for any site underlain by soft clay. Once the ground model has been set up and the development details are available, options available to improve the ground could be identified based on the model, performance criteria and the site conditions. Initially, this will be carried out by a senior experienced engineer with simple calculations if required.

Step 6: Preparation of preliminary designs and estimation of costs

After the identification of several possible ground improvement methods, it is necessary to carry out a preliminary design for each method. The design methodologies provided in this book should assist engineers to carry out this task.

It is then necessary to at least do basic costing of the different ground improvement options. At an early stage of a design this is certainly not easy, but it is the order of costs that is important rather than the exact value. SHRP2 (2014) provides guidance on costing for the various methods as applicable in USA. However, there could be many site specific issues that need a significant effort and therefore associated costs that need to be added. An example would be the occurrence of contaminated material on site, which may need significant amount of time and money for disposal.

It is also important to identify project risks as they will need to be considered when finalizing a ground improvement option.

Step 7: Comparison and selection

Each method needs to be evaluated against project constraints previously discussed including costs and risks. Selection factors should be identified based on the project and could include cost, effects on residents, effects on infrastructure including services, traffic disturbance, environmental conditions, availability of materials, availability of machinery etc. FHWA (2017) suggests that these factors should be further weighted by a weighting factor that is relevant to the project. For example, if cost is a selection factor and the project is a no-go if it exceeds a certain value, the weighting factor for this aspect would be high. The designer and the client could agree on the selection factors and the necessity for weighting factors prior to making comparisons and ultimately select the best option or options for the site.

CASE HISTORY 6.2   BRISBANE DOMESTIC AIRPORT (KILMISTER 1990)

The current domestic airport in Brisbane, Queensland, Australia is a two story structure and is built on a flood plain and underlain by weak Holocene clays. To use shallow foundations after raising the ground above the flood level, 2 to 3 m of sand was hydraulically placed across the site and preloading resulted in up to about 1 m of settlement. However, it was assessed the residual settlement on shallow footings will be still high, of the order of 300 mm. Differential settlement was expected to be greater than 50 mm over short distances because of the presence of old infilled stream channels and clay thickness variation. Rather than further ground improvement, it was decided to continue with shallow footings with a suspended floor slab but include measures in the superstructure to ensure long-term functionality of the building. The innovative measures included the provision to jack the structure that is on shallow footings back to level whenever the differential settlements exceed the 50 mm for which the floor has been designed. It is noted that the structural design catered for the amount of flexibility needed. In closing, it is understood that the superstructure has performed well over 30 years.

Mitchell and Jardine (2002) provide a simple but illustrative example of foundation options for a building to be constructed over a site underlain by soft clay. This example is presented in Figure 6.4 and Table 6.4, slightly modified with a typical subsoil profile of a shallow crust of hard clay or dense sand overlying soft clay, which in turn overlying competent materials at depth. The discussion given in Table 6.3 is also valid for embankments for road and rail projects.

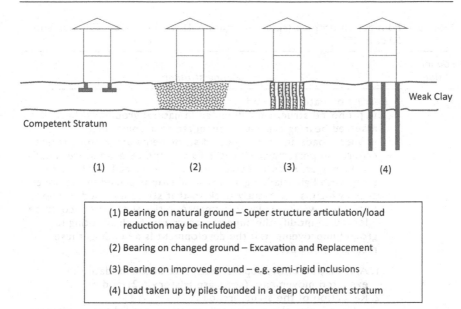

Figure 6.4 Foundation options (adapted from Mitchell and Jardine 2002).

6.4 ENVIRONMENTAL CONSIDERATIONS

The world is becoming more aware of what development could do to harm the environment unless proper steps are taken to safeguard and improve. Environmental considerations have become part of life, and more and more important in new developments. Several authors have highlighted the importance of environmental issues in general construction, which are equally valid for ground improvement projects. McGuffey et al. (1990) state that most environmental issues are addressed at the design phase and are adequately addressed in contract documents and specifications. However, they iterate that specifying how a contractor shall operate and comply with the contract and achieve the project objectives can be a very difficult and drawn-out process. They provide a detailed assessment of construction considerations, some of which are captured below.

Environmental issues are initially looked at during the very early stage of any project, and an environmental impact assessment is within the initial steps of looking at the feasibility of a project. Unfortunately, most of the time, the effects of ground improvement on the environment are not discussed or captured. This could be because limited or no geotechnical or groundwater information is available at the time and even site access may not be available. With such limited information, it would be difficult to identify whether ground improvement is required and what type is needed.

Table 6.4 Foundation options (refer to Figure 6.4) (adapted from Mitchell and Jardine 2002)

Foundation option	Bearing stratum
1	Bearing on natural ground

1   Bearing on natural ground

The proposed structure is founded in natural ground as the assessed bearing capacity is adequate to accommodate the imposed loads and the expected settlements are within project criteria on performance. This is usually the case when weak soils are absent or, when the structure loads are small, say single story, with light cladding and/or foundations placed in the upper crust, which is sufficiently thick so that stress imposed on the underlying weak clay layer is small. Additional measures could be adopted to modify the superstructure rather than opting for ground improvement. If the development is a building, these measures include:

1. Articulation of the superstructure to accommodate the expected movements (see Case History 6.2); and/or
2. Reduction of the structure loads
   - by reducing the number of stories or use lighter cladding or floors (to keep to the same office space the footprint may need to be widened); and/or
3. Changing a footing type to reduce load imparted on the underlying soft clay. For example, if the designer adopts a narrow strip footing (rather than a large spread footing) and embeds at shallow depth so that the stress bulb does not extend to the soft clay below, the expected settlements would be significantly reduced. This may not be possible unless the number of stories is reduced or lighter cladding adopted.

In a road or rail embankment problem, similar techniques could be adopted as follows to overcome stability and settlement issues:

1. Change the cross-sectional geometry
   - Flatten the batters
   - Create berms
   - Staged construction
2. Reduce load
   - In an embankment for a road/rail alignment, by working with civil designers, lower the embankment height which results in lower loads on subgrade and reduce expected settlement (Shaw et al. 2003)

2   Bearing on changed ground

In soft clay sites this refers to excavation and replacement. As the soft clay layer provides low resistance to structure loads the material could be excavated and replaced with more competent material. Depending on the loads and expected settlements, partial rather than full excavation/replacement may be sufficient although more attention may be needed to assess construction issues (see Chapter 7).

(Continued)

Table 6.4 (Continued)

Foundation option	Bearing stratum
3	Bearing on improved ground
	If the expected behavior exceeds the project performance criteria, a ground improvement technique is adopted to improve the soft clay behavior. Most of the ground improvement methods discussed in this book fall into this category. It could be a non-invasive technique such as preloading or an invasive technique such as preloading with wick drains.
4	Piled to a deep stratum
	The piles bypass the soft, weak layers and the structure loads are transmitted into a hard stratum at depth.

Not appropriately dealing with environmental issues at the design and construction stages could lead to major environmental disasters that could delay a project significantly and, in extreme cases, lead to abandonment of the project.

Environmental issues could vary widely from project to project. One of the critical issues is the occurrence of contaminated material and/or acid sulfate in the ground. In such projects some of the ground improvement techniques could become very costly as mitigation measures need to be implemented.

The following environmental issues in ground improvement projects are not different from any construction project:

- Disposal of surface water/runoff;
- Dust and noise from earthworks activities;
- Soil erosion (due to water and wind) – especially in preloading embankments;
- Traffic control in urban areas;
- Effect of earthmoving machinery in rural roads (limiting or slowing farming activities) and road pavements;
- Disposal of groundwater – wick drains;
- Disposal of soil – excavation and replacement;
- Disposal of chemicals (in chemical stabilization);
- Contaminated land/acid sulfate presence; and
- Waste/landfill sites.

6.5 TRANSITION ZONES

Part of the ground improvement design is to identify the solutions for the transition zone between the area that is improved and the adjoining area not improved, or where two types of ground improvement lie next to each other (e.g. preloaded area next to preloaded area with wick drains). Else, differential settlement could lead to highly undesirable steps on the surface or grade changes. These transition zones are not necessarily limited to ground improvement. For instance, if a building is erected on piles where the subsoils are compressible clays and the building surrounds are likely to undergo settlement due to development loads such as landscaping, pavements etc., it is necessary to figure out how to combine the two as it is not safe to allow steps in these circumstances.

In a structure scenario, the transition zone could be a structural slab attached to the building allowing it to rotate as the surrounds settle. In a road embankment situation, it is quite common to include a run-on slab or relieving slab or approach slab at the abutment, generally 3 m to 6 m in length, because it is inevitable that some settlement would occur at the edge of the bridge slab, which could be purely embankment creep or settlement due to vibrations caused by traffic. It is also known that slabs as long as 8 m have been used in certain projects.

A typical example of such a transition zone is shown in Figure 6.5.

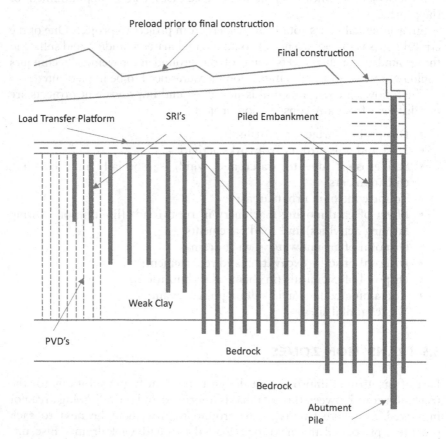

Figure 6.5 Transition zone – example.

The design of these transition zones is complex and there is no simple solution applicable to all situations. In general, 2D analysis and sometimes 3D analysis is likely to be required to design the transition zones. The use of 1D analysis maybe considered appropriate depending on the level of risk and if instrumentation is used to monitor settlements and remedial measures could easily be used if adverse behavior is observed. Usually, numerical analysis is considered a minimum for major projects because of the ramifications associated with differential settlements at critical locations.

6.6 SUMMARY

In this chapter, technologies available to improve the ground where soft soils are prevalent are summarized. The necessary steps to be taken to carry out a proper ground improvement design is discussed. It is noted that some methods carry out more than one function. An example is stone columns, which increase the stiffness of the ground but also assist pore pressure dissipation under preloading. The selection of the right type of ground improvement for any project needs to be done in a systematic manner and the steps required are listed and described. The chapter also discusses the effects of ground improvement and the necessity to assess its effect on services, residents, structures etc. in the vicinity of the project, and recommends that all available information to be taken into account when the type of ground improvement method is to be selected.

REFERENCES

Ameratunga, J., Boyle, P., De Bok, C. and Berthier, D. (2010). 'Ground improvement in Port of Brisbane (PoB) clay.' *ISSMGE Bulletin*, 4(2), 28–54.

Caltrans – California Department of Transportation (2014). *Caltrans Geotechnical Manual*. California Department of Transportation, California.

Chu, J., Varaksin, K., Klotz, U. and Menge, P. (2009). 'Construction processes – state of the art report.' *Proceedings of the 17th International Conference on Soil Mechanics and Geotechnical Engineering*, Alexandria, Egypt, 3006–30135.

FHWA (2017) *Ground Modification Methods Reference Manual – Volume I*. Publication No. FHWA-NHI-16-027. Federal Highway Administration, USDOT, Washington DC.

Han, J. (2015). *Principles and Practices of Ground Improvement*. Wiley, New Jersey.

Hausmann, M.R. (1990). *Engineering Principles of Ground Modification*. McGraw Hill, New York.

Kilmister, M.B. (1990). 'The impact of long term settlements on the structural design of major terminal facilities – a close study.' *Proceedings Airports into the 21st Century*. Hong Kong Institution of Engineers, Hong Kong.

McGuffey, V.C., Bellatty, T.A. and Haas, W.M. (1990). *Environmental Considerations. State of the Art Report 8: Environmental Considerations*. Transportation Research Board, Washington, DC.

Mitchell, J.K. and Jardine, F.M. (2002). *A Guide to Ground Treatment, CIRIA Publication C573*. Construction Industry Research and Information Association, London.

Shaw, P., Ameratunga, J. and Shipway, I. (2003). '*Port of Brisbane Motorway Alliance enhances geotechnical outcomes.*' *Proceedings of the 9th ANZ Geomechanics Conference*, Auckland, NZ.

SHRP2 (2014). *Geotech Tools: Geo-construction Information & Technology Selection Guidance for Geotechnical, Structural, & Pavement Engineers*. http:// geotechtools.org/

Ye, S.L., Han, J. and Ye, G.B. (1994). *Ground Improvement and Underpinning Technologies*, 2nd edition, China Building Industry Press, Beijing.

Chapter 7

Replacement

J. Ameratunga and N. Sivakugan

7.1 INTRODUCTION

Replacement of soft and compressible soils with better material is probably one of the oldest and simplest ground improvement techniques in the world. The intention is to remove weak soils and replace them with competent materials. Although categorized as a ground improvement technique, replacement is one method any ordinary person would call 'common sense.'

By design, Excavation and Replacement is generally considered a simple but effective ground improvement. As discussed later, the method is not appropriate if the site soils are contaminated because significant treatment prior to disposal of unsuitable material could be costly.

In construction also, Excavation and Replacement is considered a useful ground improvement technique. For example, in road and rail alignment projects, occurrence of a limited extent of soft clay, especially where culverts are located, is quite common. Unfortunately, in many instances, it is not identified during the design stage as site investigation test locations are spaced too wide apart. The result is, quite often, their presence is detected only during construction. Therefore, the engineer may need to make quick decisions on the best way forward to ensure the construction timeline and the integrity of the structure are not affected significantly. Hence, replacement of soft soils by more competent soils is quite common in construction projects.

In brief, there are two types of improvement using the replacement principle:

Type 1 Excavation and Replacement – excavation of weak material, either partially or fully, and replacement with competent material.

Type 2 Displacement and Replacement – displacement of the weak material and replacement with competent material. Displacement could also occur unintentionally when failure takes place and generally referred to as mud waving.

The extent of work for replacement could be small as in the case of a column footing below a building footprint or large, hundreds of meters along a road alignment footprint.

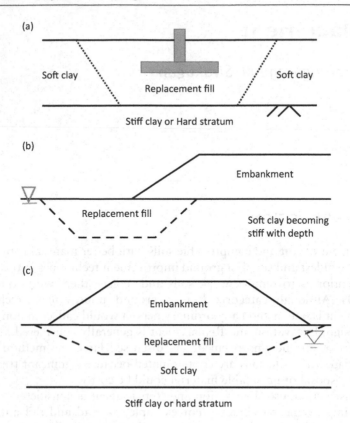

Figure 7.1 Examples of Excavation and Replacement.

There could be one or more reasons why concept of replacement is adopted (see Figure 7.1):

- Weak in bearing;
- Potential instability; and
- High settlements.

1. **Weak in bearing**: Existing soft clays on site are not strong enough to carry the design or construction loads, e.g., fail in bearing

EXAMPLE 7.1A   SHALLOW FOUNDATION

Figure 7.1a shows a typical example where a soft clay layer at shallow depth is not strong enough to carry the column load of a building. As the depth of soft clay is shallow, and local dewatering is feasible even if the water table is shallow, Type 1 Excavation and Replacement could be carried out. The depth to be replaced needs to be assessed, as discussed later in this chapter. The example shown in Figure 7.1a depicts the removal of the entire soft clay layer in the vicinity of the foundation and replaced with competent material.

2. **Slope instability:** Existing soft clay is not strong enough to accommodate construction load or design load and could lead to instability.

EXAMPLE 7.1B   EMBANKMENT EDGE STABILITY

Figure 7.1b shows an example where Excavation and Replacement has been put to good use to reduce the risk of instability in a road embankment. In this example, the road footprint is underlain by a deep deposit of weak clay with the upper layers having a soft consistency. Critical failure surfaces are found to be relatively shallow in the soft clay. In this instance, the solution is to replace some of the weak clay in the vicinity of the toe of the embankment with rock fill. Obviously, the optimal width and depth of excavation are determined by slope stability or numerical analysis. This is a typical Type 1 situation, although a Type 2 scenario cannot be ruled out.

3. **High settlements:** The settlement expected under the development load exceeds the design criteria.

EXAMPLE 7.1C   EMBANKMENT – SETTLEMENT REDUCTION

Figure 7.1c illustrates an example where Excavation and Replacement could be used to control settlement. A weak clay layer is located below the footprint of a proposed highway. Stability analysis indicates a sufficient factor of safety is available, during and after construction. However, the predicted settlement is found to be much greater than design performance criterion. Therefore, ground improvement becomes necessary.

In this particular case, say, the time for construction of this segment of the road is a major constraint. Therefore, Excavation and Replacement is considered a good option to reduce the likely future settlements. Say, the water table is near the ground surface, hence excavation and backfilling need to be carried out under water. Therefore, fill materials need to be granular, possibly rockfill if a borrow site is located nearby. Most likely this is a Type 1 case as described.

It should be noted that, in this example, we have not removed the full depth of the soft clay layer. The depth to be replaced should be assessed by settlement calculations to ensure the future final settlement complies with project performance criteria.

7.2 TYPE I – EXCAVATION AND REPLACEMENT

This type of ground improvement is mostly used in sites of limited extent. It is most popular when structures are to be constructed on shallow footings. It is also most used along linear alignments such as roads and rail when shallow soft clay paleochannels cross the alignment.

Type 1, Excavation and Replacement, could be considered one of the most reliable types of ground improvement methods especially when the work can be carried out in the dry. This is because you can clearly see what material is removed and what material would remain at the base of the excavation. In addition, the effectiveness of placement and compaction of replacement material is visible and could be tested.

One of the disadvantages is that, in soft clays with relatively low undrained shear strength, vertical cuts will not stand up like in stiff clays, even short term. Therefore, flat excavation batters may need to be adopted if slope failure is to be avoided. In confined areas such as an excavation of a pad footing (Figure 7.1a), either a strutted excavation or a trench box could be used.

Excavation and Replacement is not different to general, routine earthworks, especially when the maximum depth being improved is not high, say of the order of 3 m to 4 m, although much deeper replacement has been reported in literature. The latter needs special attention and therefore cannot be considered routine.

When the water table is shallow, it becomes more difficult for Excavation and Replacement. If the plan area is small, it may not be difficult to dewater the excavation using pump and sump techniques, depending on the permeability of the soft clay. Base heave should also be given consideration when partial Excavation and Replacement is proposed. On the other hand, excavation under water in a large site may not be difficult with earthmoving machinery such as draglines if the excavated clay does not become a slurry. A highly sensitive clay is likely to become a slurry, in which case specialist dredging machinery would be needed.

When the area to be excavated is small, similar to the situation shown in Figure 7.1a, i.e., under a pad footing or a strip footing, manual methods are likely to be needed for compaction of replacement fill, unless flowable fill or cementitious materials are used. If the excavation is dry, fill could be placed in thin layers and compacted using a plate compactor (wacker packer), or a light roller in the case of a strip footing excavation. If the water table is shallow, either local dewatering needs to be carried out or replacement fill such as sand or gravel could be used and compacted under water with the use of a vibratory poker. If rockfill is used, it is likely to be somewhat self-compacted under water, although an assessment of risks should be carried out.

7.2.1 Full removal of weak clay

When the soft clay is fully removed, replacement fill could be easily placed and compacted because the subgrade is stable.

If the excavation is wet, as previously mentioned, cohesionless materials such as sand, gravel, crushed rock or rockfill has to be used as replacement material because compaction of cohesive materials under water is not possible. When the footprint is large, especially on road/rail alignments, rockfill is frequently used as replacement material because rockfill generally needs no compaction under water, and large rollers could be used with thick layers rather than thin layers needed for clays and sands.

If the excavation is dry, cohesive fill could be used with construction carried out with appropriate layer thickness and compactive effort. However, even if the water table is currently below the excavation, there could be a possibility of it rising during its design life, especially during the wet season. Therefore, attention should be given to the characteristics of the cohesive replacement material. In an embankment situation, if the replacement material is expansive, it should be confined. It should not be located immediately below pavement subgrade. In most specifications worldwide, expansive soils are not encouraged and not allowed within the top 1m or so below the pavement. In a building situation, expansive soil should not be used as fill because ingress of moisture will give rise to swelling and force the footing upward, leading to possible building damage.

Depending on the characteristics of the exposed subgrade within the excavation and the replacement material to be used, it may be necessary to use a separation geotextile to reduce the risk of migration of finer particles from the subgrade to the replacement material.

Excavation in the wet is more popular in marine environments where the proposed extent of improvement is vast. It is easier for dredger mobilization in such areas as it allows more efficient removal and replacement. Stringent quality control procedures need to be implemented because it is not possible to observe the base of the excavation and therefore to confirm no soft clay is left behind. Invariably, some of the clays and silts will become suspended in the process of dredging and settle later. Depending on the thickness of such sedimentation, either such materials need to be removed or their presence taken into account in the design.

7.2.2 Partial removal of weak clay

Where partial removal is adopted, the base of the excavation will still be in soft clay. If the groundwater table is high or near the surface, it may not be possible to use cohesive materials as fill. Rock fill, gravel fill and sand fill are generally considered more suitable because of difficulty in compaction. If the subgrade is wet, probably the best option is to use rock fill, which will sink into the soft clay and make it stable. If the subgrade is dry and stable, any competent material could be used.

Depending on the material characteristics of the subgrade of the excavation and the replacement fill, a separation geotextile may be required to reduce the risk of finer particle migration from the soft clay into the replacement soil. In a wet subgrade scenario where you want the replacement material such as rockfill to sink into the subgrade, obviously no separation geotextile is needed.

Placement of replacement material in a partial removal situation also depends on the consistency of the material at the base of the excavation if the excavation is dry. If the soils are strong enough to accommodate the loads imparted by rollers and compactors, the placement and compaction could progress without concern. However, when the soils at the base of excavation are weak and not strong enough to accommodate the machinery loads, initial layers would be difficult to be placed and compacted. This occurs very often when the water table is just near the base of the excavation as the soil would be saturated. When construction vehicles traffic such areas the loads imparted cause excess pore pressures and heaving or pumping would occur. The situation could be resolved by several means:

- Use lighter machinery for compaction.
- Dewater the excavation to drop the water table further and allow the upper layers to become dry and make a crust. Dewatering in a road construction scenario could be a simple exercise, which may be cutting trenches along the toe of the embankment if the road is not too wide.
- In situ, stabilize the upper layers, down to a depth of the order of 0.3 m to 0.5 m using lime or another agent.
- Use crushed rock, rockfill or similar immediately above the base and allow the particles to penetrate the weak layer and provide a stable base, basically creating a working platform. Such a layer could generally be up to about 0.5 m thickness, although it depends on the consistency of the clay.

It must not be forgotten that, in most cases, especially where the total height of embankment plus the replacement depth is large, it may not be necessary to subject the initial layer of replacement fill to the same density and layer thickness requirements as for the embankment proper.

CASE HISTORY 7.1   (AFTER MITCHELL AND JARDINE 2002)

Mitchell and Jardine (2002) summarize a few projects completed using the above technique of Excavation and Replacement.

1.   Thorburn and MacVicar (1968) – In the 1960s, most developments included the excavation in the dry to fully remove weak materials and replacement with granular fill in layers and compacted using vibrating rollers.

2.   Schnabel and Martin (1983) – Soft clay, thickness varying from 3 m to 9 m, was excavated and replaced by well graded aggregate placed in layers and compacted with a vibrating roller.

3. Hilmar et al. (1984) – In a large site, 600,000 m³ was excavated and the replacement was gravel compacted by vibratory rollers. It formed a stable and competent base for the column loads of the order of 6.7 MN resulting in settlements of only 12 to 25 mm over three years.
4. Everton and Gellatley (1998) – About 4 m of peat was excavated over a length of 1.5 km railway and replaced with rockfill. One of the main risks was any disturbance to the West Coast Main Line railway, which was close to the alignment and constructed on peat. The Observational Method (see Chapter 16) was used to successfully complete the project without resorting to expensive sheet piling to support the railway.

7.3 TYPE 2 – DISPLACEMENT AND REPLACEMENT

In a Type 2 situation, the displacement occurs because of continuous failure forced at the leading edge of the replacement fill load. Usually, a fill load is pushed into the weak soil and the height of fill is increased until the soft clay below gives way leading to failure. This displaces the soft clay and is immediately replaced with replacement fill. Trial and error is needed to gauge the fill heights required to force a failure. Sometimes, the fill load may need to be kept overnight to see the failure next day. The mud waves created in front of the toe acts as a stabilizing berm and makes it more difficult for subsequent displacement. Using long arm excavators or a dredger if working under water, this leading mud wave berm needs to be removed prior to moving forward (see Figure 7.2).

Although technically it may not need to be, Type 2 usually does 'partial' replacement only as it is extremely difficult to remove the full depth of soft clay by this method. The depth of displacement is difficult to control because failure

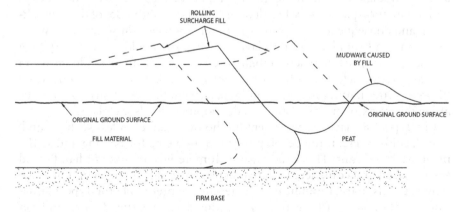

Figure 7.2 Displacement and replacement.

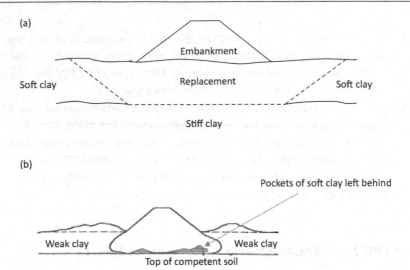

Figure 7.3 Example of a bund formation in a shallow soft clay deposit.

cannot be controlled. Therefore, it is difficult to design the ground improvement compared to Type 1. Figure 7.3 shows an example of a bund across a
shallow layer of soft soils to be improved by replacement. Figure 7.3a shows
the likely geometry when Type 1 is used and Figure 7.3b when Type 2 is used.

Even if the clay is thin and shallow, there is no guarantee the weak soil
would be completely removed by loading the soil and allowing to displace
the mud/soft clay. Whether the clay has been completely removed could only
be assessed after further site investigations are carried out. Even then, thin
layers and pockets of soft clay left behind, may not be easily detected. In a
situation such as shown in Figure 7.3b, it may not be an issue if stability
assessment shows the risk of instability is low even if a thin, soft clay layer is
left behind across the footprint. That is why Displacement and Replacement
is most popular when forming internal bunds or dikes across lagoons, bays
and reclamation paddocks where loading occurs on both sides of the embankment, and consequence of failure is not high. However, in some other situations, leaving behind a very thin layer of weak soil could create a very high
risk scenario especially where stability is concerned. In an example shown in
Figure 7.4, the leftover soft clay remains a weak thin layer sandwiched
between stiffer natural materials and replacement fill. Instability could occur
along this weak plane later on when development loads are placed.

In a Type 2 scenario, replacement fill should be cohesionless such as sand,
gravel, cobbles, etc. if the development requirements for bearing and settlement are significant. This is because such materials are easy to handle and
easily placed and compacted relative to placement and compaction of cohesive materials. Larger and heavier materials are preferable for Type 2 situations as they aid pushing out by creating local bearing failure and get
imbedded in the soft clay providing a layer of stronger material at the base
of the excavation, i.e., at the top of the remaining soft clay.

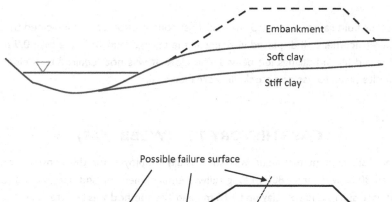

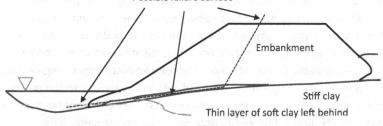

Figure 7.4 Example of Type 2 where a thin clay remains.

Compaction of the replacement material is only possible if the particle size is small, i.e., sand or gravel size. Vibroflotation or similar ground improvement will be required if the replacement layer needs to be compacted.

In general, Displacement and Replacement strategy (Type 2) could be used to improve the soils to a greater depth. There have been projects, mostly done progressing from land to water (i.e., ports or mine related), where such techniques have been employed to replace more than 10 m to 15 m, and in some cases using fill mixed with boulders. In these projects, in general, the ultimate development will be either supported on deep foundations, which would transfer the load to a competent stratum, or is a low risk project such as a car yard or similar where a maintenance regime could be adopted for future settlements. Where the project involves filling under water, collapse settlement of the replacement fill is unlikely to be a high risk in the future, although the replacement fill has not been compacted.

CASE HISTORY 7.2 (LOVE 1987)

Love (1987) describes the application of displacement technique on an 8 m deep deposit of normally consolidated mine tailings consisting predominantly of clay and silt size particles with some fine sand-sized coal in a layered formation. The extent was 35 m wide and 300 m long. The objective was for partial Displacement and Replacement, i.e., to displace mud as much as possible using a mud waving technique with any soft clay remaining beneath the replacement fill to be improved by preloading. The fill height used was of the order of 8 m and observations and assessments indicated that the mud acted more like a viscous fluid rather than a normal soil. Post construction drilling indicated that the replacement fill displaced virtually all the tailings mud with less than 0.9 m of mixed fill and tailings left behind. Therefore, it was not required to preload the site prior to use as a stockpiling area for coal.

CASE HISTORY 7.3   (WEBB 1985)

The displacement technique was successfully employed for the construction of an 80 m wide and 1800 m long railway embankment by end tipping sand to displace slurry and soft clay up to 16 m deep. The method was adopted because conventional excavation and removal based on different construction techniques clearly indicated the difficulties and complexities. Without improvement, up to 3 m of settlements were expected over ten years, due to the proposed development. Excess sand was placed to increase the settlement but had to be removed only after a few weeks because of the construction timeline. Post-construction site investigations comprising 26 boreholes and 12 Dutch Cone probes indicated that most of the soft clay had been displaced. A layer of soft clay 1 m to 3 m thick was left behind on the bed of the lagoon beneath 15 m to 20 m of sand fill where the expected settlement would be significantly less than the 3 m expected originally (see Figure 7.5).

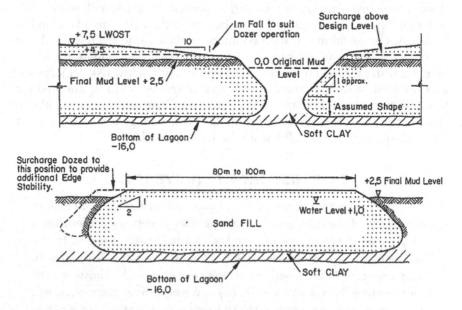

Figure 7.5 Case history 7.3 (Webb 1985).

CASE HISTORY 7.4   (BASED ON PERSONAL COMMUNICATION WITH MR. NEIL MCCLENNAN OF LYTTELTON PORT COMPANY, CHRISTCHURCH (MCLENNAN 2019))

Since 1956, Professor Lauritus Bjerrum has been a geotechnical adviser to the Harbor Board of the Port to resolve issues related with the construction of the extension of the Lyttelton Harbour. The following is based on comments he has made in reports to the Lyttelton Harbour Board.

The subsurface conditions consist of two layers of soft clay separated by a sandwich layer of sand and silt, with the top of this layer at 60 to 66 ft below so that the soft clay layer thickness is about 40 feet. The sand layer thickness varies with a minimum of about 20 ft. The lower clay is firm, and thickness varies from 20 to 65 ft. The depth of water is about 22 to 27 ft.

Professor Bjerrum states that the engineers of the Lyttelton Harbour selected the displacement method, which was described as follows:

The upper soft clay layer is displaced by inducing failures in the clay under the weight of rubble fill placed dry. The intention is to make the displacement so complete, at least under the future reclamation wall, so that the thickness of clay left behind on top of the sand layer is small as possible. To accomplish this, a considerable overweight surcharge is placed. The mud wave which is squeezed up in front of the filling is removed by dredging. A most important point is to ensure the lower clay layer is undisturbed, hence the surcharge height needs to be controlled.

Figure 7.6 Mud waving at Lyttelton Port (circa late 1950s) (McLennan 2019).

Professor Bjerrum concluded that the procedure is most adequate and feasible. Figure 7.6 shows the site where mud waving has occurred (it is likely that one of the personnel in the photograph is Professor Bjerrum).

Similar procedures have been adopted over the last few years to reclaim land at the port using rubble from the Central Business District of Christchurch, post recent earthquakes.

7.4 DESIGN PRINCIPLES

It should be remembered that design principles discussed in this book are generally sufficient for feasibility, concept or preliminary design. Further work is needed for the detailed design including numerical analysis in some cases. More details could be found in literature such as Brown (2001) and Han (2015).

With the above in mind, the design principles are discussed in terms of:

- Shallow structure foundations;
- Embankment stability; and
- Settlement.

7.4.1 Shallow structure foundations

Generally, replacement is used in shallow structure foundations with the Type 1 method, i.e., Excavation and Replacement. As previously discussed, there are two situations, viz.:

- Full replacement; and
- Partial replacement.

In both the full and partial replacement situations, the bearing capacity is significantly increased by replacement because of the strength of replacement materials, which are generally coarse-grained soils. If the replacement fill is engineered or density increased by some ground improvement, the bearing capacity further increases. Skinner (2001) states that a fill not engineered is likely to be in a loose condition, with an angle of shear resistance close to the constant volume ϕ'_{cv}, and as this is usually similar to the angle of repose, is likely to be in the range 32 to 37 degrees for sand fills and 35 to 42 rockfills. Look (2015) indicates a value range of 27 to 37 degrees for friction angle of very loose to medium dense sand. Even with lower bound values, the bearing capacity increase is likely to be significant when a footing has sufficient embedment.

Brown (2001) and Han (2015) provide detailed design steps to assess bearing capacity of partial and full replacement scenario and will not be repeated here. Das and Sivakugan (2019) also provide analysis methods for layered systems, which could be adopted to assess the bearing capacity of ground improved with replacement soils.

Han (2015) states that the design should consider the following failure modes of foundation:

- Failure in bearing within the replaced zone;
- Punching failure of the footing through replaced zone and general failure occurring below;
- Failure in bearing by the distributed load at the top of the soft clay left behind; and
- Punching failure of replaced zone.

For preliminary designs, it is considered adequate to assess the bearing capacity using a load spread of 2V:1H to the base of the replacement zone and assess the bearing capacity at that depth and compare against the reduced load due to spread of the load at that depth. Referring to Figure 7.7, if the footing load is P per meter length on a width of B (strip footing) and the thickness of replaced materials is D, the reduced load per meter length at the base of the replacement zone will be $P/(B+D)$. If the footing was rectangular with length = L and the load is P, the reduced load will be $P/(B+D)/(L+D)$.

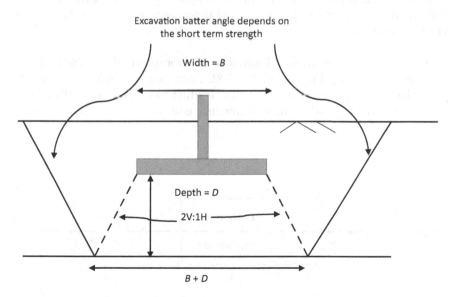

Figure 7.7 Extent of improvement – shallow foundation.

Where the subsurface conditions are more complex and especially if layered, design methods discussed by Brown (2001), Han (2015) or Das and Sivakugan (2019) have to be considered. In complex situations, numerical analysis will be preferable to obtain an accurate assessment.

In addition to bearing capacity, the designer must also carry out an assessment of the settlements. This could also be carried out using the load reduction method discussed above.

7.4.2 Embankments

When the foundation soils are weak, embankment stability becomes very important. One of the easiest methods to improve stability, apart from adopting flatter batters and/or stabilizing berms, is full or partial replacement as shown in Figure 7.1b. In some situations, for example, when the failure is non-circular and confined to the soft layer at shallow depth (similar to sliding type failure), full replacement is likely to improve the stability significantly. The best way to design is to carry out slope stability analysis using widely established software. Sometimes, as shown in Figure 7.1b, the replacement may not extend to the full footprint of the embankment. Several trial and error runs may be needed to position and/or assess the width of the replacement zone.

Partial or full replacement is a guaranteed method to reduce settlement under an embankment where soft clays are present. The assessment of the settlement one dimensionally is quite straightforward with the methods described in Chapter 3. The extent/width of the excavation is a more difficult assessment unless 2D numerical or analytical studies are carried out. At the preliminary design stage, one of the three options shown in Figure 7.8 could be assumed:

1. The base of excavation limited to the footprint of the crest of the embankment (Line 1 in Figure 7.8). There is a risk that excessive settlement could occur on the flanks leading to cracks, especially at the crest, which could create maintenance issues.

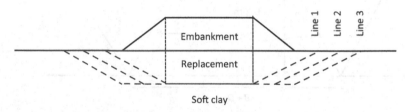

Figure 7.8 Extent of improvement – embankment.

2. The base of the excavation extends to the footprint of the base of the embankment (Line 3 in Figure 7.8). This is the most conservative way but risk of settlement becomes very low.

3. Intermediate situation. One may decide to adopt this as a safe option at the preliminary design stage (Line 2 in Figure 7.8).

7.5 ENVIRONMENTAL CONSIDERATIONS

Even with a simple technique such as replacement, there could be serious environmental issues we have to deal with. They are mostly related to any earthworks construction site and could be summarized as follows:

- Presence of acid sulfate soils;
- Presence of contaminated soils;
- Noise;
- Vibration;
- Traffic; and
- Quarry operations.

When acid sulfate soils and contaminated soils are present in the excavation, they need careful consideration prior to disposal. In the case of acid sulfate soils, space permitting, could be neutralized by the addition of lime. Then such material could be re-used in the project or disposed offsite.

In the case of contaminated soils, special arrangements may be required for disposal based on State and Country laws and requirements. Additional investigations are likely to be required depending on the hazardous nature. Contaminated soils, if present, could become a show-stopper and therefore it is expected that such an issue is comprehensively covered in the detailed design and the contractor is given clear instructions how to deal with it as part of the project requirements, design documents, drawings and specifications.

7.6 CONSTRUCTION ISSUES

Type 1 or Type 2 Excavation/Displacement and Replacement involves typical earthworks and therefore established procedures for QC/QA of fill materials, layer thicknesses and compaction standards are applicable to the replacement material as well. Such procedures include testing of borrow materials, compaction lift thickness, moisture and field density measurements, etc. Non-routine procedures may include the following:

- Geotechnical instrumentation. These are required to monitor settlement, groundwater and pore water pressure in partial replacement scenario when soft clays are left behind.

- Excavation bank stability for Type 1 (Excavation and Replacement). This could include (with or without geotechnical instrumentation) carrying out site observations by experienced geotechnical engineers/geologists to assess any adverse behavior.
- Testing of groundwater and/or soils before disposal in case acid sulfate soils and/or contaminants are present.
- Carry out site investigations where the depth of removal is uncertain, especially for Type 2 (Displacement and Replacement).
- Plate load bearing tests (or other in situ tests) if the stiffness of the replacement material is important for the pavement or ground slab design.

The construction issues would be different from site to site, depending on the depth and extent of the excavation, material characteristics of the soft clay, material below soft clay and the replacement materials. Table 7.1 and Table 7.2 list construction issues and possible actions for Type 1 (Excavation and Replacement) and Type 2 (Displacement and Replacement), respectively.

Table 7.1 Constriction issues and possible actions for Type 1 (Excavation and Replacement)

Issue	Possible actions
High water table	- Dewater. - Work under water.
Excavation base not firm	- Dewater to lower the groundwater table if it is close to the base and let a crust formed. - Place rocky fill or gravelly fill or other appropriate fill to create a stable working platform.
Bank slope stability	- Use flatter batters or berms.
Base heave	- Work under water. - Dewater to lower the groundwater table if it is close to the base.
Presence of contaminants or acid sulfate soils	- Take steps as directed by the consultant as the cost could be different from site to site based on the type and extent.
Scarcity of cohesionless borrow material	- Use the available cohesionless material to at least above the water table and then use whatever material is available. The geotechnical engineer will need to provide additional specification for compaction. - Dewater and use any fill material with conventional placement and compaction techniques. Geotechnical engineer to determine the long-term effects and remedial measures.
Nearby services/ buildings	- Slopes may need to be steep or vertical and temporary retaining walls may be needed. Whatever the solution, advanced analysis should be carried out and the Observational Approach (see Chapter 16) is desired.

Table 7.2 Constriction issues and possible actions for Type 2 (Displacement and Replacement)

Issue	Possible actions
Nearby services and buildings	- Displacement activities to be carried out perpendicular and moving away from the services or buildings - Geotechnical monitoring
Suspicion that full replacement or intended replacement not achieved	- Carry out soil investigations to make an assessment - Geotechnical engineer to advise on remedial measures
Large heave outside the footprint	- To be removed by conventional earthworks machinery - If acid sulfate or contaminated soils, consult the geotechnical and environmental engineers on the remedial measures
Base heave at the leading edge acting as a stabilizing berm	- Remove using long reach excavators, draglines or a dredger if under water - Use explosives to disturb the mudwave at the toe

7.7 ADVANTAGES AND DISADVANTAGES

The advantages and disadvantages are summarized in Table 7.3.

Table 7.3 Advantages and disadvantages for Type 1 and Type 2 improvement

	Type 1 (Excavation and Displacement)	Type 2 (Displacement and Replacement)
Time	Disadvantage – could be time consuming because of material handling and transport Advantage – immediate improvement.	Disadvantage – could be time consuming because of material handling and transport Advantage – immediate improvement.
Complexity	Advantage – very simple methodology and easily understood. Advantage – construction easily controlled.	Advantage – very simple methodology and easily understood. Disadvantage – construction difficult to control.
Reliability	Advantage – very reliable as it is easy to observe the excavation, condition of subgrade, excavated materials and replacement materials; routine testing/ inspections carried out to confirm specifications.	Disadvantage – not reliable as the extent of replacement is difficult to assess. It is never possible to remove all the soft clay because the process can never be controlled.

(Continued)

Table 7.3 (Continued)

	Type 1 (Excavation and Displacement)	Type 2 (Displacement and Replacement)
Fill volume	Disadvantage – high volumes of selected material.	Disadvantage – higher volumes than for Type 1 because of over-displacement.
Costs	Advantage – generally least expensive. Disadvantage – could be costly if replacement materials are scarce and/or borrow distances high.	Advantage – generally least expensive. Disadvantage – similar to Type 1.
Disposal Costs	Disadvantage – could be costly depending on the travel distance, occurrence of acid sulfate or contaminants.	Disadvantage – similar to Type 1.
Geotechnical instrumentation	Advantage – less geotechnical instrumentation than most other methods.	Disadvantage – the nature of construction precludes the use of some types of instruments right from the start.
Presence of sensitive clays	Disadvantage – if excavated material become a slurry, disposal is difficult unless an additive of lime or cement is used, which could be costly. Advantage – if the slurry could be mixed with an additive such as lime or cement, it may produce good, competent fill that could be used on the project.	Advantage – construction progress could increase as no stabilizing berm effect. Disadvantage – in local, small excavations, if material becomes a slurry, disposal would be difficult unless additives such as lime or cement are used, which could be costly.
Site works	Advantage – as it is a standard earthworks construction, specifications, machinery and resources are easy to implement/source.	Advantage – similar to Type 1.
Soil support	Disadvantage – for building foundation type projects, shoring may be expensive if the excavation depth is large	Not applicable.
Adjacent buildings and infrastructure	Disadvantage/advantage – could have effects that need to be investigated but controllable by analyzing at the design stage	Disadvantage – assessment of effects on adjacent buildings and infrastructure is difficult prior to construction.

7.8 SUMMARY

In this chapter, the reader has been introduced to soil replacement to increase the strength, increase stability and/or reduce settlements. Where possible, examples and case histories are presented.

Soil replacement is a common-sense technique, where the weaker soil is replaced by a more competent soil. There are two ways of proceeding: (a) Excavation and Replacement, and (b) Displacement and Replacement.

Excavation and Replacement is suitable when the volume to be replaced is small and shallow, typically up to 4 m in depth. The advantage of this technique is that one can clearly observe the soil being removed and the soil that remains at the excavation level. It is a controlled process and therefore the reliability is very high.

For larger/deeper volumes to replace, displacing the weaker soil and replacing by more competent soil is the better approach. It may not be possible to displace the entire weak soil and hence there could be only partial replacement. For example, by placing a competent granular fill at the ground level, on top of a bed of soft clay, and rolling it down using heavy machinery, the soft clay can be displaced and replaced by the granular fill. The disadvantage of this method is that the extent/depth of ground replacement is unknown.

REFERENCES

Ariema, F. and Butler, B.E. (1990). 'Embankment foundations – guide to earthwork construction.' State of the Art Report 8: Guide to Earthwork Construction, Ed. by R.D. Holtz, Transportation Research Board and National Research Council, Washington, DC.

Brown, R.W. (ed.) (2001). *Practical Foundation Engineering Handbook*, 2nd edition, McGraw-Hill, Columbus, OH.

Das, B.M. and Sivakugan, N. (2019). *Principles of Foundation Engineering*, 9th edition. Cengage, Boston.

Everton, S.J. and Gellatley, G.M. (1998). 'Innovation and cost saving through design development on the M6 DBFO,' In *The Value of Geotechnics*, Proceedings AGS Seminar at Instn Civ Engrs, London, 4 November 1998, 323–332.

Han, J. (2015). *Principles and Practices of Ground Improvement*. John Wiley & Sons, Hoboken, NJ.

Hilmar, K., Knappe, M., Antz, H. and Stark, D. (1984). '*Ground improvement by soil replacement.*' *Proceedings 8th European Conference on Soil Mechanics and Foundation Engineering*, Helsinki, AA Balkema, Rotterdam, 1, 37–43.

Look, B.G. (2015). *Handbook of Geotechnical Investigation and Design Tables*, 2nd edition. CRC Press, London.

Love, A.B. (1987). '*Removal of soft tailings by mud displacement techniques.*' In: Walker, B. and Fell, R. (eds.) *Soil Slope Instability and Stabilisation, Proceedings of an Extension Course on Soil Slope Instability and Stabilisation*. Balkema, Rotterdam.

McLennan, N. (2019). *Personnel communication.*

Mitchell, J.M. and Jardine, F.W. (2002). A guide to ground treatment. CIRIA, London, *C573.*

Schnabel, J.J. and Martin, R.E. (1983). 'Parking garage supported on fill.' *Journal Construction and Engineering Management,* 109(3), 286–296.

Skinner, H. (2001). *'Construction on Fill – Problematic Soils.' Proceedings of the Symposium Held at Nottingham Trent University School of Property and Construction.* Edited by Jefferson, I., Murray, E.J., Faragher, R. and Fleming, P.T.T.

Thorburn, S. and MacVicar, R.S.L. (1968). 'Soil stabilisation employing surface and depth vibrators.' *Structural Engineer,* 46(10), 309–316.

Webb, D.L. (1985). *'Construction of a railway embankment by displacement of deep soft clays and silts.' Proceedings 11th International Conference on Soil Mechanics and Foundation Engineering,* San Francisco, 3, 1761–1766.

Chapter 8

Preloading

S. Iyathurai and J. Ameratunga

8.1 INTRODUCTION

As discussed in Chapter 3, when a load is placed on the ground or the water table is lowered, the subsurface profile is subjected to stresses additional to the current overburden stresses. When the subsoil profile consists of saturated weak clays, these additional loads are initially taken up by the pore water pressures. These excess pore water pressures dissipate with time, rate of dissipation dependent on material properties, especially the coefficient of consolidation and the drainage conditions of the clay layers. As the excess pore water pressures dissipate, the load is gradually taken up by the soil skeleton, which leads to an increase in the effective stress of the soil, equivalent to the dissipated excess pore water pressure. This process is known as consolidation and illustrated in Figure 8.1.

The second stage of this time dependent process of consolidation is known as secondary compression or creep. Creep is a viscous process of soil particles readjusting under constant load and considered to occur over a long period of time. As discussed in Chapter 3, there are two schools of thought on when creep occurs with one group stating creep commences after primary consolidation, i.e., once excess pore pressures are completely dissipated, and the other group insisting that creep occurs simultaneously with primary consolidation. In this chapter, for simplicity, we assume the former, i.e., creep occurring at the end of primary consolidation.

Although the chapter title is 'preloading,' we need to mention a similar phenomenon known as surcharging. Although they are different in the strictest terms, the application is very similar as discussed below:

Preloading: Preloading means the application of a fill load equivalent to the development load so that the majority of settlement expected under the development load is significantly reduced prior to construction. As mentioned in Chapter 5, there is confusion as to whether compensating fill (the soil required to top up) for settlement is be included or not in estimating the settlements. The authors' opinion is that it should be included and has been adopted in this book. Preloading can be simply

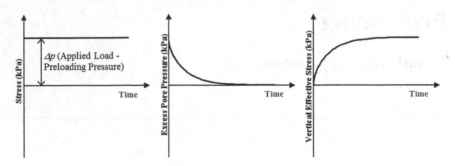

Figure 8.1 Primary consolidation process.

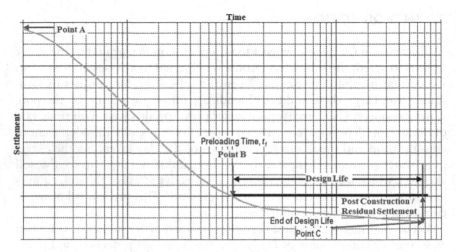

Figure 8.2 Typical illustration of preloading.

illustrated as shown in Figure 8.2 where a settlement time curve (Points A-B-C) is shown for a load applied. If the site was preloaded to time t_1 at Point B and preload is removed when the development load is applied the remaining settlement (Points B-C) is significantly less than expected under the load if preloading had not been carried out (i.e., Points A-C).

Surcharging: Surcharging is a similar process but the fill load in this instance is greater than the development load, i.e., a temporary surcharging. This process further reduces the magnitude of settlement expected under the development load. Figure 8.3 illustrates this phenomenon whereby applying a higher load than the development load, settlement is increased prior to removal of surcharge.

The main difference in technical terms is that in preloading the clay becomes normally consolidated at the end of applying the full development load whereas, in surcharging, the clay remains overconsolidated even at the end of applying the full development load.

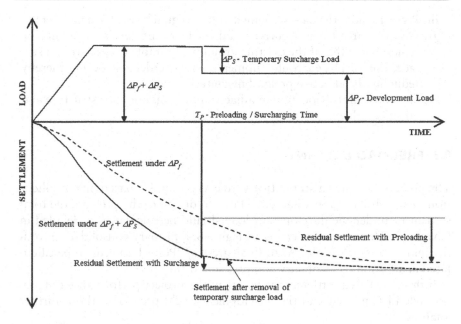

Figure 8.3 Typical illustration of surcharging.

Because of different definitions and assumptions, in this book, no differentiation is made between preloading or surcharging and only the term 'preloading' is used where some or the majority of future settlement is removed by applying an initial load.

Preloading is very popular in land development and major road projects. As discussed above, the compressibility of a clay layer could be significantly reduced by applying this method and making the clay overconsolidated. Reduction of compressibility means lowering expected future settlement as well as increasing the shear strength of the soils.

The process is dependent on the time available for construction. Where the time is limited, preloading may not be viable. This is especially the case when the subsoil profile has a thick clay layer and therefore the drainage paths are very long, leading to long primary consolidation times. In such instances, wick drains (or sand drains) could be used to accelerate the primary consolidation (see Chapter 9).

8.2 INFORMATION NEEDED FOR THE DESIGN

Preload design needs the following information and/or assessment:

   i. Ground model – includes the subsurface profile, strength profile, OCR profile, consolidation parameters, drainage conditions;
  ii. Landform details – existing and post development, and other associated loadings;

iii. Performance criteria – settlement and design life, grade changes etc.;
iv. Groundwater table – current and likely future levels. It should be remembered that if the existing groundwater table is near ground surface, the fill placed could submerge when settlement occurs thereby reducing the effective preload magnitude; and
v. Site constraints – location of adjacent buildings and infrastructure and effects on them.

8.3 PRELOAD DESIGN

The preload design is a straightforward exercise and is based on consolidation theory discussed in Chapter 3. Prior to design of the preload, the total settlement under the development load should be calculated as detailed in Chapter 3. That means, referring to Figure 8.4, primary consolidation from the initial stress σ'_{vo} (Point A) to final stress σ'_f (Point D) needs to be calculated in addition to secondary compression at the end.

If this calculated settlement is less than the project performance criteria, no preload is needed and the development could proceed without further analysis.

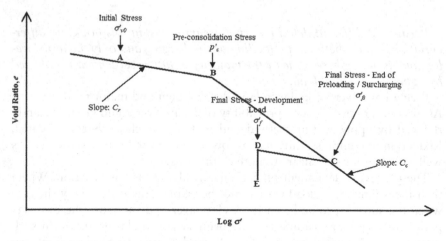

Figure 8.4 Consolidation curve for a typical load application.

EXAMPLE 8.1

Consider a 4 m high embankment (unit weight of 20 kN/m³) is to be constructed over a soft clay layer 3 m thick (i.e. $H = 3$ m) and in a NC (normally consolidated) state and the ground model is given in Figure 8.5. Assuming time is not a constraint, assess whether preloading is required if the project criterion stipulates a maximum settlement of 100 mm over ten years after the construction is completed. (Ignore secondary settlement.)

Solution:

At the center of the soft clay deposits, $\sigma'_{vo} = 1.0 \times 18.0 + 1.5 \times (16 - 9.81)$ = 27.3 kPa

The development load, $\Delta\sigma = 4.0 \times 20.0 = 80.0$ kPa

The final stress at the center of the soft clay deposits, $\sigma'_f = 27.3 + 80.0 =$ 107.3 kPa

Total primary consolidation settlement for NC clay $= H \times CR \times \log (\sigma'_f / \sigma'_{vo}) = 3.0 \times 0.25 \times \log (107.3/ 27.3) = 449$ mm

As sand layers are present at the top and bottom of the clay layer, drainage occurs vertically upward and downward. Therefore, the drainage path length, $H_{dr} = 3/2 = 1.5$ m.

From Eq. (3.14) in Chapter 3, the time factor at the end of 10 years is,

$$T = \frac{c_v t}{H_{dr}^2} = \frac{4 \times 10}{1.5^2} = 17.78$$

From Figure 3.6 in Chapter 3, the average degree of consolidation at the end of 10 years = 100%

Hence, the primary settlement at the end of 10 years = 449 mm

As the estimated primary settlement is greater than 100 mm, preloading is required to meet project criteria.

If preloading is assessed to be necessary, the following steps should be taken:

Step 1 Figure out the time (t) available for preloading – you will need to assess this based on the project information and discussion with the client and/or contractor.

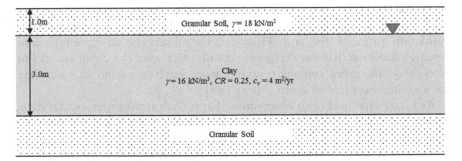

Figure 8.5 Ground model for Example 8.1.

Step 2 Find time factor, T from the expression $c_v t / H_{dr}^2$ for each layer.

Step 3 Find the degree of consolidation (U) for the above T using Figure 3.6 in Chapter 3. Assume that the time for consolidation over the OC region (i.e., from Point A to Point B in Figure 8.4) is very small, usually 1/5th to 1/10th that in the NC region.

Step 4 Assume a preload (i.e., design load and additional surcharge) slightly greater than the design load as a starting point including allowing for settlement as previously calculated. For example, in the case of a road embankment of 4 m design height (i.e. 80 kPa assuming a unit weight of 20 kN/m³), if the settlement without preloading as assessed to be 2 m, a preload height of 6 m could be adopted as an initial trial. This means the total preload/surcharge height would be 6 m which imparts a pressure of 120 kPa compared to the design load of 80 kPa.

Step 6 In primary consolidation, no pore pressure is generated in the OC region (i.e., along the line AB in Figure 8.4). Hence as pore pressure is generated when the stress is greater than the maximum past pressure p_c', subtract p_c' from the final preload stress (equivalent to 120 kPa in the above example) and multiply by U to assess the increase in the effective vertical stress over the preload period. This is added to the p_c' value, which gives the final effective stress (σ_{fs}') over the time the preload lasted as shown as Point C in Figure 8.4. Settlement up to that point can be calculated using the equations given in Chapter 3.

Step 7 Now, after the nominated preload duration is complete, part of the preload is removed. In this instance, point C in Figure 8.4 will move to a new position Point D to reflect the stress decrease. The OCR at that point can be calculated using the current stress σ_f' and the σ_{fs}' value.

Secondary settlement (Point D to Point E in Figure 8.4) can be calculated using Eq. (3.24a) or Eq. (3.24b) in Chapter 3 and Eq. (5.33) in Chapter 5.

If the estimated residual settlement is within performance criteria, surcharge height and/or preloading time could be further optimized by trial and error.

The estimation of secondary compression settlement as discussed in Chapter 3 and Chapter 5 assumes that the secondary compression commences at the end of primary consolidation (often taken as the time to reach 90% degree of consolidation) and creep strain decreases with the overconsolidation ratio of the soil. However, the observational approach and research indicate that surcharging not only decreases the creep strain but also delays the creep starting time following surcharge removal, i.e., delay for recommencement of creep (Mesri 1991).

As Mesri's method is an observational approach from experimental data, Buggy and Peters (2007) have proposed a simplified approach of estimating the starting time of creep using a direct relationship between the unloading time and the overconsolidation ratio (OCR) achieved from the surcharge.

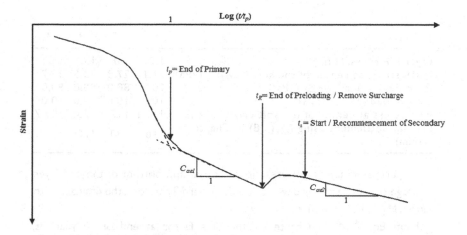

Figure 8.6 Surcharging to reduce secondary settlement (after Buggy and Peters 2007)

The relationship between recommencement of secondary consolidation (t_s) is given below and Buggy and Peters' principle (2007) is shown in Figure 8.6.

$$\log\left(\frac{t_s}{t_R}\right) = a(OCR - 1) + b \qquad (8.1)$$

where, the constants 'a' and 'b' should be based on laboratory testing. In the absence of data values use of $a = 2$ and $b = -0.075$ are reasonable (Wong 2007). Wong (2007) also recommends that if $t_p > t_s$, then the secondary settlement to be assessed using t_p, i.e., creep starting time, is the maximum of t_p or t_s.

EXAMPLE 8.2

Consider a 4 m high embankment (unit weight of 20 kN/m³) to be constructed over a normally consolidated soft clay layer 4 m thick and the ground model is given in Figure 8.7.

Estimate the residual settlement with a design load of 10 kPa and the design life of 10 years if no treatment is undertaken.

Solution:

Assessment of primary settlement:

To improve the accuracy of the settlement calculation, the soft clay is divided into four sub-layers, each of 1.0 m thickness.

Sub-layer	1	2	3	4
Layer thickness, H (m)	1.0	1.0	1.0	1.0
Initial stress at center of the soft clay, σ'_{v0} (kPa)	21.1	27.3	33.5	39.7
Embankment load, $\Delta\sigma_{emb}$ (kPa)	80.0	80.0	80.0	80.0
Design load, $\Delta\sigma_{design}$ (kPa)	10.0	10.0	10.0	10.0
Final stress at center of the soft clay σ'_f (kPa)	111.1	117.3	123.5	129.7
Primary settlement using Eq. (3.8) in Chapter 3 (mm)	216	190	170	154

As permeable layers are present at the top and bottom of the clay layer, drainage occurs vertically upward and downward. Therefore, the drainage path length, $H_{dr} = 4/2 = 2.0$ m.

From Eq. (3.14) in Chapter 3, the time factor at end of 10 years is,
$$T = \frac{c_v t}{H_{dr}^2} = \frac{4 \times 10}{2^2} = 10.$$

From Figure 3.6 in Chapter 3, the average degree of consolidation at end of 10 years = 100%.

The total primary settlement in 10 years, i.e., 100% degree of consolidation = (216 + 190 + 170 + 154) = 730 mm.

Assessment of secondary settlement:

As the ground is not preloaded, the secondary settlement will commence at t_p.

From Figure 3.6 in Chapter 3, the time factor to reach 90% degree of consolidation, $T_{90} = 0.848$.

Using Eq. (3.14), the time required to reach 90% degree of consolidation,

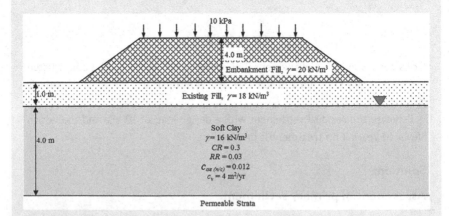

Figure 8.7 Ground model for Examples 8.2 to 8.5.

$$t_p = \frac{T_{90} H_{dr}^2}{c_v} = \frac{0.848 \times 2^2}{4} = 0.848 \text{ years} = 10.18 \text{ months}$$

Using Eq. (3.24b) in Chapter 3, the secondary settlement in 10 years = 0.012 × 4.0 × log (10/0.848) = 0.051 m = 51 mm

Total residual settlement:

Residual settlement in 10 years = 730 + 51 = 781 mm

(If the height of embankment is to be maintained at 4 m, further filling is required, which in turn will induce further settlements.)

EXAMPLE 8.3

For the problem in Example 8.2, estimate the residual settlement if the ground was improved by 9 months of preloading.

Solution:

Assessment of residual primary settlement

Sub-layer	1	2	3	4
Layer thickness, H (m)	1.0	1.0	1.0	1.0
Drainage length, H_{dr} (m)	2.0	2.0	2.0	2.0
Distance from drainage boundary to center of the layer, z (m)	0.50	1.50	1.50	0.50
Depth factor, $Z = z / H_{dr}$	0.25	0.75	0.75	0.25
Time factor, T at end of preloading – Eq. (3.11) in Chapter 3	0.75	0.75	0.75	0.75
Degree of consolidation, U (%) at end of preloading – Figure 3.5 in Chapter 3	92	82	82	92
Initial stress at center of the soft clay, σ'_{v0} (kPa)	21.1	27.3	33.5	39.7
Embankment load, $\Delta\sigma_{emb}$ (kPa)	80.0	80.0	80.0	80.0
Stress increased ($U \times \Delta\sigma_{emb}$) at end of preloading (kPa)	73.6	65.6	65.6	73.6
Vertical stress at end of preloading, σ'_{fs} (kPa)	94.7	92.9	99.1	113.3
Design load, $\Delta\sigma_{design}$ (kPa)	10.0	10.0	10.0	10.0
Final stress at center of the soft clay σ'_f (kPa)	111.1	117.3	123.5	129.7
Primary residual settlement (mm) – Eq. (3.8) in Chapter 3	21	30	29	18

Total primary residual settlement = 98 mm

Assessment of secondary settlement:

Total secondary settlement = 51 mm (from Example 8.2 – no surcharging and preloading time is smaller than t_p hence the secondary settlement will commence at t_p).

Total residual settlement:

So, the total residual settlement = 98 + 51 = 149 mm (if the height of embankment is to be maintained at 4 m further filling is required, which in turn will induce further settlements).

EXAMPLE 8.4

For the same problem in Example 8.2 and Example 8.3, estimate the residual settlement if a 2 m surcharge (i.e., total embankment height of 6.0 m) is applied and kept for 9 months before removing only the surcharge above the 4.0 m height, i.e., removal of 2 m.

Answer:

Assessment of residual primary settlement:

Sub-layer	1	2	3	4
Layer thickness, H (m)	1.0	1.0	1.0	1.0
Drainage length, H_{dr} (m)	2.0	2.0	2.0	2.0
Distance from drainage boundary to center of the layer, z (m)	0.50	1.50	1.50	0.50
Depth factor, $Z = z / H_{dr}$	0.25	0.75	0.75	0.25
Time factor, T at end of preloading – Eq. (3.14)	0.75	0.75	0.75	0.75
Degree of consolidation, U (%) at end of preloading – Figure 3.5	92	82	82	92
Initial stress at center of the soft clay, σ'_{v0} (kPa)	21.1	27.3	33.5	39.7
Embankment + surcharge load, $\Delta\sigma_{sur}$ (kPa)	120.0	120.0	120.0	120.0
Stress increased $(U \times \Delta\sigma_{sur})$ at end of surcharging (kPa)	110.4	98.4	98.4	110.4
Vertical stress at end of surcharging, σ'_{fs} (kPa)	131.5	125.7	131.9	150.1
Embankment load, $\Delta\sigma_{emb}$ (kPa)	80.0	80.0	80.0	80.0
Effective stress after surcharge removal, $\Delta\sigma_{emb} + \sigma'_{v0}$ (kPa)	101.1	107.3	113.5	119.7
Design load, $\Delta\sigma_{design}$ (kPa)	10.0	10.0	10.0	10.0
Final stress at center of the soft clay σ'_f (kPa)	111.1	117.3	123.5	129.7
Final over-consolidation ratio $OCR_f = \sigma'_{fs} / \sigma'_f$	1.184	1.072	1.068	1.157
Primary residual settlement (mm) - Eq. (3.9) in Chapter 3	1.2	1.2	1.1	1.0

Total primary residual settlement (recompression) = 4.5 mm ≈ 5 mm

Assessment of secondary settlement:

As a surcharge is applied, the secondary settlement will commence at t_p or at a later time, t_s, depending on the effectiveness of the surcharge. The recommencement of secondary time t_s is estimated using Eq. (8.1).

Sub-layer	1	2	3	4
Layer thickness, H (m)	1.0	1.0	1.0	1.0
Surcharge removal time, t_R (years)	0.75	0.75	0.75	0.75
Final overconsolidation ratio OCR_f	1.184	1.072	1.068	1.157
t_s/t_R from Eq. (8.1)	1.960	1.170	1.151	1.736
The recommencement of secondary time, t_s (years)	1.470	0.878	0.863	1.302
Overconsolidated creep strain, $C_{\alpha\varepsilon,OC}$ – Eq. (5.33) in Chapter 5	0.0048	0.0082	0.0084	0.0054
Secondary settlement (mm) – Eq. (3.24b) in Chapter 3	4.1	8.9	9.2	5.0

Total secondary settlement = 27.2 mm ≈ 27 mm

Total residual settlement:

So, the total residual settlement = 5 + 27 = 32 mm (if the height of embankment is to be maintained at 4 m the amount of surcharge removal is less than 2 m, which in turn will induce further settlements – see Example 8.5).

EXAMPLE 8.5

For the same problem in Example 8.4, estimate the residual settlement if a 4.0 m embankment height (above the original level) is to be maintained after surcharge removal, i.e., amount of surcharge removal is (surcharge height of 2 m – settlement at end of surcharging in m).

Answer:

Assessment of residual primary settlement:

Sub-layer	1	2	3	4
Layer thickness, H (m)	1.0	1.0	1.0	1.0
Drainage length, H_{dr} (m)	2.0	2.0	2.0	2.0
Distance from drainage boundary to center of the layer, z (m)	0.50	1.50	1.50	0.50
Depth factor, $Z = z / H_{dr}$	0.25	0.75	0.75	0.25
Time factor, T at end of preloading – Eq. (3.14) in Chapter 3	0.75	0.75	0.75	0.75

Sub-layer	1	2	3	4
Degree of consolidation, U (%) at end of preloading – Figure 3.5 in Chapter 3	92	82	82	92
Initial stress at center of the soft clay, σ'_{v0} (kPa)	21.1	27.3	33.5	39.7
Embankment + surcharge load, $\Delta\sigma_{sur}$ (kPa)	120.0	120.0	120.0	120.0
Stress increased ($U \times \Delta\sigma_{sur}$) at end of surcharging (kPa)	110.4	98.4	98.4	110.4
Vertical stress at end of surcharging, σ'_{fs} (kPa)	131.5	125.7	131.9	150.1
Primary settlement at end of surcharging, ρ_{prim} (mm) – Eq. (3.8) in Chapter 3	238	199	179	174
Additional load, $\Delta\sigma_{add}$ due to ρ_{prim} (kPa)	15.8	15.8	15.8	15.8
Embankment load, $\Delta\sigma_{emb}$ (kPa)	80.0	80.0	80.0	80.0
Effective stress after surcharge removal, $\Delta\sigma_{emb} + \Delta\sigma_{add} + \sigma'_{v0}$ (kPa)	116.9	123.1	129.3	135.5
Design load, $\Delta\sigma_{design}$ (kPa)	10.0	10.0	10.0	10.0
Final stress at center of the soft clay σ'_f (kPa)	126.9	133.1	139.3	145.5
Final overconsolidation ratio $OCR_f = \sigma'_{fs} / \sigma'_f$	1.036	1.000	1.000	1.032
Primary residual settlement (mm) – Eq. (3.10) in Chapter 3	1.1	7.7	7.4	0.9

Total primary settlement at end of surcharging = (238 + 199 + 179 + 174) mm = 790 mm

So, additional load due to settlement during surcharging = 0.790 × 20 = 15.8 kPa

Total residual primary settlement = 17.1 mm ≈ 17 mm

Assessment of secondary settlement:

As a surcharge is applied, the secondary settlement will commence at t_p or at a later time, t_s, depending on the effectiveness of surcharge. The recommencement of the secondary time, t_s, is estimated using Eq. (8.1).

Sub-layer	1	2	3	4
Layer thickness, H (m)	1.0	1.0	1.0	1.0
Surcharge removal time, t_R (years)	0.75	0.75	0.75	0.75
Final overconsolidation ratio OCR_f	1.036	1.000	1.000	1.032
t_s/t_R from Eq. (8.1)	0.994	0.631	0.631	0.730
The recommencement of secondary time, t_s (years)	0.848	0.848	0.848	0.848
Overconsolidated creep strain, $C_{\alpha\varepsilon,OC}$ – Eq. (5.33) in Chapter 5	0.010	0.012	0.012	0.010
Secondary settlement (mm) – Eq. (3.24b) in Chapter 3	11	13	13	11

Total secondary settlement = (11 + 13 + 13 + 11) = 48 mm

Total residual settlement:

So, the total residual settlement = 17 + 48 = 65 mm

8.4 GEOTECHNICAL INSTRUMENTATION

Preloading is one of the ground improvement methods that relies significantly on geotechnical instrumentation. Most frequently used instrumentation for preloading projects are settlement plates/points, piezometers, standpipes, inclinometers, lateral stakes and hydraulic profile gauges. A typical embankment cross section with geotechnical instrumentation is shown in Figure 8.8.

Geotechnical instrumentation for ground improvement is discussed in detail in Chapter 16.

8.5 ASSESSMENT OF PRELOAD PERFORMANCE

Geotechnical instruments and observations are used primarily for the following tasks:

1. Stability – indicate behavior that could lead to instability – mostly during the construction of the preload.
2. Settlement – indicate the settlement behavior and how it stacks up with predictions. While monitoring will progress from day one of construction, back calculation to ascertain the deviations from the predictions will be done after preload construction is complete. It is generally agreed that no useful calibration/back analysis could be carried out until 60% or so of the primary consolidation has occurred.

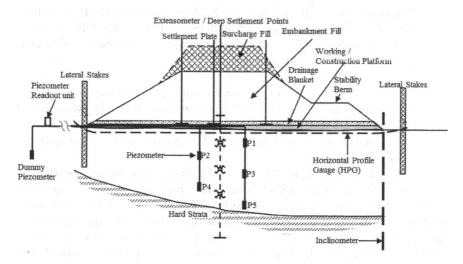

Figure 8.8 Typical embankment cross section with instrumentation.

Another method to assess the preload performance is to carry out soil investigations to assess the strength gain and therefore the dissipation of pore pressures. For example, as discussed later, a CPT could be used to measure the current pore pressure thereby the dissipated amount could be assessed.

8.5.1 Back analysis of preload performance for stability

Assessment of preload performance for stability is discussed in Chapter 16.

8.5.2 Back analysis of preload performance for settlement

Preload designs are usually designed to achieve 90 to 95% of the primary consolidation because, to obtain closer to 100% could take a relatively much longer time. For instance, if the time for 90% dissipation is only one year, it would take more than two years to obtain 99%. Theoretically, dissipation of pore pressure and settlement should follow the same pattern but often it does not. This is why there are two schools of thought, one group advocating pore pressure dissipation to be the main tool to be used in back analysis whereas the other group states that settlement should be the main tool. There have been cases that appear to advance each hypothesis.

Hsi (2016) refers to the recognition of time lag in pore pressure dissipation based on the findings by Gibson (1963), Fell et al. (2005) and Schultheiss (1990). Hsi (2016) gives the following reasons for the time lag in pore pressure dissipation:

- Blockage of water flow due to air bubbles and gas in the piezometer;
- Slow flow rate into the measuring device in low permeability soils;
- Compressibility leading to volume change in soil which could cause changes to the pore water pressure.

Schaefer et al. (1997) also advise caution concerning the use of pore pressure devices because there have been many projects where the two types of observations, viz., pore pressure and settlement, did not match. Schaefer et al. (1997) state that settlement data should be given priority as indicators of the rate of consolidation.

Another phenomenon is the effect of construction on the permanent groundwater table. When a preload is placed pore pressures develop and it is assumed that when the pore pressures dissipate 100%, the groundwater table will be the same as prior to preloading (long-term). This may not be reality because the equilibrium groundwater level after the development loads are placed could be higher than the initial groundwater table. If you

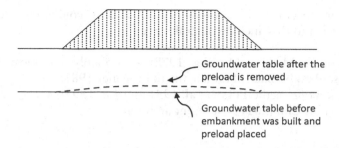

Figure 8.9 Typical groundwater profile prior and post preloading.

measure the groundwater table at the center of the preload footprint and also away from the toe of the preload before construction, and compare with the results in the long term, the water table at the center of the embankment is most likely to be higher as illustrated in Figure 8.9. Assuming the long-term groundwater table to be the same as the pre-existing is conservative.

Notwithstanding the above, CPT could also be used to assess the degree of consolidation under a preload. As Bo et al. (1999) discuss, when a CPT is pushed into the soft soil, excess pore pressure will be generated due to the cone penetration into the soil. If the cone is now held stationary, excess pore pressure generated by the CPT insertion will dissipate to attain an equilibrium pore pressure. The latter will be equal to the pore pressure in the soil at the time of testing. This allows the estimation of pore pressure that had dissipated since the application of the load.

Vane shear test could also be used to assess the state of the soil mass. The vane shear test provides an undrained shear strength value for the soil at the depth of test, which could be used in Eq. 5.38 to assess whether the soil has reached its end of primary consolidation and attained an almost normally consolidated state. Alternatively, CPT could be used to assess the shear strength as discussed in other chapters.

Back analysis of monitoring results is the main tool used to assess whether the behavior under the preload is as per design intent and to take the decision when to remove the preload. The greatest uncertainty in back analysis is the rate of consolidation, which is based on c_v. Back analysis of c_v from a trial embankment would be the preferred method because of shortcomings in field and laboratory testing to find c_v.

In the back analysis, one should first identify likely range of values for the important parameters for magnitude, i.e., CR and OCR. c_v should be allowed a wider range and back calculation must be done in using trial and error methods. It is convenient to use industry software or set up a simple spreadsheet based on Terzaghi's theory. The latter is not a huge task if the loading is simple and construction is straightforward.

There are several other methods of back analysis commonly adopted in the industry and they include the following:

- Asaoka's Method (Asaoka 1978) – probably the most popular (described by Balasubramaniam and Brenner 1981);
- Hyperbolic method (Tan et al. 1991 and Tan 1995); and
- Guo and Chu (2017) – a recent addition.

8.6 DESIGN AND CONSTRUCTION ISSUES

The following design issues are considered important but often neglected.

1. Unit weight of the preload. It is a frequent practice that preload is not compacted to the same vigor as the general fill required for the development because of its temporary nature. The preload is quite often placed in thicker layers and compacted only using site vehicles such as dozers, etc. Therefore, the unit weight of the materials is likely to be less than one would expect from a compacted fill. If, for example, the fill unit weight is 20 kN/m³ and the preload unit weight is 18 kN/m³, for a preload of 3 m it will have a difference of 6 kPa and therefore be under-preloaded.
2. Under the crest of the preload, the vertical stress imparted into the soil will be less than that under the central part of the preload. This is unlikely to be an issue for roads and highways because the permanent loads will also have a similar stress distribution under the crest. However, in land development projects, the extent of the preload should extend beyond the building/development area as shown in Figure 8.10.

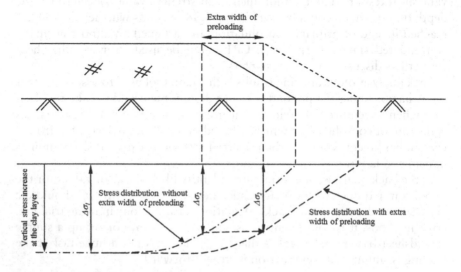

Figure 8.10 Preload edge effects.

The extra width required could be easily calculated using stress distributions discussed in Chapter 3. If this is not possible because of boundaries, the designer should take that into account in its design of the development. This could mean setbacks from the boundary for the development, use of retaining walls for the preload, etc.

3. Where the groundwater table is at or close to the ground, the surface needs special attention. This is because fill placed could go below the water table when settlement occurs. Fill under water is buoyant and therefore the imparted stress on the ground will be less and should be taken into account by the designer.

4. A possible mistake that could happen is to ignore the settlement under preload when it is time to remove the surcharge. What you will be removing is a thickness less than the initial preload height because settlement has already occurred. This needs to be considered in the design.

5. When carrying out a preload design, it is customary to nominate the preload as a reduced level as earthworks are usually based on reduced levels. However, it must be remembered that settlement occurs during construction and by the time the final layers are placed to the nominated reduced level (i.e., RL) the preload height could be significantly higher than the design intent. This is obviously conservative, and it is possible to use the fill height rather than RL as the measure for preload placement. The fill height could be easily monitored using marks on the pipes/rods of settlement plates.

6. The existing water table is likely to be elevated when the fill and preload is placed. Therefore, a higher water table should be adopted for the duration of the preload as the stress will be less than that when the groundwater table had not risen. Especially when a dredger is used for a large reclamation project, the short-term water table during the preload period could be significantly high. It is possible that initially there could be an artificial water table in the fill because of the water pumped in with reclamation materials. Gradually, the water will travel downward and connect with the pre-existing groundwater table and make it elevated. Depending on the characteristics of the materials above the natural water table, attaining groundwater equilibrium could take time. Once the water table in the fill is connected with the existing groundwater table, there will only be a single water table but highly elevated than the pre-existing water table. This means, although initially the preload had a heavy weight imposed on the ground because of the saturated materials having a higher bulk density, now as the water table is elevated, the stress increase felt by the soft clays below will be significantly less because of the lesser effective stress imposed.

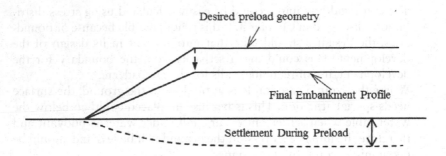

Figure 8.11 Embankment crest width and settlement.

The designer should also take into account that the new water table is temporary, i.e., only during the preload period. After the preload is removed, in the long-term, the water table will gradually recede and come to an equilibrium state, probably at a slightly higher elevation than the pre-existing water table. When the groundwater table drops, additional stresses are imposed that should be considered by the designer. If no information is available, a conservative design could be carried out by assuming the long-term water table to be the same as the original water table.

7. If the designed side slope of the embankment is to be maintained, the slope of the preloaded embankment needs to be steeper to allow for settlement effects as shown in Figure 8.11.

8. Some designers use the SHANSEP approach to assess the shear strength increase when staged construction is adopted. One must remember that strength gain in the OC state is very minimal and therefore should be neglected. This means, until the preload attains a minimum height that takes the in situ stresses past the maximum past pressure, no strength gain should be assumed.

The following construction issues are also considered important.

• Where preloading is to be constructed, usually there will also be engineered fill below to raise the existing ground surface to the development level or, in the case of a road embankment, to the subgrade level above which the temporary preload will be placed. While the same compaction requirements as for the engineered fill could be applied for the preload fill, there is no harm in relaxing the compaction requirements for the preload (subject to stability considerations) as it will be removed later anyway. But, as discussed above, the lesser unit weight of the materials should be considered in the design.

• It is imperative that monitoring instruments are installed prior to commencement of earthworks as discussed in detail in Chapter 16.

- While instruments play a major role, site observations are essential to ensure the works do not create instability. Daily walkovers by an experienced geotechnical engineer may allow the observation of undesirable movements that can be interrogated and remedial measures taken if necessary.

8.7 NUMERICAL ANALYSIS

Numerical modelling based on the finite element or finite difference methods are widely used to assess the performance or simulate the behavior of preloading due to complex soil-structure interaction. Nowadays, numerical analysis is strongly recommended, especially for detailed designs of complex situations due to the ability of such programs to model stress dependent and time dependent soil characteristics. One of the commonly used commercial software is PLAXIS (e.g., PLAXIS 2D 2018, PLAXIS 3D 2018), a finite element program for plane strain and axisymmetric geotechnical applications in which various constitutive models are used to simulate soil behavior. Structural elements (e.g., geotextiles, piles, beams, etc.) can be incorporated into the models to facilitate soil-structure interaction analyses. The widely adopted constitutive models for embankments on soft clay is either the Soft Soil (SS) model or Hardening Soil (HS) model, which allows the calculation of deformation due to primary consolidation only. If secondary settlement is to be assessed using finite element methods then a different constitutive model, like Soft Soil Creep (SSC) model, can be used. If the SSC model is used in PLAXIS to assess the secondary settlement, it is often found that this is not consistent with conventional approaches due to the following reasons:

- PLAXIS assumes that creep occurs concurrently with primary consolidation.
- Aging is crucial when clay is normally consolidated.
- If a clay is showing creep behavior, it cannot be normally consolidated clay because the pre-consolidation pressure should have increased due to creep. The consequence of using low OCR values (<1.5) with the SSC model is that it can give a significant settlement under its own weight.

Numerical modelling is most useful in the following scenarios:

- To consider complex ground/geological conditions;
- To assess complex soil-structure interaction issues;
- To assess effects on adjacent structures or services;
- To assess the rate of construction/fill placement and construction staging;

- To assist the effects on services crossing the site at depth;
- To assess effects at the boundaries;
- To assess grade changes, for example near bridge abutments or where preload heights change.

In numerical modelling, the following needs to be considered:

- Type of analysis, e.g., deformation (plastic/consolidation), stability, coupled;
- Type of behavior, e.g., drained or undrained;
- Extent of boundaries, i.e., the boundaries should not influence the results;
- Constitutive soil model, e.g., Mohr-Coulomb, SS, SSC, HS, etc.;
- Type of elements, e.g., six node elements for plane strain, working load analysis and 15 node elements for failure loads, stability assessment, axisymmetric;
- Drainage boundaries.

Some tips for practicing engineers modelling embankments on soft clay are as follows:

- Use the SS model for estimating the primary consolidation and include secondary consolidation using conventional methods.
- Assign appropriate consolidation parameters based on in situ and laboratory results or well-documented correlations, or from back calculated parameters from trials.
- Geotechnical engineers often work with the coefficient of consolidation but finite element programs use hydraulic conductivity/permeability. Hence, an appropriate conversion needs to be done based on oedometer modulus (E_{oed}). This is stress dependent and can be estimated using the following theoretical relationship (refer to Chapter 5 for derivation):

For NC soil:

$$E_{oed(NC)} = \frac{2.3\sigma_v'}{CR} \tag{8.2a}$$

For OC soil

$$E_{oed(OC)} = \frac{2.3\sigma_v'}{RR} \tag{8.2b}$$

Here, σ_v' is the average effective normal stress during consolidation.

For state change from OC to NC: $E_{oed} = \dfrac{\Delta\sigma}{\left[RRlog\left(\dfrac{\sigma'_p}{\sigma'_{vo}}\right) + CRlog\left(\dfrac{\sigma'_f}{\sigma'_p}\right)\right]}$

$$(8.2c)$$

- As the effective stress increases with time during consolidation, the oedometer modulus will increase; hence the coefficient of consolidation will increase with time if a constant permeability is used. Hence, it is essential that a change in permeability due to consolidation is considered in the assessment. PLAXIS recommends a permeability variation with void ratio as given in Eq. (8.3) where C_k is the permeability change index and shall be taken as either $0.5e_0$ or C_c

$$\log\left(\frac{k}{k_0}\right) = \frac{\Delta e}{C_k} \qquad (8.3)$$

  where, k_0 – permeability at initial void ratio e_0; k – permeability at any void ratio e; and Δe – change in void ratio.
- Make sure that the permeability ratio between two layers is less than 10^5.

8.8 ENVIRONMENTAL CONSIDERATIONS

Environmental considerations are related to earthworks construction and could be summarized as follows:

- Vibration and noise; and
- Traffic.

Vibration and noise are closely connected with earthworks equipment and machinery. If the development site is close to a noise or vibration sensitive site such as a hospital, measures should be taken to overcome the potential issues. Some of the measures include the use of static rather than vibratory rollers, use of lighter equipment at the expense of thinner fill layers, operations limited to certain times of the day and access roads moved to the far side of the site, etc.

Traffic could be affected if the site is in an urban area or adjacent to another road (for example, for a widening project). Traffic control should be studied in detail and measures taken to reduce this risk. This could mean the construction of temporary roads as a last resort to help construction without delay.

8.9 ADVANTAGES AND DISADVANTAGES

The advantages and disadvantages are summarized in Table 8.1.

Table 8.1 Advantages and disadvantages – preloading

Time	Disadvantage – Could be time consuming as proper construction cannot commence until the preload is removed.
	Disadvantage – It is extremely difficult for the designer to accurately predict preload time because the assessment of c_v is difficult from in situ and/or laboratory testing.
Complexity	Advantage – Very simple methodology and easily understood.
	Advantage – Construction easily controlled.
	Disadvantage – The elevation of water table during preloading and afterwards may be difficult to assess and therefore conservative assumptions need to be made.
Reliability	Advantage – Generally very reliable if proper instrumentation is used.
	Disadvantage – There is a risk that secondary compression could exceed predictions in the long term. Where critical, such as near bridge abutments, a conservative design should be adopted, or more elaborate testing should be conducted to study secondary compression behavior especially if the clay is to be in an OC state throughout its design life.
Fill volume	Disadvantage – High volumes of borrow materials, part of which has to be carted away after preloading. This could be a significant cost depending on the distance to borrow sites unless the material could be used on the same project.
Cost	Advantage – Relatively cheap compared to many other ground improvement techniques.
	Disadvantage – Could be costly if borrow/disposal distances are high.
Geotechnical instrumentation	Advantage – Simple geotechnical instrumentation such as settlement plates and lateral markers suffice for most projects with more complex instrumentation needed only at critical locations.
Site works	Advantage – As it is a standard earthworks construction, specifications, machinery and resources are easy to implement/source.
Adjacent buildings and infrastructure	Disadvantage – Could have effects that need to be investigated including settlement, lateral movement and vibration.

8.10 SUMMARY

This chapter introduced preloading as a ground improvement technique. It is one of the widely used methods around the world to improve soft soils. The design is discussed in detail with examples to illustrate the procedures to be adopted.

Preloading is a simple method but depends heavily on monitoring and using monitored results to back-calculate parameters and assess whether the design intent could be realized. If not, usually the preload has to be increased and/or removal is delayed. Methods to assess the performance of the preload are described as well as risks associated with adopting preloading as a ground improvement technique.

Finally, the advantages and disadvantages of preloading are summarized.

REFERENCES

Arulrajah, A., Nikraz, H. and Bo, M.W. (2003). 'Factors affecting settlement assessment and back-analysis by the Asaoka and Hyperbolic methods.' *Australian Geomechanics Journal*, 38(2), 29–37.

Asaoka, A. (1978). 'Observational procedure of settlement prediction.' *Soils and Foundations*, 18(4), 87–101.

Balasubramaniam, A.S. and Brenner, R.P. (1981). 'Consolidation and settlement of soft clay.' In Brand E.W. & Brenner, R.P. (eds.) *Soft Clay Engineering*. Elsevier, Amsterdam, 481–566.

Bjerrum, L. (1967). 'Engineering geology of Norwegian normally-consolidated marine clays as related to settlement of buildings.' Seventh Rankine Lecture. *Geotechnique*, 17, 81–118.

Bo, M.W., Chu, J. and Choa, V. (1999). 'Factors affecting the assessment of degree of consolidation.' *International Conference on Field Measurements in Geomechanics, 1999*, Leung, C.F., Tan, S.A. and Phoon, K.K. (Eds) Balkema, Rotterdam. 481–486.

Buggy, F. and Peters, M. (2007). 'Site investigation and characterisation of soft alluvium for Limerick Southern Ring Road – Phase II, Ireland.' *Soft Ground Engineering Symposium, Ireland Geotechnical Engineering 2007*, paper 1.6.

Fell, R.L., MacGregor, P., Stapledon, D. and Bell, G. (2005). *Geotechnical Engineering of Dams*. A.A. Balkema Publishers, London.

Gibson, R.E. (1963). 'An analysis of system flexibility and its effect on time-lag in pore water pressure measurements.' *Geotechnique*, 13(1), 1–11.

Guo, W. and Chu, J. (2017). 'New observational method for prediction of one-dimensional consolidation settlement.' *Geotechnique*, 67(6), 516–522.

Hsi, J. (2016). '*Soft soil engineering in practice*.' *Australian Geomechanics Society Symposium*, Sydney, 31–54.

Mesri, G. (1991) 'Prediction and performance of earth structures on soft clay.' *Proceedings of The International Conference on Geotechnical Engineering on Coastal Development, Yokohama*, 2, G2.1–G2.16.

PLAXIS (2018) (PLAXIS 2D 2018, PLAXIS 3D 2018) *User Manual*. Bentley Systems, Incorporated.

Schaefer, V.R., Abramson, L.W., Drumheller, J.C. and Sharp, K.D. (1997). *Ground Improvement, Ground Reinforcement and Ground Treatment: Developments 1987–1997*. ASCE.

Schultheiss, P.J. (1990). 'Pore pressure in marine sediments: An overview of measurement techniques and some geological and engineering applications.' *Marine Geophysical Researches*, 153–168.

Tan, S.A. (1995). 'Validation of hyperbolic method for settlement in clays with vertical drains.' *Soils and Foundations*, 35(1), 101–113.

Tan, S.A., Inoue, T. and Lee, S.L. (1991). 'sHyperbolic method for consolidation analysis.' *Journal of Geotechnical and Geoenvironmental Engineering*, 117(1), 1723–1737.

Wong, P. (2007). Preload design to reduce post-construction creep settlement. *ANZ Geomechanics Soft Soil Workshop* (Brisbane, 25 October 2007).

Chapter 9

Preloading with wick drains

S. Iyathurai

9.1 INTRODUCTION

As discussed in Chapter 8, when preloading is used as a ground improvement technique, we try to remove a majority of consolidation settlement expected under the development loads during the preloading process. Usually, 90% or more of the primary consolidation settlement is removed, although lower values are sometimes targeted when time is a critical factor. The time needed for completion of primary consolidation is dependent on the "square" of the drainage path in a clay layer (Eq. (3.14) in Chapter 3). Hence, when the clay layer thickness is large, the time for primary consolidation becomes very large especially when single drainage conditions prevail. This makes preloading not viable for many sites because it would be years and not months before preload could be removed.

The concept of using artificial drainage paths to accelerate excess pore pressure dissipation was first introduced in 1925 in the USA. Since then, many developments have taken place, in theory and materials used to facilitate excess pore water pressure dissipation. Sand drains were the initial development to speed up the consolidation process and are still in use in some parts of the world, especially where sand is in bulk supply. Wick drains first came to light in the late 1940s and were initially made of cardboard, before moving to geosynthetics. Wick drains are also known as prefabricated vertical drains (PVDs). For the purposes of this book, all types of PVDs will be referred to as wick drains.

Wick drains are generally flat, about 100 mm wide, and about 2 to 5 mm in thickness (see a typical wick drain in Figure 9.1). It has a core wrapped in a separation geotextile. The core allows the water to flow to the surface and the geotextile functions as a filter to ensure fine particles do not block the movement of water from the surrounding soil. Different types of core are available, and some are shown in Figure 9.1. There are circular types of wick drains too, generally the diameter varying from 35 mm to 50 mm, which are mostly used in vacuum consolidation applications. The wick drains allow the pore water to flow laterally for short distances to the closest

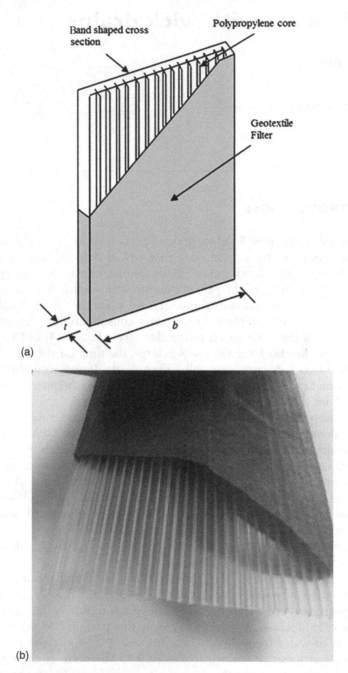

(a)

(b)

Figure 9.1 Typical wick drain ([a] adapted from Sathananthan (2005); [b] courtesy Zen Ng;

(*Continued*)

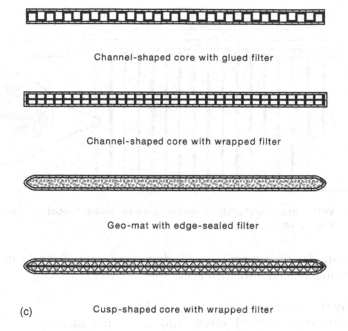

Channel-shaped core with glued filter

Channel-shaped core with wrapped filter

Geo-mat with edge-sealed filter

(c)   Cusp-shaped core with wrapped filter

Figure 9.1 (continued) [c] with permission from Standards Australia – AS 8700 (2011) Figure A3).

wick drain rather than travel vertically over a long distance, thereby reducing the consolidation time significantly.

Wick drains provide additional drainage paths, ensuring quicker dissipation of excess pore water pressure. They are usually installed vertically from the ground level to the base of the soft, compressible clay and spaced generally between 1 m to 3 m in a rectangular or triangular pattern. This means pore water has to travel only short distances, of the order of 1 m to 2 m laterally, rather than several meters vertically to a drainage boundary if there were no wick drains. A similar outcome could be achieved using sand drains instead of wick drains, although they are likely to be more expensive for the majority of sites because of the relative speed of construction as well as the relative cost of the wick drains. A typical cross section of an embankment with wick drains and preloading is presented in Figure 9.2. Here, a drainage blanket is placed between the preload embankment and the soft clay deposit, which is a drainage layer facilitating water flow when consolidation is taking place and water is dispelled from the soil via wick drains.

The acceleration of pore pressure dissipation and therefore the resulting benefits are presented schematically in Figure 9.3 using a simple example. In this example, a soft clay layer is preloaded with wick drains in place.

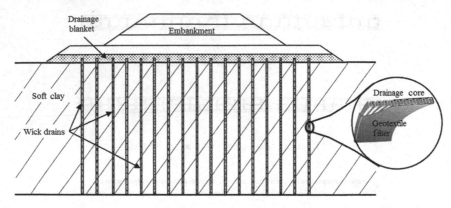

Figure 9.2 Wick drain installation under a road embankment (adapted from Rixner et al. 1986).

To illustrate the effectiveness of wick drains and keep the problem simple, the following assumptions are made:

- Clay thickness = 10 m (two-way drainage);
- Coefficient of vertical consolidation, c_v = 3 m²/year;
- Coefficient of horizontal consolidation, c_h = 6 m²/year; and
- Wick drains (ignore smear) are installed in a triangular pattern.

Figure 9.3 shows the degree of consolidation achieved against time for several wick drain spacings. It shows clearly how wick drains reduce the time for primary consolidation, from years if there are no wick drains to four months with wick drains at 1.0 m spacing.

One should remember the following:

- Wick drains greatly increase the rate of consolidation and settlement, but the magnitude of primary consolidation is not reduced.
- If it is assumed that secondary consolidation starts at the end of primary consolidation, when wick drains are used, secondary compression commences early, and therefore, over the design period, slightly higher secondary compression occurs.

9.2 INFORMATION NEEDED FOR THE DESIGN

Design parameters needed for preloading as discussed in Chapter 8 are equally important for analysis related to preloading with wick drains. In addition, the following parameters or information are needed:

- Wick drain size, pattern and spacing;
- c_h – coefficient of consolidation of the clay in the horizontal direction;

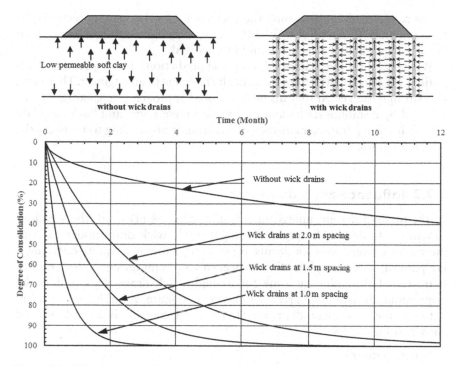

Figure 9.3 Effectiveness of wick drains – example.

- Smear zone radius and permeability in the smear zone; and
- well resistance.

9.2.1 Coefficients of consolidation – c_h and c_v

As previously mentioned in Chapter 8, the rate of consolidation is not easy to predict compared to the magnitude of consolidation. This is because of the inherent nature of soil deposits where the presence of lenses and the variable fabric across a site would make it extremely difficult to predict dissipation rates unless a trial, such as a trial embankment, is carried out on the development site itself.

As the primary direction of pore water movement surrounding a wick drain is in the horizontal direction, c_h becomes a very important parameter. Typically, in situ tests such as CPT/piezocone dissipation tests or laboratory oedometer tests are performed to find c_v or c_h. It is generally accepted that a CPT dissipation test provides c_h while a routine laboratory test provides c_v. Therefore, unless sophisticated laboratory tests are conducted, a ratio of c_h/c_v needs to be assumed to obtain both c_h and c_v. It is common practice for practicing engineers to adopt a c_h/c_v ratio of 1 to 4 based on published literature. This ratio is approximately the same as the ratio of the permeabilities in the two directions, assuming the coefficient of volume compressibility is the same in both directions. Where layering dominates the subsurface profile, significantly higher ratios could be expected.

Whatever the value finally adopted, the prediction of pore pressure dissipation will remain an extremely difficult challenge to an engineer. Therefore, it is strongly suggested that an observational approach (see Chapter 16) be adopted by installing settlement plates, piezometers, etc. and back calculate the dissipation parameters as the construction is proceeding to re-assess the adopted parameters.

9.2.2 Influence zone diameter

In the preload design, unless numerical analysis is involved, the standard practice is to design the preload using a single wick drain under axisymmetric conditions. As wick drains are laid out, either in a square or triangular pattern, some simplification is needed to obtain an equivalent radius/diameter of the influence zone for a single wick drain. This is done by approximately equating areas as shown in Figure 9.4. The relationship between the equivalent diameter (D_e) and spacing (S) is as follows and shown in Figure 9.4.

Square pattern:

$$D_e = 1.128S \tag{9.1a}$$

Triangular pattern:

$$D_e = 1.05S \tag{9.1b}$$

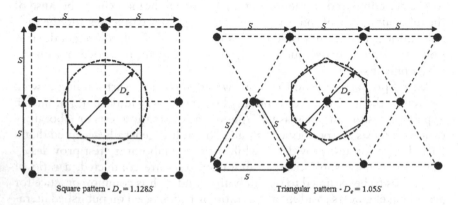

Square pattern - $D_e = 1.128S$ Triangular pattern - $D_e = 1.05S$

Figure 9.4 Relationship between drain spacing to drain influence zone.

9.2.3 Conversion of rectangular shape of the wick drain cross section to an equivalent circular area

Most wick drains are rectangular in section with a width of about 100 mm and a thickness generally between 2 mm and 5 mm. A schematic diagram of a typical wick drain is shown in Figure 9.1 where b is the breadth/width and t is the thickness of the wick drain.

Although wick drains are rectangular in cross section, as previously mentioned, design analysis is generally carried out under axisymmetric conditions. Hence, the cross section needs to be converted to an equivalent circle, say with a diameter d_w. As the dissipation occurs on the perimeter of the wick drain, Hansbo (1979) proposed that the perimeter of the cross section be made equivalent to a circle and obtain the equivalent diameter, i.e.,

$$d_w = 2(b+t)/\pi \qquad\qquad (9.2)$$

where d_w is the equivalent diameter (see Figure 9.5).

Numerical analysis carried out by Rixner et al. (1986) proposed the following modification:

$$d_w = (b+t)/2 \qquad\qquad (9.3)$$

This was supported by Hansbo in 1987 and is routinely used in practice. The difference between the d_w values obtained from Eq. (9.2) and Eq. (9.3) is about 20%. However, the estimated degree of consolidation using the drain diameter from Eq. (9.2) and Eq. (9.3) is negligible, less than 5%.

9.2.4 Smear zone radius and permeability

Wick drains are usually pushed into the ground using an open mandrel which is rectangular in cross section. The wick drain is connected to an

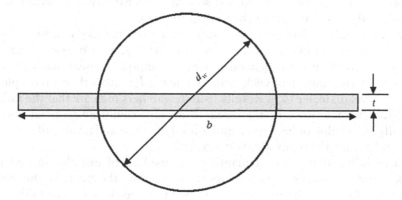

Figure 9.5 Equivalent diameter of a wick drain.

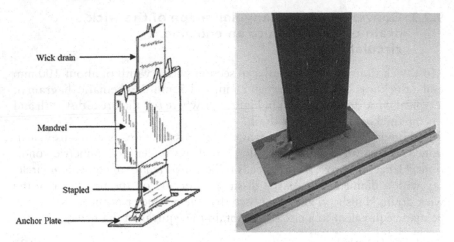

Figure 9.6 (a) Anchor plate (after Rixner, et al. (1986)); (b) Mandrel (courtesy of Zen Ng and Jarreau Alinur).

anchor plate that keeps the soil not being pushed into the space inside the mandrel. Figure 9.6 shows a typical mandrel and an anchor plate. The wick drain is slipped through the handle of the anchor plate and stapled. Mandrel is stopped when it pushes into the harder stratum where the toe should be and the wick is anchored. Then the wick drain is cut at the top, mandrel withdrawn and the machine commences the next wick drain installation.

One of the disadvantages of wick drain installation is the disturbance it makes in the form of smear around the hole created by the mandrel pushing the soil laterally in the process of advancing. Although the hole created and the smear zone are expected to be rectangular because the mandrel cross section is rectangular, in analysis, a circular hole and a circular smear zone are assumed. The diameter of the smear zone (d_s) depends not only on the mandrel cross section but the soil type too. In the smear zone, which comprises basically disturbed/remolded soil, the permeability is expected to be significantly less than that of the undisturbed soil.

Various authors have provided dimensions for the likely radius of the smear zone and c_h of the smear zone materials (e.g. Hansbo 1979; Sharma and Xiao 2000). It is the general practice to impose a smear diameter as a multiple of the equivalent wick drain diameter (d_w) rather than a multiple of the mandrel diameter (d_m). Results from researchers indicate that the radius and the characteristics of the smear zone vary significantly. Practitioners usually use a value of between 2 and 7 for d_s/d_w ratio and a value of between 1.5 and 5.0 for the permeability ratio, $k_h/k_{h,smear}$.

In addition to the size of mandrel, the method of installation and the experience of installers are also important to reduce the extent of the smear zone. For instance, the mandrel could be driven using static methods (i.e., pushing), dynamic methods or using vibration. The pushing type creates the least disturbance and therefore recommended especially when the compressible clays are sensitive. In such soils vibratory methods should be avoided. Dynamic or vibratory methods may need to be used to push through the upper crust and fill, which could be dense/hard, before reverting to ordinary pushing. Wick drain machinery now has the capability to switch on and off using just a switch and therefore it is easy to switch off the vibratory mode soon after passing through a dense layer or the crust.

9.2.5 Well resistance

Well resistance (i.e., limited discharge capacity) refers to the finite permeability of the vertical drain with respect to that of soil. Head loss occurs when water flows along the drain and delays radial consolidation to a certain extent. It should be pointed out that well resistance is controlled by the drain discharge capacity, soil permeability, the maximum discharge length and any geometric deficiencies (bending, kinks, etc.) in the drains.

Excess water trying to flow radially is expected to be resisted by the vertical drain. which has a finite permeability/discharge capacity. However, the discharge capacity of a wick drain is generally very high (i.e., >150 m^3/year) and only a little resistance is expected. Unless the depths of the drains are high, say 30 m or more, well resistance has little effect and therefore can be ignored in the analysis.

9.2.6 Discharge capacity

The discharge capacity controls the amount of water that would be dispelled and therefore directly relates to the performance of a wick drain. To maintain free flow, it is necessary to clearly specify the size of opening of the geotextile and the characteristics of the geotextile, which will ensure adequate discharge capacity and prevent significant loss of that capacity due to clogging.

As discussed in Australian Standard AS8700, the filter jacket of a wick drain has to meet two basic but contrasting requirements, viz.,

(a) Retain soil particles; and
(b) Allow the pore water to pass through without hindrance.

Indraratna and Bamunawita (2002) proposed requirements that should satisfy the above expectations, which includes suggestions by Bergado et al. (1996) as well.

Kinking is another issue that needs to be considered when high settlements are expected in a compressible clay layer. Kinking could be avoided by using a drain type that has the right stiffness and high flexibility. The wick drain selection should consider this, although it is not a major concern for most projects. Deformed drains with small kinks hardly reduce the flow rate but this can be assessed by carrying out tests in large triaxial setups. Ameratunga et al. (2017) mentions a project where wick drains were trialed and there was no indication of blocking, kinking or breaking due to high settlements, which were of the order of 1.5 m to 3 m. Prior to the field trials, large-scale laboratory tests have also been conducted on the proposed wick drain types to be used at the site to provide confidence that there would be no adverse behavior.

9.3 ASSESSMENT OF DEGREE OF CONSOLIDATION

Hansbo (1981) presented an approximate solution for vertical drains by considering both smear and well resistance based on the 'equal strain' hypothesis. The average degree of consolidation by radial flow, $\overline{U}_h$, presented by Hansbo (1981) is shown in Figure 9.7 and expressed as:

$$\overline{U}_h = 1 - \exp\left(\frac{-8T_h}{\mu}\right) \tag{9.4}$$

where, horizontal time factor, $T_h = \dfrac{c_h t}{D_e^2}$;

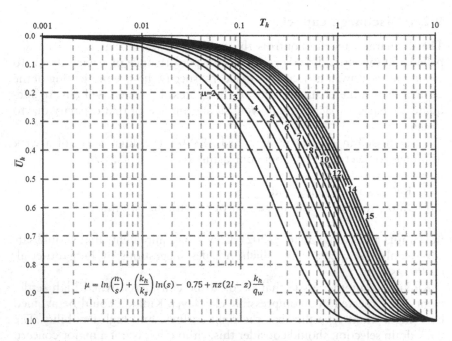

Figure 9.7 Variation of degree of consolidation $(\overline{U}_h)$ with time factor, T_h.

smear and well resistance factor,

$$\mu = \ln\left(\frac{n}{s}\right) + \left(\frac{k_h}{k_s}\right)\ln(s) - 0.75 + \pi z(2l - z)\frac{k_h}{q_w}$$

where

wick drain spacing ratio, $n = \dfrac{D_e}{d_w}$;

smear zone ratio, $s = \dfrac{d_s}{d_w}$;

k_s – permeability of disturbed/smear zone material;

z – depth; and

q_w – wick drain discharge capacity at the unit hydraulic gradient.

It is noted that the rate of consolidation is dependent on both the coefficient of consolidation and μ. The rate of consolidation will decrease with increasing μ as shown in Figure 9.7.

By ignoring well resistance, $\mu = \ln\left(\dfrac{n}{s}\right) + \left(\dfrac{k_h}{k_s}\right)\ln(s) - 0.75$ for a perfect drain (no well or no smear effect where $d_s = d_w$) $\mu = F(n) = \ln(n) - 0.75$

The average degree of consolidation due to both horizontal and vertical drainage $\overline{U}_{vh}$ can be assessed using the following equation and shown in Figure 9.8:

$$\overline{U}_{vh} = 1 - \left(1 - \overline{U}_v\right)\left(1 - \overline{U}_h\right) = 1 - \left(1 - \overline{U}_v\right)\exp\left(-\lambda T_v\right) \text{ where } \lambda = \frac{8}{\mu}\frac{T_h}{T_v} \quad (9.5)$$

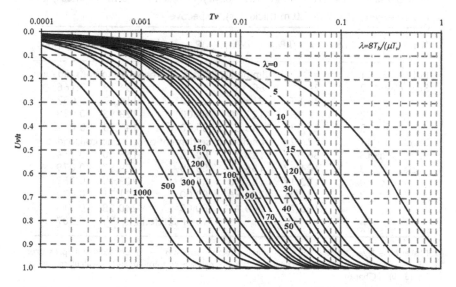

Figure 9.8 Variation of Average Degree of Consolidation ($\overline{U}_{vh}$) with Time Factor, T_v.

EXAMPLE 9.1

A cross section of a typical road embankment is shown in Figure 9.9 along with soil parameters. Ignoring creep behavior (purely for simplification) and assuming the construction is instantaneous, estimate the following:

a. Total primary settlement with embankment loading only;
b. Residual settlement (i.e., the settlement that is yet to occur after preloading) with a preloading time of 6 months with no wick drains;
c. Residual settlement with a preloading time of 6 months and 100 mm × 5 mm wick drains installed at 1.5 m spacing in a triangular pattern. Assume $c_h = 2c_v$; ignore vertical drainage; and neglect smear and well resistance of the wick drains (for simplification).

Note: As the embankment settles, the final embankment height would be slightly lower than 4.0 m. If a final height of 4.0 m is required additional fill (a surcharge) would be required to allow for such settlement. The effect of this compensation fill has not been taken into account in this example to keep the problem simple.

Answer:

a. *Estimation of total settlement*:

To improve the accuracy of the settlement calculation, the soft clay and soft to firm clay layers are divided into four sub-layers (each of 2.0 m thickness) and five sub-layers (each of 3.0 m thickness), respectively.

Layer	Hs (m)	σ'_{vo} (kPa)	σ'_p(kPa)	ΔP_{emb} (kPa)	σ'_f (kPa)	ρ_{prim} (m)*
Fill	1	9.00	Immediate settlement only, i.e., no consolidation			
Soft clay	2	24.19	60.48	80.00	104.19	0.166
	2	36.57	91.43	80.00	116.57	0.087
	2	48.95	122.38	80.00	128.95	0.038
	2	61.33	153.33	80.00	141.33	0.022
Sand	1	72.12	Immediate settlement only, i.e., no consolidation			
Soft to firm	3	86.00	128.99	80.00	166.00	0.114
clay	3	104.57	156.85	80.00	184.57	0.079
	3	123.14	184.70	80.00	203.14	0.053
	3	141.71	212.56	80.00	221.71	0.032
	3	160.28	240.41	80.00	240.28	0.016
				Total settlement $\sum \rho_{prim}$		0.607

Note: Hs – layer thickness; σ'_{vo} – initial effective stress; σ'_p – preconsolidation pressure; ΔP_{emb} – embankment load; σ'_f – final stress; ρ_{prim} – primary settlement; and * – calculated using Eq. (3.10) in Chapter 3

Total primary settlement = 0.607 m = 607 mm

b. Estimation of residual settlement with 6 month preloading:

Time factor $T_v = \dfrac{c_v t}{H_{dr}^2} = \dfrac{6 \times 0.5}{4^2} = 0.1875$ for the soft clay layer; and

$T_v = \dfrac{2.5 \times 0.5}{7.5^2} = 0.0222$ for the soft to firm clay layer

Layer	Hs (m)	σ'_{vo} (kPa)	σ'_p (kPa)	ΔP_{emb} (kPa)	Z	U_v	$\dfrac{U_v}{\Delta P_{emb}}$ (kPa)	$\sigma'_v (EPL)$ (kPa)	σ'_f (kPa)	ρ_{prim} (m)
Fill	1	9.00	Immediate settlement only, i.e., no consolidation							
Soft	2	24.19	60.48	80.00	0.25	0.687	54.96	79.15	104.19	0.072
clay	2	36.57	91.43	80.00	0.75	0.262	20.95	57.52	116.57	0.075
	2	48.95	122.38	80.00	1.25	0.262	20.95	69.90	128.95	0.028
	2	61.33	153.33	80.00	1.75	0.687	54.96	116.29	141.33	0.005
Sand	1	72.12	Immediate settlement only, i.e., no consolidation							
Soft to	3	86.00	128.99	80.00	0.20	0.343	27.42	113.41	166.00	0.104
firm	3	104.57	156.85	80.00	0.60	0.004	0.35	104.92	184.57	0.079
clay	3	123.14	184.70	80.00	1.00	0.000	0.00	123.14	203.14	0.053
	3	141.71	212.56	80.00	1.40	0.004	0.35	142.06	221.71	0.032
	3	160.28	240.41	80.00	1.80	0.343	27.42	187.69	240.28	0.010
Total settlement $\sum \rho_{prim}$										0.458

Note: Z – depth ratio; U_v – degree of consolidation due to vertical drainage; and $\sigma'_v (EPL)$ = $\sigma_{vo} + U_v \, \Delta P_{emb}$ – effective stress at end of preloading

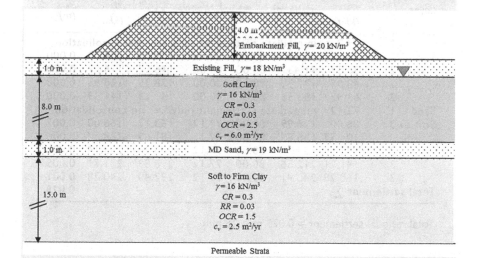

Figure 9.9 Ground model for Example 9.1.

Total residual settlement = 0.458 m = 458 mm

c. Estimation of residual settlement with six months of preloading with wick drains:

Equivalent wick drain diameter using Eq. (9.2b), $d_w = 2 \times (100 + 5) / \pi = 66.8$ mm = 0.0668 m

Diameter of influence zone using Eq. (9.1b), $D_e = 1.05 \times 1.5 = 1.575$ m

$$n = \frac{D_e}{d_w} = \frac{1.575}{0.0668} = 23.562;$$

$$F(n) = \ln(n) - 0.75 = \ln(23.562) - 0.75 = 2.41$$

Soft clay:

$$T_h = \frac{c_h t}{D_e^2} = \frac{12 \times 0.5}{1.575^2} = 2.42; U_h = 1 - \exp\left(\frac{-8T_h}{F(n)}\right) = 1 - \exp\left(\frac{-8 \times 2.42}{2.41}\right) = 1.000$$

S – F Clay:

$$T_h = \frac{5 \times 0.5}{1.575^2} = 1.01; U_h = 1 - \exp\left(\frac{-8T_h}{F(n)}\right) = 1 - \exp\left(\frac{-8 \times 1.01}{2.41}\right) = 0.964$$

Layer	Hs (m)	σ'_{vo} (kPa)	σ'_p (kPa)	ΔP_{emb} (kPa)	$U\Delta P_{emb}$ (kPa)	σ'_v (EPL) (kPa)	σ'_f (kPa)	ρ_{prim} (m)
Fill	1	9.00	Immediate settlement only, i.e., no consolidation					
Soft	2	24.19	60.48	80.00	80.00	104.19	104.19	0.000
clay	2	36.57	91.43	80.00	80.00	116.57	116.57	0.000
	2	48.95	122.38	80.00	80.00	128.95	128.95	0.000
	2	61.33	153.33	80.00	80.00	141.33	141.33	0.000
Sand	1	72.12	Immediate settlement only, i.e., no consolidation					
Soft to	3	86.00	128.99	80.00	77.12	163.12	166.00	0.007
firm	3	104.57	156.85	80.00	77.12	181.69	184.57	0.006
clay	3	123.14	184.70	80.00	77.12	200.26	203.14	0.006
	3	141.71	212.56	80.00	77.12	218.83	221.71	0.005
	3	160.28	240.41	80.00	77.12	237.40	240.28	0.001
Total settlement $\sum \rho_{prim}$								0.025

Total residual settlement = 0.025 m = 25 mm

9.4 GEOTECHNICAL INSTRUMENTATION

Similar to preloading discussed in Chapter 8, geotechnical instrumentation and monitoring are essential to assess ground improvement under a preload with wick drains, and similar instruments are used. A typical cross section with geotechnical instrumentation for a highway project is shown in Figure 9.10.

Geotechnical instrumentation for ground improvement is discussed in detail in Chapter 16.

9.5 ASSESSMENT OF PRELOAD PERFORMANCE

The tools used for the assessment of preload behavior with wick drains are not different to the approach discussed in Chapter 8.

9.5.1 Back analysis of preload performance for stability

Assessment of preload performance for stability is discussed in Chapter 16.

9.5.2 Back analysis of preload performance for settlement

Back analysis of preload performance for settlement is carried out similar to that under preloading without wick drains, i.e., using trial and error. The

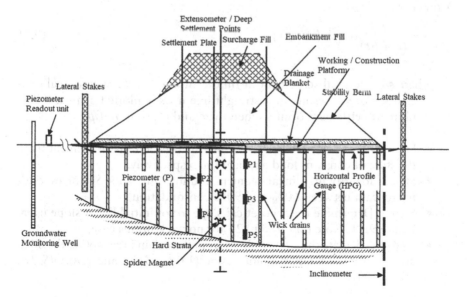

Figure 9.10 Typical instrumented cross section.

issues are more complex because of additional variables as discussed in Section 9.2.

The most used graphical methods of back analysis are:

- Asaoka method (Asaoka 1978) – probably the most popular (described by Balasubramaniam and Brenner (1981)). Although the procedure is similar to the case of preloading without wick drains, there is a modification to cater for horizontal dissipation and obtain c_h (see Magnan and Deroy 1980);

$$c_h = -\frac{\dfrac{\ln(\beta_1)}{\Delta t}}{\dfrac{8}{d_e^2\,\mu} + \dfrac{\pi^2}{4H_{dr}^2}\dfrac{c_v}{c_h}} \tag{9.6}$$

- Hyperbolic method (Tan et al. 1991 and Tan 1995).

The use of the Asaoka method was discussed in detail in Chapter 8; the hyperbolic method is discussed in detail here.

9.5.2.1 Hyperbolic method

The hyperbolic method was introduced by Tan et al. (1991; Tan 1995) by proposing the following hyperbolic equation to fit the Terzaghi consolidation curve.

$$s = \frac{t}{(\alpha + \beta t)} \ or \ \frac{t}{s} = (\alpha + \beta t) \tag{9.7}$$

where s is the total settlement at time 't' and α and β are empirical constants. Eq. 9.7 results in a straight line if t/s is plotted against t. The ultimate settlement occurs when $t = \infty$ and is given by $1/\beta$.

- Step 1: Plot t/δ vs t where t and δ are the time and the settlement from the start of constant load application, respectively;
- Step 2: Identify first linear segment and measure slope S_i (corresponding to data between 60% and 90% consolidation);
- Step 3: Determine the theoretical slope (α) of initial linear slope using Figure 9.11. Note that $\alpha = 0.824$ for no wick drains case;
- Step 4: Calculate the slope of lines from the origin intercepting the 60% and 90% consolidation points using $(1/0.6)(S_i/\alpha)$ and $(1/0.9)(S_i/\alpha)$, respectively;

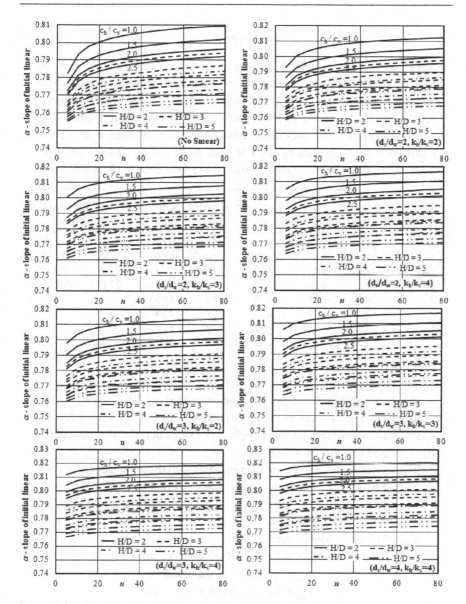

Figure 9.11 Relationship initial linear slope (α) and drainage parameters.

- Step 5: Construct these lines in the plot (Step 1) and locate settlement corresponding to 60% and 90% consolidation (i.e., δ_{60} and δ_{90});
- Step 6: The total primary settlement shall be estimated from α/S_i, $\delta_{60}/0.6$ or $\delta_{90}/0.9$. Note that all three values should be close to each other.

EXAMPLE 9.2

Settlement monitoring record of an embankment over a 10.0 m thick soft clay layer sandwiched between two permeable layers, is given below. The construction fill history indicates that the full height is reached by day 60. Estimate the total primary settlement using the hyperbolic method considering the following: smear zone radius = 4 × wick drain radius, permeability in smear zone = 1/4 of undisturbed zone permeability, c_h = 1.5c_v, wick drains are installed at 1.3 m spacing in a triangular pattern and the dimension of a wick drain is 100 mm × 3 mm.

Time (Days)	Settlement (mm)	Time (Days)	Settlement (mm)	Time (Days)	Settlement (mm)	Time (Days)	Settlement (mm)
0	0	54	332	121	548	212	776
10	27	59	357	128	589	218	778
16	35	80	393	135	598	226	790
26	67	93	449	144	620	232	792
33	141	101	479	163	680		
42	162	108	496	194	739		
50	262	114	524	207	768		

Solution:

Step 1: The plot of t/δ vs t is presented in Figure 9.12.

Step 2: Identify first linear segment and measure slope S_i = 0.00081.

Step 3: Theoretical slope of initial linear slope using Figure 9.11, α = 0.78.

Step 4: Draw two lines through origin with a slope of $(1/0.6)(S_i/\alpha)$ = 0.00174 and $(1/0.9)(S_i/\alpha)$ = 0.00116.

Step 5: Locate the intersection of above lines with the initial slope line. From Figure 9.12 the intersections are (t_{60} = 117, t_{60}/δ_{60} = 0.203) and (t_{90} = 313, t_{90}/δ_{90} = 0.362). Hence, δ_{60} = 117 / 0.203 = 576 mm and δ_{90} = 313 / 0.362 = 865 mm

Step 6: The total primary settlement, δ_{100} is:

$$\delta_{100} = \alpha/S_i = 0.78 /0.00081 = 963 \text{ mm}$$
$$\delta_{100} = \delta_{60}/0.6 = 576/0.6 = 960 \text{ mm}$$
$$\delta_{100} = \delta_{90}/0.9 = 865/0.9 = 961 \text{ mm}$$

9.6 NUMERICAL ANALYSIS

Numerical analysis of embankments with wick drains is slightly different to those without wick drains (refer to Section 8.7) because wick drains and smear zones have to be included in the modelling. The influence of wick drains could be simulated in numerical modelling as follows:

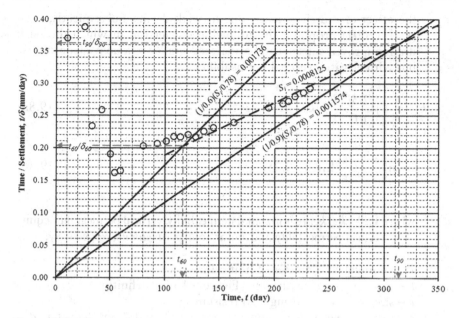

Figure 9.12 Hyperbolic plot.

- Either using simple mass permeability method; or
- Using zero excess pore pressure element to simulate the drain (i.e., drain element) and adjusting the permeability to cater for the drainage boundary and smear.

The disadvantage of using the simple mass permeability method is that the excess pore pressure at a particular point cannot be matched.

The available methods are discussed below.

9.6.1 Equivalent (mass) vertical permeability method

In this method, the wick drain is not modelled but the vertical permeability is increased to cater for wick drains and smear. This is ideal if we are interested in a settlement profile or average degree of consolidation. There are a number of methods available and some of them are:

- Chai et al. (2001):
  Chai et al. (2001) discussed that in a macro point of view, wick drains increase the mass hydraulic conductivity of subsoil and hence it is logical to establish an equivalent value of vertical hydraulic conductivity to represent both the effect of vertical drainage of natural subsoil and

the effect of radial drainage due to wick drains. Subsequently, Chai et al. (2001) proposed the following conversion in which the wick drain improved subsoil can be analyzed similar to a no wick drain case but with elevated vertical permeability.

$$k_{ve} = \left(1 + \frac{2.5l^2}{\mu D_e^2} \frac{k_h}{k_v}\right) k_v \qquad (9.8a)$$

$$\mu = \ln\left(\frac{n}{s}\right) + \frac{k_h}{k_s} \ln(s) - 0.75 \qquad (9.8b)$$

where k_h, k_v, k_{ve}, k_s are horizontal permeability, vertical permeability, equivalent vertical permeability and smear zone permeability, respectively;

D_e – equivalent diameter of unit cell;
l – drainage path length (refer Figure 9.13 for definition);
μ – smear factor ignoring well resistance;
s – ratio between smear zone diameter (d_s), and drain diameter (d_w), i.e. d_s/d_w; and
$n = D_e/d_w$

- **CUR 191 (1997):**
  This method is very similar to Chai et al. (2001) but a factor of $32/\pi^2$ (= 3.24) is recommended instead of 2.5 in Eq. (9.8a), i.e., the equation should be modified as:

$$k_{ve} = \left(1 + \frac{32}{\pi^2} \frac{l^2}{\mu D_e^2} \frac{k_h}{k_v}\right) k_v \qquad (9.8c)$$

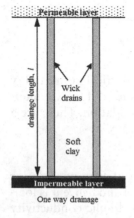
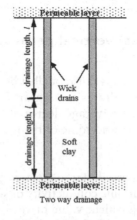
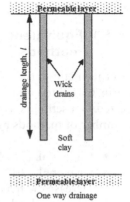

Figure 9.13 Illustration of drainage path length.

9.6.2 Equivalent (plane strain) horizontal permeability method

In this method, the wick drain is simulated using a drain element with zero excess pore water pressure along the drains. However, in modelling the wick drains, it is a general industry practice to ignore the smear zone due to complications with meshing (slender element) or to reduce the modelling/calculation time. Hence, it is necessary to amend the horizontal permeability to cater for smear and flow directions, i.e., radial flow vs horizontal flow. The equivalent horizontal permeability depends on the model used as discussed below:

- CUR 191 (1997):
  In this method, the horizontal conductivity (i.e., equivalent plane strain permeability, k_{hp}) is amended to simulate the wick drains as given in Eq. (9.9). However, this method is rarely used as k_{hp} is a function of degree of consolidation, i.e., the equivalent permeability is time-dependent.

$$k_{hp} = \alpha \frac{D_e^2}{\mu l^2} k_h \tag{9.9}$$

  where α is a function of the degree of consolidation:

$$\alpha = 3.24 \frac{\ln(1-U) + 0.21}{\ln(1-U)} \tag{9.10}$$

- 3D or axisymmetric:
  In 3D or axisymmetric modelling, the actual condition could be modelled. However, one can expect mesh complications due to the thin smear zone around the wick drains and require finer elements, which will significantly increase the computation time. Hence, it is often the industry practice to use an idealized model of having no smear zone as shown in Figure 9.14 but accounting for the smear zone by the use of an amended horizontal permeability. The equivalent horizontal permeability (k_{he}) could be estimated using the following equation:

$$k_{he} = \left(\frac{\ln(n) - 0.75}{\ln\left(\dfrac{n}{s}\right) + \dfrac{k_h}{k_s} \ln(s) - 0.75} \right) k_h \tag{9.11}$$

- Plane strain:
  Although actual consolidation around wick drains is axisymmetric, most of the finite element analyses are generally conducted based on a plane strain model due to high computational/modelling time or, in the

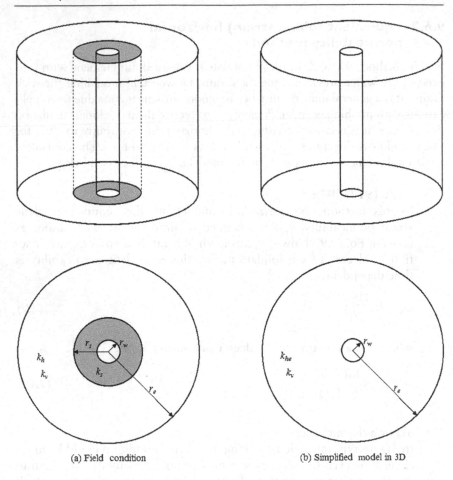

(a) Field condition   (b) Simplified model in 3D

Figure 9.14 Idealization of wick drains in 3D modelling – simplified model.

case of 3D modelling, requiring high speed computers. Hence, it is essential to use an idealized model with the horizontal permeability amended to cater for flow direction and the smear zone (if the smear zone is not modelled due to mesh complication). Refer to Figure 9.15 for an illustration of model conversion. The equivalent plain strain horizontal permeability in the undisturbed zone (k_{hp}) and smear zone (k_{sp}) could be estimated using the following equations (Indraratna et al. 2005):

$$k_{hp} = \left(\frac{\frac{2}{3}\left(1-\frac{1}{n}\right)^2}{\ln(n)-0.75} \right) k_h \qquad (9.12)$$

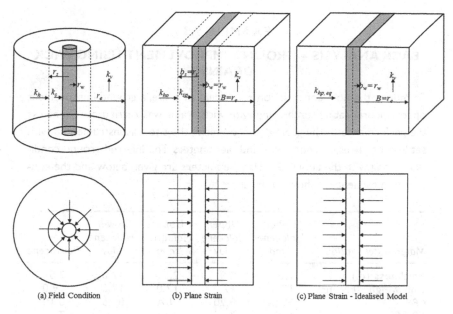

(a) Field Condition   (b) Plane Strain   (c) Plane Strain - Idealised Model

Figure 9.15 Conversion of axisymmetric into plane strain wall.

$$k_{sp} = \frac{\beta k_{hp}}{\dfrac{k_{hp}}{k_h}\left(\ln\left(\dfrac{n}{s}\right) + \dfrac{k_h}{k_s}\ln(s) - 0.75\right) - \alpha} \tag{9.13}$$

where $\alpha = \dfrac{2}{3}\dfrac{(n-s)^3}{n^2(n-1)}$ and $\beta = \dfrac{2}{3}\left(1 - \dfrac{1}{n}\right)^2 - \alpha$

If the smear zone is not modelled due to mesh complications, the equivalent plane strain permeability, $k_{hp,eq}$ could be estimated by combining Eq. (9.12) and Eq. (9.13) as shown below:

$$k_{hp,eq} = \left(\frac{\dfrac{2}{3}\left(1 - \dfrac{1}{n}\right)^2}{\ln\left(\dfrac{n}{s}\right) + \dfrac{k_h}{k_s}\ln(s) - 0.75}\right) k_h \tag{9.14}$$

EXAMPLE 9.3

BACK ANALYSIS – GROUND IMPROVEMENT USING WICK DRAINS

To reduce the post construction settlement of a site, a ground improvement option incorporating surcharging with wick drains was carried out. To assess the performance of the ground improvement, the site was instrumented with settlement plates, inclinometers and piezometers. The inferred ground conditions along with the adopted design parameters are given below and the construction sequence is shown in Figure 9.16.

Material description	Upper Holocene sand	Upper Holocene clay	Upper Holocene sand	Lower Holocene clay	Pleistocene
Thickness (m)	1.5	2.0	2.5	17.0	7.0
Unit weight (kN/m³)	17.0	17.0	17.0	17.0	19.0
CR	N/A	0.20	N/A	0.25	0.10
CR / RR	N/A	7	N/A	7	7
OCR	N/A	1.5	N/A	1.0	3.0
c_v (m²/yr)	N/A	6.0	N/A	3.0	5.0

The design parameters used in the wick drain design are given below:

Wick drain size = 100 mm wide × 3 mm thickness

Extent of smear zone, s = 3

Smear zone permeability, k_h/k_s = 5

Wick drain spacing = 1.25 m center to center, square pattern

Wick drain installation = installed from natural ground over a period of 30 days

The ratio between horizontal to vertical coefficient of consolidation = 2

Ground water table = 1.0 m below ground level

The performance of the embankment was analyzed using a finite element program, PLAXIS, based on the permeability conversion as discussed in Section 9.6.

Numerical assessment:

Step 1: Estimate design permeability of the clay materials. Most of the finite element programs require the coefficient of permeability (k) and hence it is essential to estimate /assess the permeability from the coefficient of vertical consolidation (c_v) using the following equations:

$$k_v = c_v m_v \gamma_w$$

where $m_v = \dfrac{1}{\Delta\sigma}\left(CR\log\left(\dfrac{\sigma'_{v0}+\Delta\sigma}{p'_c}\right) + RR\log\left(\dfrac{p'_c}{\sigma'_{v0}}\right)\right)$

The estimated coefficient of vertical permeability of the upper and lower Holocene clay is 1.610×10^{-4} m/day and 5.691×10^{-5} m/day, respectively.

Step 2: Estimate equivalent axisymmetric / plane strain permeability using Eq. (9.11) / Eq. (9.14) for modelling based on design parameters.

Upper Holocene clay: equivalent axisymmetric and plane strain permeability is 1.112×10^{-4} m/day and 2.908×10^{-5} m/day, respectively.

Lower Holocene clay: equivalent axisymmetric and plane strain permeability is 3.931×10^{-5} m/day and 1.028×10^{-5} m/day, respectively.

Step 3: Undertake modelling (initial run) with the design parameters. The monitored settlement along with numerical results based on design parameters are shown in Figure 9.17.

Step 4: The numerical analysis indicates that the actual rate of consolidation is smaller than design rate of consolidation. Hence, a re-assessment is undertaken using a smaller coefficient of consolidation. Figure 9.18 shows a comparison of monitored and predicted settlement based on a coefficient of consolidation of 2/3 of the design coefficient of consolidation. This re-assessment indicates that the actual rate of consolidation with modified coefficient of consolidation is comparable.

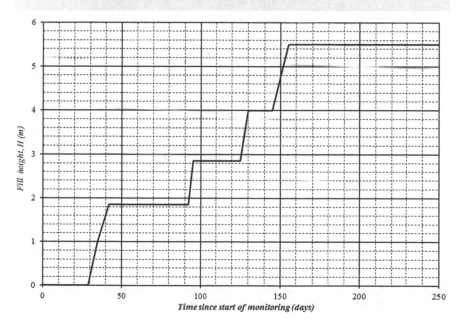

Figure 9.16 Construction history.

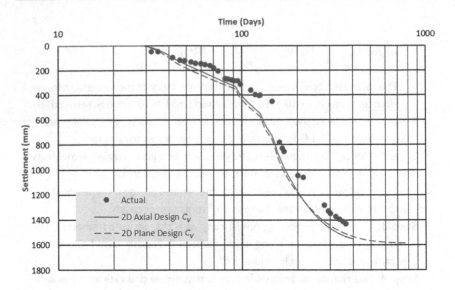

Figure 9.17 Comparison of monitored settlement with prediction based on design parameters.

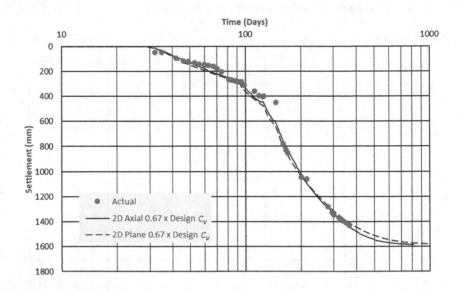

Figure 9.18 Comparison of monitored settlement with prediction based on 2/3 of design coefficient of consolidation.

9.7 ENVIRONMENTAL CONSIDERATIONS

Environmental considerations related to earthworks construction and dis-cussed in Chapter 6 are equally valid for wick drains and preloading. The use of a large machine such as the wick drain machine, which could have a mast as high as 40 m or more for routine wick drain work, needs special attention related to adequate working platforms and effects on services, not only underground but above ground.

Wick drains accelerate dissipation of excess pore water pressure and that means a significant amount of water is likely to come out of the ground. A settlement of 1 m means the volume of water removed from the soil strata is 1 m^3 per square meter on the plan area of the site. Hence, proper surface drainage should be maintained throughout the preload period to ensure that water, that comes out of the ground, has no impediment to flow. In addition, if acid sulfate soils are present, how the water could be treated should be planned prior to commencing work. It could be simply spreading hydrated lime along surface drains where the water is allowed to run. The amount of hydrated lime needed should be assessed by an experienced engineer.

More attention should be given if contaminants are found in the soil as to containment and processing of water coming out.

9.8 DESIGN AND CONSTRUCTION ISSUES

Design and construction issues related to preloading as discussed in Chapter 8 apply for preloading with wick drains also. Additional issues related to wick drains are given below.

- If the mandrel has to pass through a fill that has obstacles such as concrete rubble, wick draining will not be possible. In these instances, preboring, pre-augering or use of a drop hammer may be needed to penetrate this layer.
- If the subsurface profile includes layers of gravel or dense sand, push-ing the mandrel may be difficult. If the layer is limited to the overbur-den above the soft clay, the wick drain machine could be switched to vibratory mode to push through. Else, pre-boring will be needed.
- If sufficient anchoring is not achieved in a stiffer/dense layer, the wick drain could rise up with the mandrel withdrawal.
- If wick drains are found to be shorter than intended, i.e., total penetra-tion has not been achieved, analysis must be carried out to assess the effects. Depending on the depth of soft clay and untreated thickness of clay, and if the base is a free draining material, it is possible that no remedial measures may be needed except perhaps slightly longer dura-tion of preloading. Else, additional surcharging and/or time may be needed.

9.9 ADVANTAGES AND DISADVANTAGES

Most of the advantages and disadvantages, except time constraints, discussed under Chapter 8 apply for preloading with wick drains also. Additional points are summarized in Table 9.1.

Table 9.1 Additional advantages and disadvantages of preloading with wick drains

Relative time	Advantage – Time for preloading is significantly reduced by the incorporation of wick drains.
Spacing	Installing wick drains at too closer spacing may not be effective because of disturbance effects and the extent of the smear zone. The general practice is not to reduce the spacing below 1 m (Walker and Indraratna, 2007; Rankine et al., 2008), although spacing as low as 0.5 m has been reported.
Complexity of analysis	Disadvantage – Analysis is complex compared to preloading because of the introduction of wick drains. However, advanced but simple techniques to assess the results are now available.
Cost	Disadvantage – Relatively expensive compared to some other ground improvement methods including preloading without wick drains. Advantage – Still less expensive compared to some other ground improvement techniques such as semi-rigid inclusions.
Preload height	Advantage – Generally, the height of preload could be reduced when wick drains are introduced if available time for preloading is limited.
Parameters	Disadvantage – Too many design parameters, which makes it difficult to back analyse.
Edge effects	Disadvantage – Could create differential settlement issues adjoining a non-wick drain area and therefore transition should be taken into account in the design. In residential or building blocks, it is strongly suggested not to partially wick drain the footprint of the house.
Stability	Advantage – The rate of strength gain in soft clay is increased, which allows faster construction and/or less high strength geotextiles to manage stability.
Sensitive clays	Disadvantage – Wick drains may not be effective in very sensitive clay because of remoulding due to disturbance. There are several case histories where the performance was ineffective when the soils were sensitive (Liquidity index greater than 1 or more) – Chu et al. (2014); Wijeyakulasuriya et al. (1999).
Drained water	Disadvantage – Similar to preloading without wick drains, if the water coming out of the ground is contaminated, containment and processing would be needed.
Remedial measures if wick drains not properly installed	Disadvantage – If wick drains have not been executed properly as per design, say they are significantly short, depending on the clay properties, the first remedial action one would think of is the use of additional wick drains. However, installing additional wick drains may not be very effective because of extent of smear zones of closely spaced wick drains.

9.10 SAND DRAINS

Sand drains preceded wick drains but are still used as a form of ground improvement. A sand drain is basically a small hole, generally of the order of 300 mm or so, filled with sand, and acts in a similar manner to wick drains to facilitate dissipation of excess pore water pressure. Some sand drains use a mandrel with a closed end pushed through the soft soils and then filled with sand while withdrawing the mandrel. Others use augering or water jetting to make the hole before filling with sand.

Always, clean sand is used for sand drains to provide high permeability for the drains to act. One of the concerns is the effect of squeezing ground reducing the cross-sectional area of the drain and, in an extreme case, complete blockage.

Methods of analysis described for wick drains apply for sand drains also.

9.11 VACUUM CONSOLIDATION

The ground improvement with or without wick drains with conventional preloading is generally cost effective compared to many other ground improvement techniques. However, sometimes it requires a significant height of preload when,

- the clay thickness is high; and/or
- the target time for preloading is low; and/or
- stringent settlement criteria apply.

High surcharges create issues of instability and excessive lateral displacements. Hence, some other means of treatment is required to mitigate such issues. One of the methods that could be adopted is vacuum consolidation in conjunction with reduced surcharge. The main difference between the conventional preloading and vacuum preloading is that the former will increase the total stress while the latter will reduce the pore pressure while maintaining a constant total stress as illustrated in Figure 9.19. The main advantages of the vacuum consolidation technique are:

- Increased stability;
- Significantly reduced lateral displacements;
- Reduction in the required surcharge by 3.5 m to 4.0 m if a vacuum pressure of 70 to 80 kPa could be maintained thereby reducing the fill transport costs significantly.

The fundamental procedure of vacuum consolidation process consists of removing atmospheric pressure from a confined sealed (membranes/cut of wall) medium of soft soils to be consolidated and maintain the vacuum

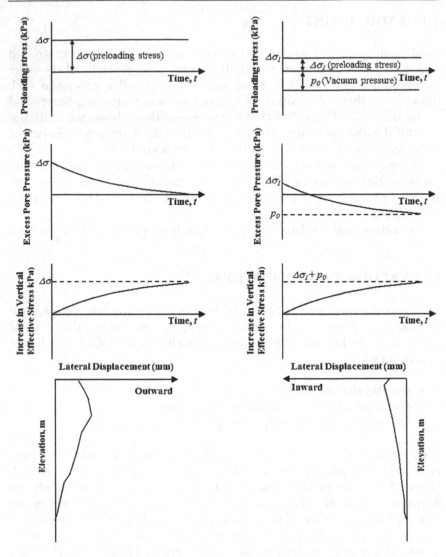

Figure 9.19. Consolidation process and typical lateral displacement at the boundary (a) conventional loading, (b) vacuum preloading.

pressure for a predetermined preloading time. A typical vacuum system is
illustrated in Figure 9.20.

The main setback in vacuum preloading is that the efficiency of the entire
system depends on the ability of the membrane to prevent any air leaks to
withstand a sufficient suction pressure over an extended period of time.
Another setback is that if there are sand lenses present within vacuum consolidation area, the vacuum consolidation process will be ineffective and would
require a cut of wall along the perimeter to reduce the lateral seepage.

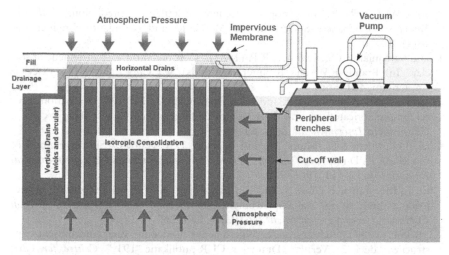

Figure 9.20 Typical vacuum consolidation set up (Menard Vacuum™).

To obtain an assessment at the pre-feasibility stage of a project, the design could be simplified to regular preloading with wick drains considering the vacuum as an additional load equivalent to the vacuum pressure. If the vacuum pressure to be maintained is 80 kPa, the equivalent preload height would be 4 m assuming a unit weight of 20 kN/m³ for the fill material.

A proper design of vacuum consolidation is possible using published literature. The reader is referred to Indraratna et al. (2005, 2012, 2013) and Indraratna (2017) for detailed descriptions and analytical methods.

9.12 SUMMARY

This chapter may be considered as an extension to Chapter 8 with the inclusion of wick drains to preloading. The design methodologies are described in detail and simple examples have been provided to comprehend easily. The advantages and disadvantages are also listed. Illustrations have been provided to allow the reader to visualize complex behavior and analysis.

REFERENCES

Ameratunga, J., Honeyfield, N., Cheah, B. and Ng, Z. (2017). *"Lessons learnt from geotechnical trials at Port of Brisbane."* *Proceedings of the 19th International Conference on Soil Mechanics and Geotechnical Engineering*, Seoul, South Korea, 1267–1270.

AS 8700 (2011). *Execution of Prefabricated Vertical Drains*. Standards Australia, Sydney.

Asaoka, A. (1978). "Observational procedure of settlement prediction." *Soils and Foundations*, Japanese Society of Soil Mechanics and Foundation Engineering, 18(4), 87–101.

Balasubramaniam, A.S., Brenner, R.P. (1981). 'Consolidation and settlement of soft clay.' In: Brand, E.W. and Brenner, R.P. (eds) *Soft Clay Engineering*. Elsevier, Amsterdam, 481–566.

Bergado, D.T., Manivannan, R. and Balasubramaniam, A.S. (1996). "Filtration criteria for prefabricated vertical drain geotextile filter jackets in soft Bangkok clay." *Geosynthetics International*, 3(1), 63–83.

Chai, J.C., Shen, S.L., Miura, N. and Bergado, D.T. (2001). "Simple method of modelling PVD improved subsoil." *Journal of Geotechnical and Geoenvironmental Engineering*, 127(11), 965–972.

Chu, J., Indraratna, B., Yan, S. and Rujikiatkamjorn, C. (2014). "Overview of preloading methods for soil improvement." *Proceedings of the Institution of Civil Engineers, Ground Improvement*, 167(GI3), 259–267.

CUR 191. (1997). "Achtergronden bij numierike modellung van geotechnische constructies, deel 2. Vertikal Drainage CUR-publikatie 191." *Centre for Civil Engineering Research Codes*, Gouda, Netherlands.

Hansbo, S. (1979). "Consolidation of clay by band-shaped prefabricated drains." *Ground Engineering*, 12(5), 16–25.

Hansbo, S. (1981). "*Consolidation of fine-grained soils by prefabricated drains.*" *Proceedings, 10th International Conference on Soil Mechanics and Foundation Engineering*, Stockholm, 3, 667–682.

Hansbo, S. (1987). "Design aspects of vertical drains and lime column installation." *Proceedings 9th Southeast Asian Geotechnical Conference*, 2(8), 1–12.

Indraratna, B. (2017). "*Recent advances in vertical drains and vacuum preloading for soft ground stabilization.*" *Proceedings of the 19th International Conference on Soil Mechanics and Geotechnical Engineering*, Seoul, Honours Lectures, 141–166.

Indraratna, B., Balasubramaniam, A., Poulos, H., Rujikiatkamjorn, C. and Ameratunga, J. (2013). "Performance and prediction of marine clay treated with vacuum and surcharge consolidation at Port of Brisbane." *Australian Geomechanics Journal*, 48(4), 161–180.

Indraratna, B. and Bamunawita, C. (2002). "Soft clay stabilisation with mandrel driven geosynthetic vertical drains." *ISSMGE-TC36 International Workshop*, Mexico City, Mexico. Published in *Australian Geomechanics Journal*, 37(5), 57–86.

Indraratna, B., Rujikiatkamjorn, C., Kelly, R. and Buys, H. (2012). "Soft soil foundation improved by vacuum and surcharge loading." *Proceedings of the Institution of Civil Engineers, Ground Improvement*, 165(GI), 87–96.

Indraratna, B., Sathananthan, I., Rujikiatkamjorn, C. and Balasubramaniam, A.S. (2005). "Analytical and numerical modelling of soft soil stabilized by prefabricated vertical drains incorporating vacuum preloading." *International Journal of Geomechanics, ASCE*, 5(2), 114–124.

Magnan, J.P. and Deroy, J.M. (1980). "The graphic analysis of observed settlements under structures." *Bulletin de Liaison des laboratories des Ponts et Chaussées*, 109, 45–52.

Rankine, B., Indraratna, B., Sivakugan, N., Wijeyakulasuriya, V. and Rujikiatkamjorn, C. (2008). "Foundation behaviour below an embankment on soft soils." *Proceedings of the Institution of Civil Engineers, Geotechnical Engineering*, 161(GE5), 259–267.

Rixner, J.J., Kraemer, S.R. and Smith, A.D. (1986). "Prefabricated vertical drains." *Summary of Research Report – Final Report, Federal Highway Admin, Report No. FHWA-RD-86/169*, Washington D.C, 433p.

Sathananthan, I. (2005). *"Modelling of vertical drains with smear installed in soft clay."* PhD thesis, University of Wollongong.

Sharma, J.S. and Xiao, D. (2000). "Characterisation of a smear zone around vertical drains by large-scale laboratory tests." *Canadian Geotechnical Journal*, 37(6), 1265–1271.

Tan, S.A. (1995). "Validation of hyperbolic method for settlement in clays with vertical drains." *Soils and Foundations*, 35(1), 101–113.

Tan, S.A. Inoue, T. and Lee, S.L. (1991). "Hyperbolic method for consolidation analysis". *Journal of Geotechnical and Geoenvironmental Engineering*, 117(1), 1723–1737.

Walker, R. and Indraratna, B. (2007). "Vertical drain consolidation with overlapping smear zones". *Geotechnique*, 57(5), 463–467.

Wijeyakulasuriya, V., Hobbs, G. and Brandon, A. (1999). *"Some experiences with performance monitoring of embankments on soft clays."* *Proceedings 8th Australia New Zealand Conference on Geomechanics*, Hobart, 783–788.

Chapter 10

Stone columns

K. Chan and B. Poon

10.1 INTRODUCTION

Stone columns have a long history as a ground improvement method to improve clayey soils and silty sands. It is said that for the construction of the Taj Mahal in India in the 1650s, as well as for work on military sites in France in 1830, this technique was used (Chan and Poon 2015). The stone column system is based on a simple concept to improve the mechanical characteristics of the soft soils, mainly shear strength and compressibility, by inserting columns comprising stone/gravel.

In reality, the concept cannot be any simpler since the installation method consists of holes being created in the ground and then filled with well-compacted gravel or stones. As compacted gravel has significantly higher strength and lower compressibility than those of the untreated surrounding soil, the columns improve the bearing capacity of the soft soils, increase shear resistance, reduce magnitude of settlement and increase the rate of settlement by providing better drainage. Where settlement control is the main objective, stone columns are generally used in combination with pre-loading and/or surcharging.

The stone column technique is frequently used on major roads and highways, wharves, buildings, warehouses, storage tanks as well as to support utilities. It can also be adopted to increase slope stability and reduce liquefaction potential of sand.

It should be noted that stone columns can yield and should not be treated as structural members, i.e., they have very limited bending capacity and no tensile capacity. It is generally accepted that stone columns cannot be effectively used in very soft clays because of bulging failure due to lack of confinement, which the designer should assess during the concept evaluation stage of a project. Design and analysis, as will be discussed later, can be carried out based on theoretical approaches although empirical methods are very popular due to their simplicity. During the last two to three decades, numerical analysis has also become popular because of the availability of suitable industry software.

Stone columns are generally limited to about 20 m to 25 m depth, but deeper columns have been successfully implemented. The maximum depth of installation is usually controlled by the height of the installation rig and probe. The stone columns usually range in diameter from 800 mm to 1200 mm and the load carried by each column could be up to about 300 kN. As the column spacing, in either a square or triangular grid, varies from about 1.5 m to 3.5 m, the percentage of area replacement generally varies between 15% and 40%.

Where the soil layer requiring treatment is thicker than the maximum length of probe, the columns will act as floating columns. In such cases, it is possible to adopt a solution in combination with preloading and prefabricated vertical drains.

10.2 VARIATIONS OF STONE COLUMN SYSTEMS

There are many stone column systems contractors use in the field. Some of the more frequently used stone column installation techniques are briefly described below.

10.2.1 Vibro-replacement (wet) method

The vibro-replacement method involves using water jetting to displace the soil being treated. A probe is progressed to the design depth during the water jetting operation. After flushing out the hole, gravel is added usually from the ground surface and compacted using a vibrator attached to the probe. This method of backfilling the hole with feeding gravel from the ground surface is called the 'top feed' method. The vibro-replacement method is best suited for sites where the groundwater table is high and soils are relatively soft and therefore borehole stability is doubtful. Columns created using this method are generally larger than those created using the dry method. Hence, relatively large load per column can be accommodated.

10.2.2 Vibro-displacement (dry) method

This method of stone column installation involves a construction process without introducing water. The stone column hole is created by relying on the weight of the probe and vibration. The soils being treated are therefore displaced laterally rather than brought up to the ground surface. The gravel is introduced into the hole by either using the 'top feed' method or the 'bottom feed' method whereby the gravel is discharged at the tip of the probe. Since the side wall of the hole needs to stand without collapsing and no support, it is most suited for sites where the water table is relatively deep and the undrained shear strength is 40 to 60 kPa (FHWA 1983).

10.2.3 Geotextile encased columns

Both the vibro-replacement and vibro-displacement methods require the in situ soils to have adequate undrained shear strength such that some lateral restraints can be provided to the stone columns. Otherwise, the installed stone columns could fail by bulging. One possible solution to overcome the bulging issue is to adopt geotextile encased columns. These columns comprise high strength geotextile tubes installed into the soft soil using a mandrel. Both the replacement and displacement methods can be used to create the column hole and to install the geotextile tubes. The geotextile tube is then filled with fine gravel or other granular materials to form the stone column. These columns are typically 750 mm in diameter.

10.2.4 Dynamic replacement method

Using the dynamic replacement method, the stone columns are installed by compaction using high energy (see Figure 10.1). A 15 to 30-tonne pounder is dropped from heights as high as 40 m in some projects. Craters are formed that are filled with gravel or rock. It is a continuous process of pounding to create a hole, filling and pounding again to achieve compaction. Another difference to other types of stone columns is the size of the hole. The column diameter formed by the dynamic replacement method is generally greater than those formed by vibro-displacement methods. The diameter of an individual column is usually greater than the size of the pounder because

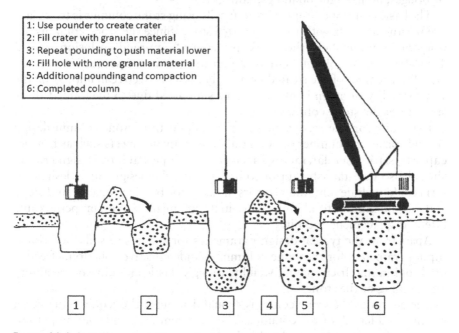

1: Use pounder to create crater
2: Fill crater with granular material
3: Repeat pounding to push material lower
4: Fill hole with more granular material
5: Additional pounding and compaction
6: Completed column

Figure 10.1 Installing dynamic replacement columns.

pounding pushes soil laterally. One advantage of dynamic replacement is the speed of installation. It appears to be a crude technique to improve the ground but certainly yields results to the designer. However, several limitations of dynamic replacement need to be highlighted. They are:

- The depth of installation is usually limited to about 6 m. Once a column reaches its maximum depth, further pounding does not make the hole deeper but moves laterally and upward, creating heave.
- The process of pounding gives rise to significant vibration and may not be suitable where sensitive structures and/or utilities are present.
- The pounding also gives rise to significant noise, which may not be acceptable to occupants in some buildings such as hospitals or aged care.

The stone columns formed by dynamic replacement method can be used in combination with preloading when settlement control is a major objective. In addition, wick drains can be used to accelerate the consolidation process or to facilitate pore pressure dissipation of lower layers when the depth of weak clay extends below 6 m depth.

10.3 DESIGN CONSIDERATIONS

Some typical applications of stone columns include foundation treatment for the construction of road and rail embankments, approach embankments of bridges, lightweight buildings, tanks, etc.

The basic concept of improving soft soils using stone columns is to replace the in situ soft soils with the compacted stone. Since the stone columns are not structural members, they are considered a 'semi-rigid' inclusion system. Typically, a stone column system can greatly reduce the settlement and lateral displacement of the treated ground, but is less effective in improving the slope stability of an embankment being supported due to the limited shear strength of the stone columns.

Depending on the requirements of the application, stone column design should consider a number of potential improvement aspects such as bearing capacity, settlement, lateral displacement and slope stability. Design criteria should also be established prior to undertaking the design. Such design criteria could include minimum factors of safety for bearing capacity and slope stability, maximum total and differential settlement, maximum permissible lateral displacement and tilt.

Apart from the typical design parameters for the in situ soils, additional input parameters for the stone columns include effective cohesion, effective angle of internal friction, unit weight, Young's elastic modulus, permeability and grading of the stones.

The design of the stone columns should determine the required diameter, spacing and length of the columns. The treatment area and column pattern (rectangular or triangular) should also be specified.

10.4 ULTIMATE BEARING CAPACITY OF SINGLE STONE COLUMNS

10.4.1 Failure mechanisms

The failure mechanisms of single stone columns are outlined in Figure 10.2. For a long column with a length to diameter ratio greater than about 3, a bulging failure mechanism will usually develop under excessive loading as shown in Figure 10.2a. A short column less than about 3 diameters in length may undergo a general or local bearing failure mechanism near the surface as shown in Figure 10.2b, or a punching failure when founded on underlying weak layer as shown in Figure 10.2c.

The determination of the bearing capacity of single, isolated stone columns can be based on several approaches:

- General bearing capacity and punching failure of short columns (Madhav and Vitkar 1978);
- Cavity expansion theory of long columns (Hughes and Withers 1974);
- Undrained shear strength method for long columns (Mitchell 1981).

Since most stone columns have length to diameter ratios equal to or greater than 3, bulging is usually the controlling failure mechanism under heavy loads.

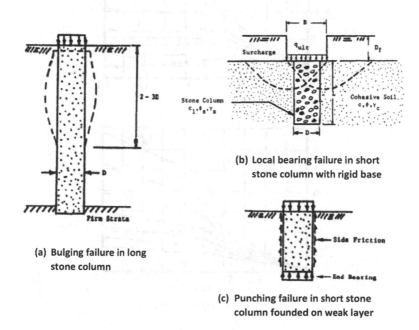

(a) Bulging failure in long stone column

(b) Local bearing failure in short stone column with rigid base

(c) Punching failure in short stone column founded on weak layer

Figure 10.2 Failure mechanisms of single stone columns.

10.4.2 Bearing capacity of short stone columns

A short stone column may undergo a general bearing capacity failure of the stone and surrounding soil, or a punching failure into the underlying soft soil.

Barksdale and Bachus (1983) showed that the ultimate capacity for punching failure can be determined by summing the end bearing capacity of the stone column using conventional bearing capacity theories and the skin friction along the shaft of the column.

By considering the plane strain solution developed by Madhav and Vitkar (1978) for the general bearing capacity of a trench filled with granular material constructed in a cohesive soil, Barksdale and Bachus (1983) showed that the general bearing capacity of a short stone column can be given by:

$$q_{ult} = \frac{\gamma_c B}{2} N_\gamma + c N_c + D_f \gamma_c N_q \tag{10.1}$$

where N_γ, N_c and N_q are bearing capacity factors given in Figure 10.3, and the other terms used in the equation are defined in Figure 10.2b. γ_c is the unit weight of the clay. ϕ_s in Figure 10.3 is the effective friction angle of the stones.

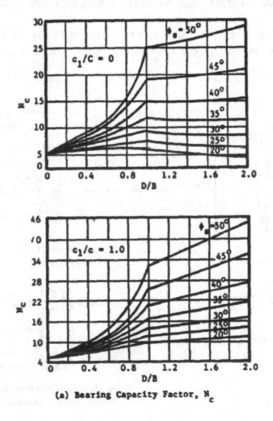

(a) Bearing Capacity Factor, N_c

Figure 10.3 Bearing capacity factors.

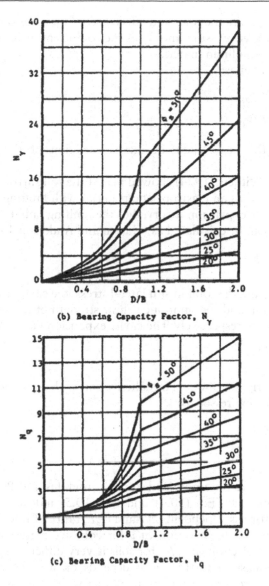

(b) Bearing Capacity Factor, N_γ

(c) Bearing Capacity Factor, N_q

Figure 10.3 (Continued) Bearing capacity factors.

10.4.3 Cavity expansion theory for long stone column undergoing bulging failure

By considering plastic equilibrium within the stone column material, Hughes and Withers (1974) showed that the ultimate capacity of a single column undergoing bulging failure can be given by:

$$q_{ult} = K_{ps} \times \sigma'_{rL} = K_{ps} \times (\sigma_{rL} - u) \tag{10.2}$$

where σ_{rL} and σ'_{rL} are the total and effective confining pressures respectively, u is the pore water pressure and K_{ps} is the coefficient of passive earth pressure of the stone column material.

K_{ps} can be taken as:

$$K_{ps} = \frac{1+\sin\phi_p}{1-\sin\phi_p} \tag{10.3}$$

where ϕ_p is the peak angle of internal friction of the stone column aggregate.

Hughes and Withers (1974) showed that, if the soil surrounding the stone column is treated as an elasto-plastic material with limiting stress given by cylindrical cavity expansion theory, then the limiting radial stress, σ_{rL}, of the stone column can be calculated as (Gibson and Anderson 1961):

$$\sigma_{rL} = \sigma_{ro} + \beta \times c_u \tag{10.4}$$

where σ_{ro}, β, c_u, are the total in situ lateral stress, the cavity expansion factor for stone column and the undrained shear strength of the soil surrounding the stone column, respectively. The cavity expansion factor, β, is given by:

$$\beta = 1 + \ln(I_r) \tag{10.5}$$

where I_r is the rigidity index. For cohesive soil with zero undrained friction angle, I_r may be expressed as:

$$I_r = \frac{E_u}{2c_u(1+v)} \tag{10.6}$$

where E_u and v are the undrained Young's modulus and Poisson's ratio of the clay, respectively. Eq. (10.4) and Eq. (10.5) indicate that the cavity expansion factor β is dependent on v and E_u/c_u ratio. Although there is much higher certainty in the selection of v for cohesive materials in undrained conditions, the selection of E_u/c_u for soils is very difficult as pointed out by Frikha and Bouassida (2015), as it depends on a wide range of factors including the degree of consolidation, plasticity, shear stress ratio and strain level. Figure 10.4 shows a plot of E_u/c_u versus OCR as given by Ladd et al. (1977), where E_u is the Young's modulus of soils in undrained condition, and OCR is the overconsolidation ratio. It can be seen that there is a wide range of E_u/c_u values that also depends on the index properties of the soils. The plot was developed based on the applied shear stress level being about one-third of the soil shear strength. As the applied shear stress increases further, the ranges of the E_u/c_u as a function of OCR will reduce.

Figure 10.4 indicates that E_u/c_u ranges from 100 to 1500 for clays of high to low plasticity. Therefore, for undrained deformation of saturated clay

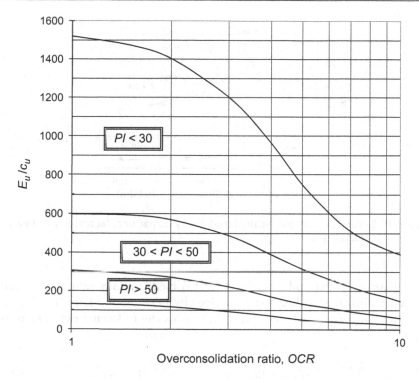

Figure 10.4 Plot of E_u/c_u versus OCR.

with $\nu = 0.5$, the β values are expected to be within the range of about 4.5 to 7.0. From a detailed examination of many field records of quick expansion pressuremeter tests, Hughes and Withers (1974) indicated that the β value could be less than the above expected range and $\beta = 4$ was adopted. Using the recorded in situ data from load tests performed on different stone column configurations, Frikha and Bouassida (2015) calibrated the cavity expansion model and indicated that $\beta = 3.1$ gives the trend line that best fits the data set.

Based on a database of high-quality load test data of spread footings supported on stone column reinforced clay, Stuedlein and Holtz (2013) back-calculated β representative of the large strains associated with passive failure of the matrix soil mass. It was found to vary as a function of c_u as shown in Figure 10.5 and with a best fit relationship of:

$$\beta = -1.45\ln(c_u) + 8.52 \tag{10.7}$$

By equating Eq. (10.5) and Eq. (10.7) and using $\nu = 0.5$ for the calculation of I_r using Eq. (10.6), the back-calculated mean value of the secant undrained Young's modulus, E_u, at large strains near the failure state of the soil mass is about 30 times c_u. The back-calculated E_u/c_u mean ratio is consistent

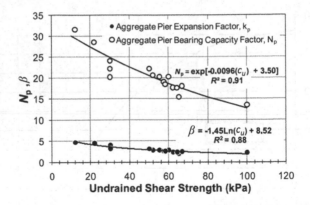

Figure 10.5 Variation of back-calculated cavity expansion factor, β and bearing capacity factor N_p.

with that of Frikha and Bouassida (2015) based on a β value of 3.1 calibrated against the loading test data given by Bergado and Lam (1987).

Combining Eq. (10.1) and Eq. (10.3) gives the following expression for q_{ult}:

$$q_{ult} = \frac{1+\sin\phi_p}{1-\sin\phi_p} \times (\sigma_{ro} + \beta c_u - u) = \frac{1+\sin\phi_p}{1-\sin\phi_p} \times (\sigma'_{ro} + \beta c_u) \qquad (10.8)$$

where σ'_{ro} is the effective in situ lateral stress, $\sigma'_{ro} = \sigma_{ro} - u$.

10.4.4 Undrained shear strength method for long stone column

A simple empirical method was proposed by Mitchell (1981) to compute the ultimate bearing capacity of a single stone column as:

$$q_{ult} = c_u \times N_p \qquad (10.9)$$

where N_p is the bearing capacity factor. Mitchell (1981) recommended that $N_p = 25$. Barksdale and Bachus (1983) recommended $18 \leq N_p \leq 22$ for vibro stone columns based on back-calculations from field tests. Bergado and Lam (1987) recommended N_p values of 15 to 18 for dynamic replacement columns within Bangkok clay (i.e. crushed rocks rammed into the soft ground using falling dead weights to form a relatively large diameter stone column).

Using a database of high-quality load test data of spread footings supported on stone column reinforced clay, Stuedlein and Holtz (2013) back-calculated the N_p factor as a function of c_u as shown in Figure 10.5, which has a best fit relationship of:

$$N_p = e^{(-0.0096c_u + 3.5)} \tag{10.10}$$

It is interesting to note that similar to pile foundation in clay, the end bearing capacity of a long stone column is approximately equal to $9c_u$. A corollary of this is that the end bearing of the stone column contributes to roughly half of the bearing capacity reported in the literature.

EXAMPLE 10.1

This example illustrates the consistency between the cavity expansion theory and the undrained shear strength method.

Consider a stone column installed in a normally consolidated clay soil as shown in Figure 10.6.

For the adopted peak friction angle of the stone column, ϕ_p of $38°$, $q_{ult} = 4.2\sigma'_{rL}$ based on Eq. (10.1). According to the cavity expansion theory for an elasto-plastic material,

$$\sigma'_{rL} = \sigma'_{ro} + \beta \times c_u$$

For normally consolidated soft clay (with c_u <20 kPa), β is expected to be about 4.5 as indicated in Figure 10.5. σ'_{ro} can be expressed in terms of c_u as follows:

 i. The coefficient of lateral earth pressure at rest for the normally consolidated clay is $K_0 = 1 - \sin\phi_{soil} = 0.58$ by considering a typical internal friction angle of $\phi_{soil} = 25°$ for clayey soil.
 ii. The vertical effective stress, σ'_{vo}, for the normally consolidated clayey soil can be calculated from the SHANSEP model proposed by Ladd and Foott (1974), in which $c_u/\sigma'_{vo} \approx 0.24$ for high plasticity clay. Therefore, $\sigma'_{vo} \approx c_u/0.24 = 4.2\,c_u$
 iii. $\sigma'_{ro} = K_0\,\sigma'_{vo} = 0.58 \times 4.2\,c_u = 2.4\,c_u$

Subsequently,

$$q_{ult} = K_{ps}\sigma'_{rL} = K_{ps}\left(\sigma'_{ro} + \beta \times c_u\right) = 4.2 \times \left(2.4c_u + 4.5c_u\right) \approx 29c_u$$

This is consistent with the q_{ult} assessed using the undrained shear strength method, for which the N_p value for a normally consolidated clay is about 29 assessed from Eq. (10.9) as shown in Figure 10.5 above.

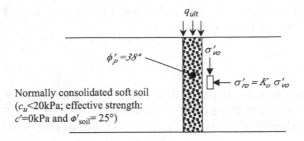

Figure 10.6 Single stone column in normally consolidated soft soil.

10.5 FUNDAMENTAL CONCEPTS FOR STONE COLUMN DESIGN

A group of stone columns is commonly used to increase bearing capacity and shear resistance, or to reduce settlement in soft soils. A single stone column is rarely installed in isolation in engineering practice. The bearing capacity of a single stone column outlined in the previous section, however, provides a fundamental understanding of the failure mechanism that is essential in the design of stone column groups. This section covers additional concepts pertinent to ground treatment design using stone columns, viz:

- Unit cell concept;
- Area replacement ratio;
- Load transfer mechanism and stress concentration ratio.

10.5.1 Unit cell concept

Stone columns are generally installed at a regular center-to-center spacing in either a square or equilateral triangular pattern as shown in Figure 10.7. The tributary area of each stone column within a group can be approximated by an equivalent circle having the same total area. For an equilateral triangular pattern of stone columns, the equivalent circle has an effective diameter, D_e, of

$$D_e = 1.05 \times s \tag{10.11}$$

And for a square pattern:

$$D_e = 1.13 \times s \tag{10.12}$$

where s is the center-to-center spacing of stone columns. The cylinder of soil materials within the tributary area that encompasses one stone column is

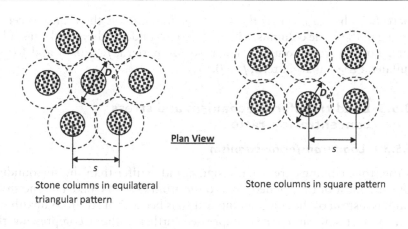

Plan View

Stone columns in equilateral triangular pattern

Stone columns in square pattern

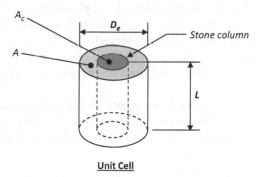

Unit Cell

Figure 10.7 Square or equilateral triangular pattern of stone columns.

known as a 'unit cell.' Each column within a large group of stone columns can be considered as a unit cell. Because of symmetry of load and geometry, the unit cell can be considered as a cylinder having a frictionless, rigid exterior wall concentrically located around the stone column.

10.5.2 Area replacement ratio

The area replacement ratio, a_s, is defined as the ratio of the cross-sectional area of the stone column, A_c, to the total area of a unit cell, A. Therefore,

$$a_s = \frac{A_c}{A} \tag{10.13}$$

The area replacement ratio, a_s, can be expressed in terms of the diameter and spacing of the stone columns by Eq. (10.13) as follows:

$$a_s = f_1 \left(\frac{D}{s} \right)^2 \tag{10.14}$$

where D is the diameter of the stone column and s is the center-to-center spacing of the stone columns. The f_1 is a constant depending on the installed pattern of the stone columns. For a square pattern, $f_1 = 0.78$ and for an equilateral triangular pattern, $f_1 = 0.91$.

10.5.3 Load transfer mechanism and stress concentration ratio

10.5.3.1 Load transfer mechanism

As the stone columns are usually significantly stiffer than the surrounding soft soils, most of the applied load to the ground will be taken by the stone columns instead of the surrounding soft soils because of strain compatibility (i.e., the soft soil cannot be compressed further without compressing the stone column). If the super-imposed load is applied via a relatively rigid slab, then the settlement of the rigid slab and the soft soils should be the same, i.e., under an equal strain condition (Figure 10.8a). In this instance, the ratio of the load carried by the columns to the load carried by the soft soils should be roughly proportional to the stiffness ratio between the stone columns and the soft soils.

If the stone columns are used to support a fill embankment, the settlement in the columns and the soil are not the same. However, an equal stress condition as shown in Figure 10.8b cannot be assumed as the columns still carry

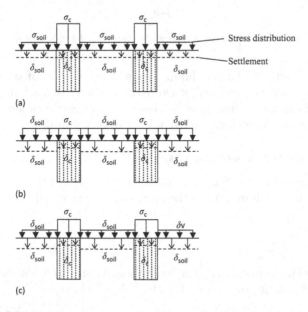

Figure 10.8 Schematic stress strain distribution for (a) equal strain under rigid footing; (b) equal stress under flexible loading; (c) unequal strain and unequal stress conditions under fill embankment.

more load than the soil due to soil arching as shown in Figure 10.8c. The amount of soil arching depends on a number of factors including the soil-column stiffness ratio, the column spacing and the yield strength of the columns.

The stress distribution between the columns and the soil can be quantified by the stress concentration ratio, n, which is defined as the ratio of the vertical stress within a stone column, σ_c, to the vertical stress of the surrounding soil, σ_{soil}, at the same depth level:

$$n = \frac{\sigma_c}{\sigma_{soil}} \tag{10.15}$$

where n varies along the length of the column, depending on the stress and strain levels and the yield strength of the column. The variation of n value can be used to describe the load transfer mechanism between the columns and the surrounding soils. In general, an increase in n indicates load transfer from soil to column, whereas a decrease of n signifies load transfer from column to soil. For example, when the stone columns are subject to increasing fill embankment load, the columns initially exhibit a relatively high stress concentration ratio (or n value) that is commensurate with the elastic range of the columns. The n value increases to a maximum when the column stresses reach the yield strengths, particularly at the top sections of the columns where the column stresses are higher. After the yielding of the columns, a greater proportion of the applied load is transferred to the soils. The additional load applied to the soils will cause consolidation settlement. As the soils consolidate, some load transfer to the columns will occur, which in turn will cause further settlement and yielding of the stone columns, leading to subsequent load transfer back to the soils. This process will continue until an equilibrium is reached and the final n value becomes a constant less than the peak stress concentration ratio.

For an equilibrium of vertical force to exist within a unit cell, the applied stress, $\Delta\sigma$, over the unit cell area at a given depth can be calculated by:

$$\Delta\sigma = \sigma_c a_s + \sigma_{soil}\left(1 - a_s\right) \tag{10.16}$$

Solving Eq. (10.16) using Eq. (10.15), the stresses in the soil and stone column may be expressed by the following equations:

$$\sigma_{soil} = \mu_{soil}\Delta\sigma = \frac{1}{\left[1+\left(n-1\right)a_s\right]}\Delta\sigma \tag{10.17a}$$

$$\sigma_c = \mu_c\Delta\sigma = \frac{n}{\left[1+\left(n-1\right)a_s\right]}\Delta\sigma \tag{10.18a}$$

where

$$\mu_{soil} = \frac{1}{\left[1+(n-1)a_s\right]} \tag{10.17b}$$

$$\mu_c = \frac{n}{\left[1+(n-1)a_s\right]} \tag{10.18b}$$

From Eq. (10.17b) and Eq. (10.18b), it can be seen that $\mu_c = \mu_{soil} \times n$.

The value of n is a function of the modular ratio, E_{soil}/E_c, of the stone column and the soil. There are several approaches to appraise the stress concentration ratio, n, viz.:

- Typical values from past measurements;
- Elastic finite element solution;
- Elastic-plastic finite element solution accounting for the yielding of stone columns;
- Analytical method using cavity expansion theory or undrained shear strength method.

Details of each approach are discussed in the following sections.

10.5.3.2 Typical n value from past measurements

FHWA (2017) indicates that the measured values of stress concentration ratio, n, are generally between 2 and 5. The n value may be towards the upper range (3 to 5) if the in situ soil is weak and the column spacing is small. For stronger in situ soils and large column spacing, lower range of the stress ratio (2 to 2.5) may be used. For preliminary design, FHWA (2017) recommends a ratio of 2.5 to be used.

Plate load test data on stone columns given in Han and Ye (1991) indicates that the stress concentration ratio of the loaded stone columns ranges from 1 to 5, as shown in Figure 10.9.

10.5.3.3 Elastic finite element solution given by FHWA (1983)

The stress concentration ratio, n, is a function of the modular ratio, E_{soil}/E_c, of the stone column and the soil. By using finite element method with linear elastic soil models, FHWA (1983) provided a plot of n versus modular ratio for different area replacement ratios, a_s, under embankment fill loading as shown in Figure 10.10. It can be seen that n increases almost linearly with increasing modular ratio. The influence of a_s on n for a particular modular ratio is, however, less significant. For a modular ratio of 10, n is approximately equal to 10. FHWA (1983) indicated that for modular ratio greater

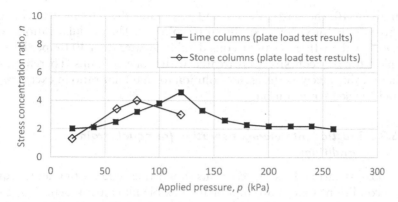

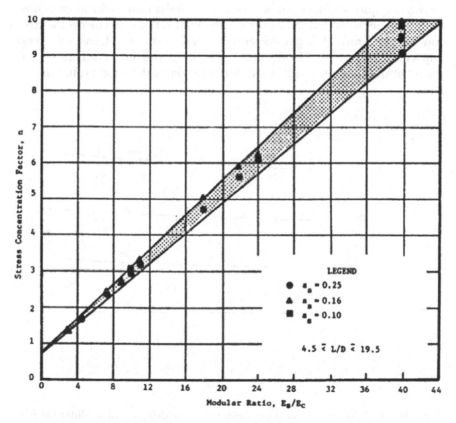

Figure 10.9 Stress concentration ratio of loaded stone columns as given by Han and Ye (1991).

Figure 10.10 Plot of **n** versus modular ratio for different area replacement ratios as under embankment fill loading (FHWA 1983).

than about 10, the elastic theory underestimates the settlements primarily due to bulging (or yielding) of the stone columns. The modular ratio is usually greater than 10 and can be considerably greater than 40 (upper limit of the plot) for soft cohesive soils treated with stone columns. Therefore, the higher n value given by the elastic solution for modular ratio in excess of 10 should be used with caution.

10.5.3.4 Elastic finite element solution for equal strain conditions

More load is carried by stone columns supporting rigid footing under equal strain conditions than those supporting flexible fill embankment. Using the elastic finite element method, Balaam and Poulos (1982) carried out a parametric study to assess the effect of modular ratio on the vertical stress σ_c in the stone column for the different unit cell effective diameter to column diameter ratios, d_e/d_c. The results given by Balaam and Poulos can be converted to an equivalent plot of n versus d_e/d_c for different modular ratios as shown in Figure 10.11. It can be seen that modular ratio is the most influential parameter such that the higher the modular ratio, the higher is the stress concentration ratio. The parameter d_e/d_c, on the other hand, has only minor impact on the n value. The n values for rigid footing given in Figure 10.11 are insightful, but need to be used with caution as the elastic solution does

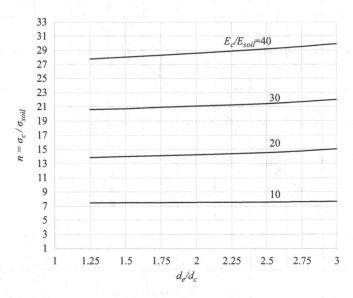

Figure 10.11 Variation of stress concentration n with d_e/d_c and modular ratio for rigid footing with equal strain condition (based on elastic finite element solution).

not consider the yielding of stone columns. The elastic solution may overestimate the stress concertation of the stone columns especially for the n values corresponding to a high modular ratio.

10.5.3.5 Elastic finite element solution for stone columns supporting fill embankment (stone columns on rigid base)

The elastic solution for stone columns supporting fill embankment was obtained based on axisymmetric finite element analysis of a 'unit cell' consisting of stone column and the surrounding soil within a column's tributary area. The embankment fill was modelled as soil elements in order to capture the arching stresses developed above the columns.

The elastic solution obtained from the finite element analysis, as shown in Figure 10.12 for the case of $d_e/d_c = 2$, indicates that the stress concentration ratio, n, is not constant along the length of the column. At the top of the column, n is consistent with that given by FHWA (1983). The n value then increases with depth until reaching a maximum threshold that is commensurate with the equal strain solution given by Balaam and Poulos (1982).

The location along the column depth at which equal strain condition occurs can be considered as being the point of maximum curvature on the n versus depth curve as shown in Figure 10.12. The critical depths, Z_t (see Figure 10.13) of the maximum curvatures for the different combinations of d_e/d_c and E_c/E_{soil} were obtained from the elastic finite element solution and the results are summarized in a normalized plot given in Figure 10.13.

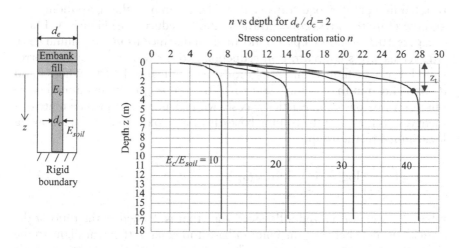

Figure 10.12 Variation of stress concentration n along column depth for the case of $d_e/d_c = 2$ (based on elastic finite element solution of stone columns supporting fill embankment).

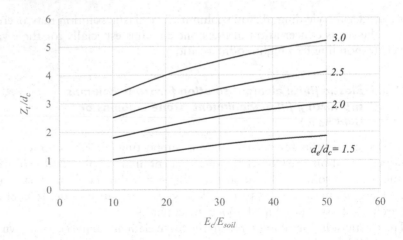

Figure 10.13 Normalized Z_t/d_c values for different E_c/E_{soil} and d_e/d_c ratios.

10.5.3.6 Elastic finite element solution for stone columns supporting fill embankment (stone columns founded on compressible soil)

The elastic solution outlined in Section 10.5.3.5 is for stone columns founded on a rigid base, whereby the surrounding soil generally settles more than the stiffer stone columns. For stone columns founded on compressible soil, the finite element solution based on an elastic model has indicated that there exists a lower equal settlement plane, below which the columns move more than the soil to mobilize positive skin resistance of the soil. This has resulted in a greater load transfer from the column to the surrounding soil and therefore the stress concentration ratio, n reduces (see Figure 10.14a).

Figure 10.14b shows a plot of normalized distance from the column base, y/d_c versus stress concentration reduction ratio, r (= n/n_{max}) for the corresponding elastic finite element solutions given in Figure 10.14a, where y and d_c are defined in the inset in Figure 10.14b; and n_{max} is the maximum computed n value based on elasticity as shown in Figure 10.14a. The finite element results for r near the column base can be approximated by the following logarithmic relationship.

$$r = 1 - \frac{1}{m}\left[\log\left(\frac{\xi}{y/d_c}\right)\right] \text{(for } y/d_c \leq \xi \tag{10.19}$$

where ξ is the normalized influenced zone that is defined as the ratio of the critical distance between the column base and equal settlement plane to the diameter of the stone column d_c. The parameter m in Eq. (10.19) controls the rate of reduction of r with y/d_c. The higher the magnitude of m the more rapid reduction of r would be towards the column tip.

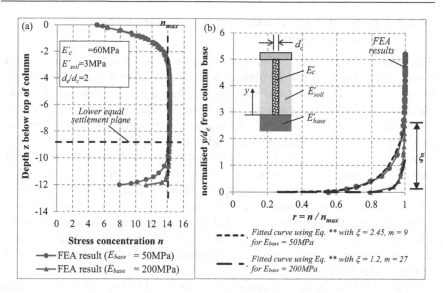

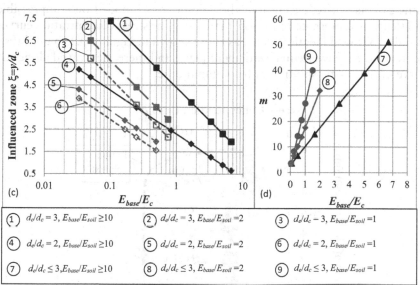

Figure 10.14 Stone column with compressible base – elastic solution.

Figure 10.14b indicates that as the Young's modulus, E_{base} of the soil beneath the columns increases, the extent of ξ reduces. In addition, the ratio r reduces more rapidly towards the tip of the column (i.e., m increases) as E_{base} increases. Figure 10.14c and 10.14d present the computed ξ and m for the different E_{base}/E_c and E_{base}/E_{soil} ratios based on linear elasticity finite element analyses. Detailed discussion of the behavior of stone columns on compressible soils can be found in Chan and Poon (2015).

10.5.3.7 Finite element solution accounting for yielding of stone columns

This section is devoted to the discussion of stress concentration of stone column that accounts for the plastic yielding and internal friction within the stone columns. The analytical solution was based on an axisymmetric finite element analysis of a 'unit cell" consisting of stone column and the surrounding soil within a column's tributary area. Details of the finite element analysis also can be found in Chan and Poon (2015).

When the column and soil were modelled as elastic-plastic Mohr-Coulomb materials, yielding elements began to form at the column top after a small load (~20 kPa) was applied, leading to a reduction in stress concentration. The yielding of the column (hence the reduction of n) progressed downwards through the column as the applied load level increased, as shown by the solid curves in Figure 10.15. Below the yielding zone, the n values follow the elastic solution under equal strain condition, which are consistent with that given by Figure 10.11.

Figure 10.16a shows the computed n under different fill loads for the same case as above, except that the column was founded on compressible soil that was represented by $c' - \phi'$ materials. The stress concentration curves initially followed identical paths as those shown in Figure 10.16b until they intercepted the lower equal settlement plane and thereafter traced along the curve of the elastic solution at the column base.

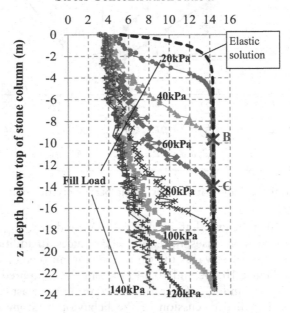

Figure 10.15 Stone column with rigid base (elastic-plastic solution).

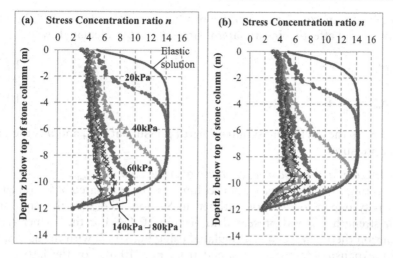

Figure 10.16 Stone column on compressible (a) $c' - \phi'$ and (b) c_u soils

Figure 10.16b presents the results for the case where $c_u = 30$ kPa has been adopted for the soils surrounding and below the column. Some yielding occurs in the soils, which has altered the shape of the stress concentration curves towards the column base as compared to that of the $c' - \phi'$ soils. However, the differences are not significant and for the purpose of assessing n, the problem can be idealized by assuming that there is no failure in the surrounding soil so that its behavior is essentially elastic.

10.5.3.8 Design procedure using existing analytical methods to account for yielding of stone columns

While the effect of yielding on stone column stress concentration can be assessed in detail by means of numerical analyses such as those outlined in Section 10.5.3.7, it can be practically assessed using simple analytical methods such as the cavity expansion theory and the undrained shear strength method outlined in Sections 10.4.3 and 10.4.4 respectively.

This section describes a simple approximate procedure to calculate the stress concentration of stone columns supporting fill embankment. The design procedure has utilized the aforementioned analytical methods in conjunction with the elastic finite element solutions described in Section 10.5.3.3 to Section 10.5.3.6, and can be described as follows:

Step 1: Calculate effective stress of in situ soil profile (σ'_{vo}).
The effective vertical stress, σ'_{vo}, of the in situ soil with depth prior to stone column ground treatment can be calculated based on the assessed bulk unit weight of the soil, γ_{bulk}, and the assessed water level.

Step 2: Calculate total and effective in situ lateral stress (σ_{ro} and σ'_{ro}).
The total and effective in situ lateral stresses can be calculated by:

$$\sigma'_{ro} = K_0 \times \sigma'_{vo} \tag{10.20}$$

$$\sigma_{ro} = \sigma'_{ro} + u \tag{10.21}$$

where K_0 is the coefficient of lateral earth pressure at rest, and u is the pore water pressure.

Step 3: Calculate the bearing capacity of stone column (q_{ult}).
The bearing capacity of each stone column can be calculated based on either the cavity expansion theory outlined in Section 10.4.3 or the undrained shear strength method presented in Section 10.4.4. For the former cavity expansion theory, q_{ult} can be calculated using Eq. (10.7) with the associated cavity expansion factor β, obtained from Eq. (10.6). For the latter, undrained shear strength method, q_{ult}, can be calculated using Eq. (10.8) with the associated bearing capacity factor N_p obtained from Eq. (10.9).

Step 4: Derive an approximate stress concentration factor n with depth based on elastic finite element solutions.
The nonlinear stress concentration factor n of elasticity with depth presented in Sections 10.4.3 and 10.4.4 for stone columns founded on rigid base and compressible soil can be approximated using piecewise linear functions as follows:

- The first linear segment consists of n increasing from the non-equal strain solution given by FHWA (1983) at the top of the column (i.e., Figure 10.10) to that of equal strain solution given by Balaam and Poulos (1982) (i.e., Figure 10.11) at a critical depth z_t obtained from Figure 10.13.
- The second linear segment consists of a constant n obtained from equal strain solution (i.e., Figure 10.11), starting from a critical depth z_t obtained from Figure 10.13 and ending before the compressible influence zone ξ obtained from Figure 10.14c.
- For stone columns founded on compressible soil, a third linear segment representing n values within the compressible influence zone ξ can be considered. This linear segment consists of n reducing from that of equal strain solution (say n_{max}) at the end of the second linear segment to a reduced n value equal to $r \times n_{max}$ at the stone column tip, where r is the stress concentration reduction ratio and can be assessed based on Eq. (10.19) (i.e., Figure 10.11), starting from a critical depth z_t obtained from Figure 10.13 and by considering $y/d_c = 0.01 \times \xi$, therefore:

$$r = 1 - \frac{2}{m} \text{ (for } r \text{ at } y/d_c = 0.01 \times \xi) \tag{10.22}$$

Step 5: Calculate the limiting vertical stress of stone column, σ_c, based on elastic solution.
Of the vertical stress applied over a unit cell area, the limiting stress along the depth of stone column based on elastic solution, σ_c, can be calculated using Eq. (10.18a) in conjunction with the elastic n values with column depth obtained from Step 4.

Step 6: Determine the final vertical stress of stone column, σ_c^*.
The final vertical stress of the stone column, σ_c^*, is the lower of the q_{ult} obtained from Step 3 and the elastic σ_c obtained from Step 5.

Step 7: Calculate the soil stress, σ_{soil}^*.
The vertical stress of the soil surrounding the stone column, σ_{soil}^*, can be calculated by rearranging Eq. (10.17a) and by substituting σ_c with σ_c^* as follows:

$$\sigma_{soil}^* = \frac{\Delta\sigma - \sigma_c^* \times a_s}{(1 - a_s)} \tag{10.23}$$

Step 8: Calculate stress concentration, n.
The stress concentration, n, is calculated based on $n = \sigma_c^* / \sigma_{soil}^*$ (i.e., Eq. (10.15)).

EXAMPLE 10.2

Consider an example outlined in Figure 10.17a where stone columns are installed within the 17 m thick Holocene soft soil underlain by stiff to very stiff Pleistocene clay. The design configuration of the stone columns is shown in Figure 10.17b. Estimate the column stress σ_c^* and the stress concentration ratio n with depth for the different applied embankment fill load of $\Delta\sigma = 80$ kPa, 110 kPa and 160 kPa.

Solution

Equivalent diameter d_e and area replacement ratio a_s
For the 0.8 m diameter stone columns installed at 2.3 m equilateral triangular spacing, the effective diameter, d_e, of the unit cell is about 2.4 m. The area replacement ratio, a_s, is 11%.

Calculation of q_{ult} of stone column
The q_{ult} of individual stone column can be calculated using cavity expansion theory (Eq. (10.7)), in which the cavity expansion factor, β, can be assessed from c_u using Eq. (10.4) and is equal to 3.85. The K_{ps} of the stone column is 4.62 based on $\phi_p = 40°$. Figure 10.18 presents the assessed σ'_{vo}, σ'_{ro} ($= K_0 \times \sigma'_{vo}$) and q_{ult} with depth.

Elastic n profile

From Figure 10.12 and without the consideration of yielding of stone column, the n value under non-equal strain condition at the top of the stone column is 5 for the E_c/E_{soil} ratio of 20. The corresponding n value under equal strain condition is 15 based on Figure 10.11. From Figure 10.13, z_t/d_c ratio = 4 for E_c/E_{soil} and d_e/d_c ratios of 20 and 3.0, respectively. Hence, z_t = 3.2 m. Figure 10.14c, ξ = 5 under the conditions of d_e/d_c = 3.0, E_{base}/E_{soil} = 4 and E_{base}/E_c = 0.2. Hence, the compressible influence zone is equal to y = 4 m measured from the tip of stone column. From Figure 10.14d and Eq. (10.19), m = 5 and r = 0.6. Therefore, the n value at stone column tip is equal to $r \times n_{max}$ = 0.6 × 15 = 9, where n_{max} is the elastic n value under equal strain condition and is equal to 15.

From the conditions outlined above, piecewise linear functions for n of elasticity with depth can be derived and presented in Figure 10.19.

Final column stress σ_c^* and stress concentration ratio n with depth

Using the piecewise linear functions of n, the limiting elastic column stresses, σ_c, can be calculated for the different applied stress levels, $\Delta\sigma$, based on Eq. (10.18). Figure 10.20a shows that when $\Delta\sigma$ = 80 kPa, the calculated elastic column stresses σ_c are less than the q_{ult} of the stone column. Therefore, the final column stresses σ_c^* are equal to the σ_c with depth, and the final n values with depth are the same as that of elasticity. When $\Delta\sigma$ is increased to 110 kPa (Figure 10.20b), yielding of the stone column occurs within the upper 13 m thickness as the elastic column stresses σ_c exceed the q_{ult} of the stone column. The final n values within the top 13 m thickness are also assessed to be lower than those of the elastic values. When $\Delta\sigma$ is increased to 160 kPa (Figure 10.20c), the stone column is completely yielded as the elastic column stresses σ_c exceed the q_{ult} along the entire depth of the stone column. The final n values with depth are also assessed to be reduced further below those of the elastic values.

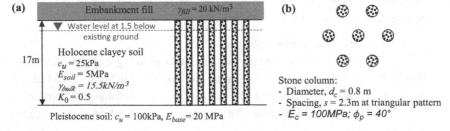

Figure 10.17 Stone column treated soft soil supporting fill embankment.

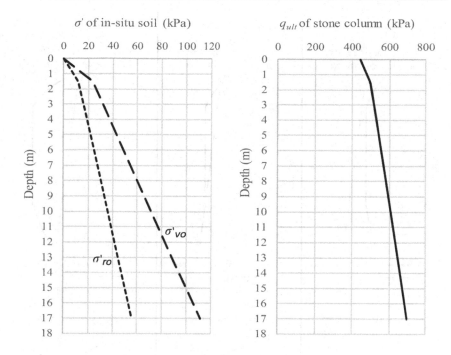

Figure 10.18 Calculated σ'_{vo}, σ'_{ro} and q_{ult} with depth.

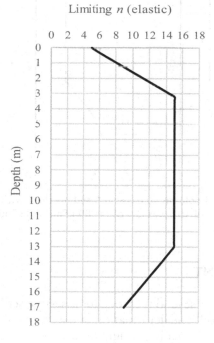

Figure 10.19 Assessed n values of elasticity with depth.

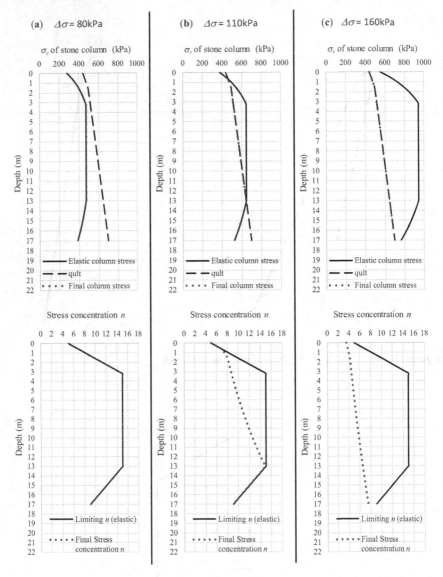

Figure 10.20 Assessed column stresses $\sigma_c{}^*$ and n values with depth for the different applied stresses $\Delta\sigma$ of (a) 80 kPa, (b) 110 kPa, and (c) 160 kPa.

10.6 SETTLEMENT

This section describes several methods for assessing settlement of stone columns treated soft soil. The deformation of the stone column group depends on whether the type of loading is rigid or flexible.

10.6.1 Balaam and Booker method for rigid foundations

Figure 10.21 shows a rigid foundation founded on a column-clay unit cell that was used by Balaam and Booker (1981) to compute settlement of the stone column group using finite element method.

By assuming that the unit cell remains in an elastic state, Balaam and Booker provide a plot of settlement reduction factor (i.e., ratio of settlement of column-clay unit cell to settlement of the clay without any column) with respect to different d_c/d_e and E_c/E_{soil} (see Figure 10.22).

If the stone column yields, then the settlement of the columns can be computed by treating the stone column as an elastic-perfectly plastic material. Balaam and Booker provide correction factors that can be applied to the elastic solution of Figure 10.22 and are presented in Figure 10.23 for the

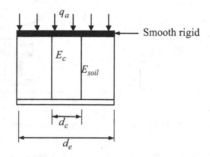

Figure 10.21 Column-clay unit cell adopted in Balaam and Booker (1981).

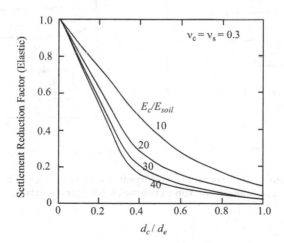

Figure 10.22 Effect of spacing ratio (d_e/d_c) on the settlement reduction factor from elastic finite element solution.

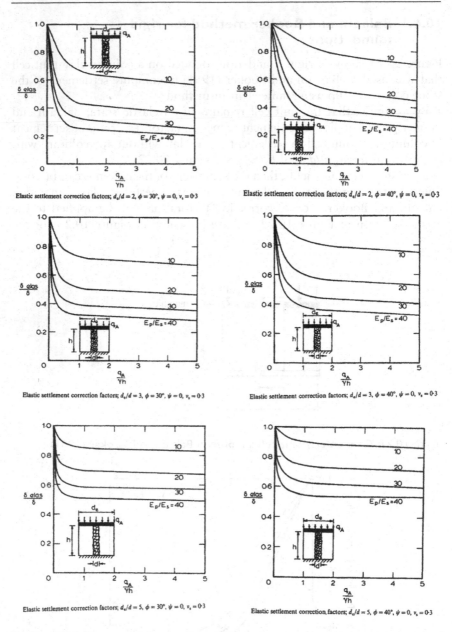

Elastic settlement correction factors; $d_e/d = 2$, $\phi = 30°$, $\psi = 0$, $v_s = 0.3$

Elastic settlement correction factors; $d_e/d = 2$, $\phi = 40°$, $\psi = 0$, $v_s = 0.3$

Elastic settlement correction factors; $d_e/d = 3$, $\phi = 30°$, $\psi = 0$, $v_s = 0.3$

Elastic settlement correction factors; $d_e/d = 3$, $\phi = 40°$, $\psi = 0$, $v_s = 0.3$

Elastic settlement correction factors; $d_e/d = 5$, $\phi = 30°$, $\psi = 0$, $v_s = 0.3$

Elastic settlement correction, factors; $d_e/d = 5$, $\phi = 40°$, $\psi = 0$, $v_s = 0.3$

Figure 10.23 Correction factors for different combinations of d_e/d_c ratio, internal friction angle of stone column and Poisson's ratios.

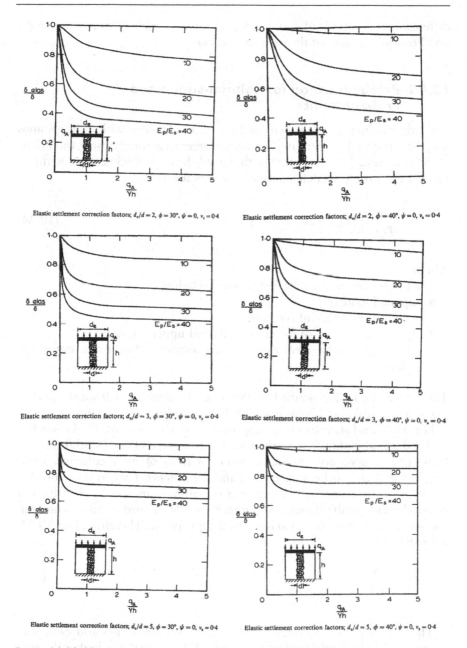

Figure 10.23 (Continued) Correction factors for different combinations of d_e/d_c ratio, internal friction angle of stone column and Poisson's ratio.

different combinations of d_e/d_c ratio, peak friction angle of stone column ϕ_p, and Poisson's ratio v_s of the surrounding clayey soil.

10.6.2 Priebe method for column-supported embankments

The reduction in primary settlement due to ground treated with stone columns may be estimated by deriving the Improvement Factor, n_i, using the Priebe (1995) method. The settlement within a sub-layer of thickness, Δs within a layered soil profile, as defined by Priebe is as follows:

$$\Delta s = \frac{p}{D_s n_2}\left[(s/s_\infty)_l\, d_l - (s/s_\infty)_u\, d_u\right] \tag{10.24}$$

where
n_2 = Improvement factor value (discussed below)
p = applied loading
D_s = constrained modulus of soil
d_l, d_u = depth to base of lower layer, d_l, and upper layer, d_u
$(s/s_\infty)_l$, $(s/s_\infty)_u$ = settlement ratio for the respective lower and upper soil layers.

The settlement s_∞ is described by Priebe as the settlement from the performance of an unlimited column grid below an unlimited load area.

For single and strip footings supported by stone columns, Priebe has formulated graphs showing Settlement Ratio (s/s_∞) versus Depth/Diameter Ratio (d/D) curves for varying numbers or rows of stone columns. These graphs show s/s_∞ to be always ≤ 1, and approaches 1 when the ratio d/D approaches zero, or when the number of columns approaches infinity. For relatively wide embankments supported on a large grid of stone columns, the assumption that $s/s_\infty = 1.0$ is deemed appropriate. Therefore, Eq. (10.24) simplifies to:

$$\Delta s = \frac{p\Delta z}{D_s n_2} \tag{10.25}$$

The Improvement Factor, n_i, takes on several subscripts and each subscript defines the level of refinement applied. In general, the higher the subscript number, the higher the level of refinement in the improvement factor.

- n_0 is the basic improvement factor based on the area ratio, A_c/A, and the coefficient of active earth pressure of the column material, K_{aC}. [$K_{aC} = \tan^2(45-\phi_c/2)$, where ϕ_c equals to the column friction angle.]

For a Poisson's ratio, v, of 1/3, n_0 may be calculated using the following equation:

$$n_o = 1 + \frac{A_c}{A}\left[\frac{5 - A_c/A}{4K_{aC}\left(1 - A_C/A\right)} - 1\right] \tag{10.26}$$

where:

A_c = cross section area of stone column

A = gross area treated by stone column = $A_c + A_s$

K_{aC} = coefficient of active earth pressure of stone column material

Values of n_o for various values of the reciprocal area ratio A/A_c are presented by Priebe (1995) graphically as reproduced in Figure 10.24.

- n_1 is a further refinement on the basic improvement factor, n_0, to account for column compressibility. This is related to the ratio of the constrained moduli of the column material and the soil (D_c/D_s). The basis behind this refinement is that the theoretical equation for calculating n_o results in infinity for the reciprocal Area Replacement Ratio $(A/A_c) = 1$. Due to column compressibility, this theoretical result is not considered valid. Hence, the n_1 improvement aims to provide a limiting value on the Improvement Factor based on the ratio of D_c/D_s where D_c and D_s are the constrained moduli of the column and soil, respectively. The value n_1 may be approximated as follows:

$$n_1 = 1 + \overline{\left(\frac{A_c}{A}\right)}\left[\frac{5 - \overline{\left(A_c/A\right)}}{4K_{aC}\left(1 - \overline{A_c/A}\right)} - 1\right] \tag{10.27}$$

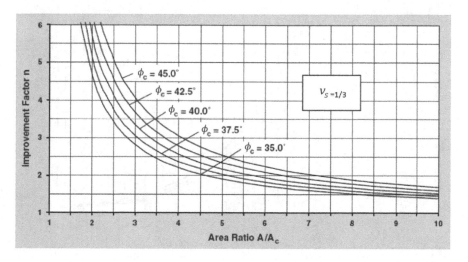

Figure 10.24 Design chart for vibro-replacement.

$$\left(\overline{\frac{A_c}{A}}\right) = \frac{1}{A/A_c + \Delta\left(A/A_c\right)} \tag{10.28}$$

$$\Delta\left(A/A_c\right) = \frac{1}{\left(A_c/A\right)_1} - 1 \tag{10.29}$$

$$\left(\frac{A_c}{A}\right)_1 = -\frac{4K_{aC}\left(D_c/D_s - 2\right) + 5}{2\left(4K_{aC} - 1\right)}$$
$$\pm\frac{1}{2}\sqrt{\left[\frac{4K_{aC}\left(D_c/D_s - 2\right) + 5}{4K_{aC} - 1}\right]^2 + \left[\frac{16K_{aC}\left(D_c/D_s - 1\right)}{4K_{aC} - 1}\right]} \tag{10.30}$$

Alternatively, the quantity $\Delta(A/A_c)$ may be read from Figure 10.25 reproduced from Priebe (1995). The value $\Delta(A/A_c)$ may then be added to A/A_c and Figure 10.24 used to assess the new n_1.

- n_2 is a higher level of refinement on the improvement factor, n_1, to account for overburden and compatibility controls.

Stone columns subjected to loading tend to bulge but due to soil resistance offered by the overburden pressure, the bulging is reduced. The soil resistance is considered to increase with increasing overburden pressure. As overburden pressure is a function of depth, the depth factor, f_d, is derived.

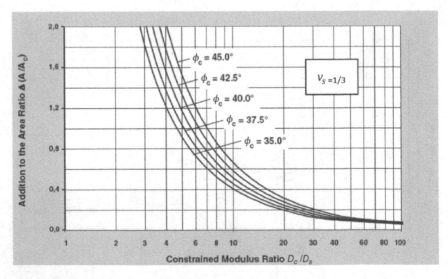

Figure 10.25 Consideration of column compressibility.

An increase in depth factor f_d, corresponds to an increase in the improvement factor.

n_2 is calculated using the following equations:

$$n_2 = f_d \times n_1 \text{ but not greater than } n_{max} \qquad (10.31)$$

where

$$f_d = \frac{1}{1 + \dfrac{(K_{oC} - 1)}{K_{oC}}\dfrac{\sigma'_v}{P_c}} \qquad (10.32)$$

$$K_{oC} = 1 - \sin\phi_c \qquad (10.33)$$

σ'_v = effective stress at mid-point of the sub-layer using the in situ soil stress

$$p_c = \frac{p}{\dfrac{A_c}{A} + \dfrac{1 - A_c/A}{p_c/p_s}} \qquad (10.34)$$

$$\frac{p_c}{p_s} = \frac{1/2 + f\left(v_s, \overline{A_c/A}\right)}{K_{aC} f\left(v_s, \overline{A_c/A}\right)} \qquad (10.35)$$

$$f\left(v_s, \overline{A_c/A}\right) = \frac{(1 - v_s)\left(1 - \overline{A_c/A}\right)}{1 - 2v_s + \overline{A_c/A}} \qquad (10.36)$$

where v_s is the Poisson's ratio.

Alternatively, f_d may be assessed from the equation in Figure 10.26 using the influence factor shown in the same figure reproduced from Priebe (1995).

The following conditions should also be satisfied such that n_2 will not be greater than n_m:

Compatibility controls – due to the theoretical nature of the Priebe method associated with simplifications and assumptions, the result obtained may not always 'guarantee that no more load is assigned to the columns than they can bear in accordance with their compressibility.' Compatibility controls are therefore introduced to limit:

- Depth Factor – This compatibility control seeks to limit the Depth Factor, f_d, to ensure that the settlement of the loaded column is not greater than that of the composite system. The maximum value of f_d is given in Eq. (10.37) with Eqs. (10.31) and (10.32).

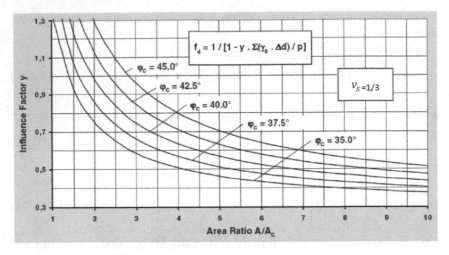

Figure 10.26 Determination of the depth factor.

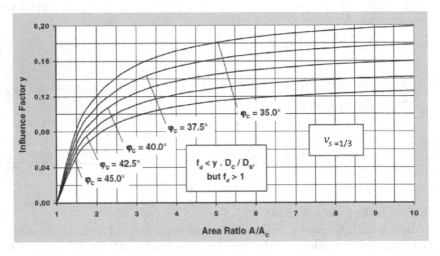

Figure 10.27 Limit value of the depth factor.

$$f_d \leq \frac{D_c/D_s}{p_c/p_s} \text{ but not less than 1} \tag{10.37}$$

Alternatively, the maximum value of f_d may be assessed using the equation in Figure 10.27 and the influence factor shown in the same figure reproduced from Priebe (1995).

- Maximum Improvement Factor, n_m – This factor is introduced to ensure that the settlement of the columns resulting from their inherent compressibility does not exceed the settlement of the surrounding soil resulting from its compressibility by the loads which are assigned to each other. An equation is assigned for this limiting value:

$$n_m = 1 + \frac{A_c}{A}\left[\frac{D_c}{D_s} - 1\right]$$

(10.38)

where
n_m = limiting Improvement Factor value
A = total area = $A_c + A_s$
A_c, A_s = area of column and soil, respectively
D_c, D_s = constrained modulus of column and soil, respectively.

10.6.3 Alternative equilibrium method accounting for yielding of stone column

The Priebe method is based on a combination of elasticity and the Rankine earth pressure theory, and it takes account of limited bulging of the columns by limiting the improvement factor. However, Priebe's improvement factor does not reduce with high applied load. The equilibrium method also does not appear to consider stone column yield unless the designer judiciously limits the stress ratio between the stone columns and the surrounding soil. Further, Priebe (1995) does not provide advice on the calculation of second-ary (creep) settlement for stone column improved soil.

Assuming settlement is approximately proportional to the applied stress, the efficiency of stone columns to reduce settlement compared to that of the untreated ground may be expressed as a settlement reduction ratio equal to $\sigma_{soil}/\Delta\sigma$, which is the same as μ_{soil} defined in Eq. (10.17b). However, if the columns reach their limiting stress at a particular depth, the balance of the applied stress must be taken by the soil surrounding the columns, and the stress taken by the soil will increase from σ_{soil} to σ_{soil}^*. In this situation, the settlement reduction ratio, μ_{soil}^*, is equal to:

$$\mu_{soil}^* = \sigma_{soil}^*/\Delta\sigma$$

(10.39)

where σ_{soil}^* is determined using the design procedure outlined in Section 10.5.3.8.

The primary settlement of the stone column improved soft soil can be calculated using the following consolidation settlement equation in conjunction with the improved compression ratio CR^* and recompression ratio RR^*:

$$S_{primary} = RR^* \log\left(\frac{\sigma_p'}{\sigma_0'}\right)H + CR^* \log\left(\frac{\sigma_0' + \Delta\sigma}{\sigma_p'}\right)H$$

(10.40)

where

$RR^* = RR \times \mu_{soil}^*$

$CR^* = CR \times \mu_{soil}^*$

RR = recompression ratio = $C_r/(1+e_0)$

CR = compression ratio = $C_c/(1+e_0)$

C_r = recompression index

C_c = compression index

σ'_p = preconsolidation pressure

σ'_0 = initial vertical stress

$\Delta\sigma$ = applied stress

H = thickness of compressible layer treated with stone column

FHWA (1983) suggests that secondary settlement of stone column treated ground may be estimated using the usual log-time relationship as shown in Eq. (10.41), with the coefficient of secondary consolidation selected at the appropriate stress level (i.e., at a lower stress level due to the presence of the stone columns). FHWA (1983) also indicates that, because of the relatively short time usually required for primary consolidation to take place in stone column (act as drains) treated ground, secondary settlement is even more important than where stone columns are not used.

$$S_{creep} = HC_{\alpha\varepsilon}^* \cdot \log\left[\frac{t_2}{t_1}\right]$$ (10.41)

where:

$C_{\alpha\varepsilon}^*$ = improved coefficient of secondary compresssion of the treated ground

H = thickness of compressible layer treated with stone column

t_2 = time at which the value of secondary compression (creep) settlement is required

t_1 = time at commencement of secondary compression (creep)

Using the constant C_α/C_c principle suggested by Mesri and Feng (1991), it would appear reasonable to reduce the coefficient of secondary compresssion $C_{\alpha\varepsilon}^*$ in the same way as the compression and recompression indices. That is, $C_{\alpha\varepsilon}^* = C_{\alpha\varepsilon} \times \mu_{soil}^*$, where $C_{\alpha\varepsilon}$ is the coefficient of secondary compresssion of untreated ground depending on the overconsolidation ratio of the soil at the end of consolidation.

EXAMPLE 10.3

From Example 10.2, if the CR, RR and $C_{\alpha\varepsilon}$ of the foundation soil at the in situ state are 0.3, 0.03 and 0.012, respectively, estimate the improved CR^*, RR^* and $C_{\alpha\varepsilon}^*$ after stone column treatment and under the applied fill load of $\Delta\sigma = 110$ kPa.

Solution

By considering yielding of the stone column, Figure 10.28a (reproduced from Figure 10.20b in Example 10.2) presents the assessed σ_c^* under the applied $\Delta\sigma$ = 110 kPa. Figure 10.28b shows the corresponding σ_{soil}^* calculated based on Eq. (10.23) and Figure 10.28c shows the settlement reduction ratio μ_{soil}^* with depth calculated based on Eq. (10.39). Subsequently, CR^*, RR^* and $C_{\alpha\varepsilon}^*$ are calculated based on:

$$CR^* = \mu_{soil}^* \times CR$$

$$RR^* = \mu_{soil}^* \times RR$$

$$C_{\alpha\varepsilon}^* = \mu_{soil}^* \times C_{\alpha\varepsilon}$$

Figure 10.29 shows the plots of CR^*, RR^* and $C_{\alpha\varepsilon}^*$ with depth,

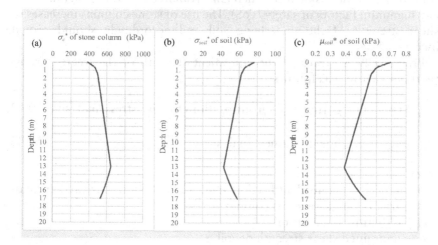

Figure 10.28 Assessed σ_c^*, σ_{soil}^* and μ_{soil}^* with depth under the applied fill load of $\Delta\sigma$ = 110 kPa.

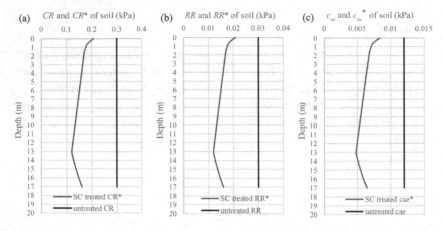

Figure 10.29 Assessment of improved CR^*, RR^* and C_{ae}^* after treatment with stone columns.

10.7 EQUIVALENT STRENGTH PARAMETERS FOR STABILITY ASSESSMENT

The underlying soft soil improved using stone columns can be assessed for rotational stability using a commercially available software program to satisfy minimum factors of safety (FoS). The use of stone columns increases the shear strength of the underlying soil and accordingly should provide increased FoS.

Priebe (1995) expressed the equivalent strength of the treated ground (without yielding) as follows:

$$\phi_{eq} = \tan^{-1}\left[\left(1-\mu_c\right)\tan\phi_{soil} + \mu_c\tan\phi_c\right] \text{(for effective stress analysis)} \quad (10.42)$$

$$c_{u(eq)} = \mu_c c_u \text{(for undrained analysis)} \quad (10.43)$$

where:
ϕ_{eq} and $c_{u\,(eq)}$ = equivalent friction angle and cohesion of treated ground
ϕ_c = friction angle of stone column
ϕ_{soil} = friction angle of soil
c_u = undrained shear strength of soil
$\mu_c = (n_1 - 1)/n_1$

Taking into consideration the yielding of the stone columns, the equivalent strengths of the treated ground become:

$$\phi_{eq}^* = \tan^{-1}\left[\left(1-\mu_c^*\right)\tan\phi_{soil} + \mu_c^*\tan\phi_c\right] \text{(for effective stress analysis)} \quad (10.44)$$

$$c^*_{u(eq)} = \mu^*_c \times c_u \text{ (for undrained analysis)} \tag{10.45}$$

where

$\mu_c^* = \sigma_c^* / \Delta\sigma$

σ_c^* = final vertical stress of the stone column determined using the design procedure outlined in Section 10.5.3.8.

It should be pointed out that the proposed method of assessing the effects of stone column yield described above is only applicable for the center of a wide embankment. Lateral yielding near the edge of the embankment has not been considered. It is essential that edge stability of stone column supported embankments be carefully assessed, particularly for high embankments. As a first approximation, the reduced equivalent parameters of the treated ground shown in Eq. (10.44) and Eq. (10.45) should also be applied to the stone column treated ground below the batters. However, to be prudent, the maximum embankment height should be adopted to allow for maximum reduction in the equivalent friction angle, rather than adopting a reduced batter height.

10.8 HORIZONTAL BEHAVIOR OF STONE COLUMNS

The stone column design outlined in the previous sections involves the prediction of settlements using either a simplistic theoretical method (e.g. Priebe 1995) or a composite material approach in which equivalent strength and deformation parameters are derived using semi-empirical correlation to represent the entire reinforced soil (e.g., Kundu et al. 1994; Madhav and Nagpure 1996). While these approaches have been accepted as reasonable methods for vertical displacement prediction, they are less certain for the prediction of horizontal displacement.

Recently, some researchers (e.g., Tan et al. 2008; Chan and Poon 2012) have started to simulate rows of stone columns using equivalent plane-strain strips in 2D FEA modelling. This approach is considered more appropriate to analyze stone columns subject to lateral loading.

For the modelling of stone columns in 2D FEA, the width of the stone column strips can be made to be equal to the side of an equivalent square for the cross-sectional area of a stone column (Figure 10.30). The spacing of the strips is equal to the actual spacing, b, for square column arrangement and $\sqrt{3}\, b/2$ for equilateral triangular arrangement. The stone column strips are modelled as Mohr-Coulomb materials with Poisson's ratio of 0.3, which was taken to be the value of the soil itself. The equivalent Young's modulus

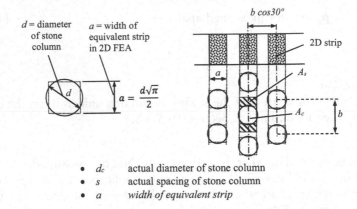

- d_c actual diameter of stone column
- s actual spacing of stone column
- a *width of equivalent strip*

Figure 10.30 2D stone column strips.

E_{eq} and the cohesion c_{eq} of the strips can be calculated based on weighted average approach as given by Eq. (10.45):

$$E_{eq}\left(or\, c_{eq}\right) = \frac{E_{soil}\left(or\, c_{soil}\right)A_s + E_c\left(or\, c_{column}\right)A_c}{A_s + A_c} \tag{10.46}$$

where A_s and A_c are the areas of the soil and column inside a unit cell within the 2D strip as shown in Figure 10.30.

The equivalent friction angle $\phi_{eq\,strip}$ of the strips can be derived based on force equilibrium approach as given by:

$$\tan\left(\phi_{eq\,strip}\right) = \frac{A_s \tan\left(\phi_{soil}\right) + nA_c \tan\left(\phi_c\right)}{A_s + nA_c} \tag{10.47}$$

The determination of $\phi_{eq\,strip}$ requires a presumption of stress concentration, n, which can be determined using the methods outlined in Section 10.5.3. The validation of the 2D stone column strip approach is detailed in Chan and Poon (2012). Comparison of the predicted lateral behavior of stone columns with measurements are presented in several case studies given in Section 10.13.

10.9 ENVIRONMENTAL CONSIDERATIONS

Environmental considerations related to earthworks construction are generally applicable to stone column sites as well. These have been listed in Chapter 6 and include disposal of surface water/runoff, dust and noise, traffic control, effects on local roads due to earthmoving machinery, disposal of groundwater etc.

If the vibro-replacement wet method is used, the amount of water circulated is significantly high. The water jetting and replacement process can generate a large amount of effluents that will escape to the ground surface. These effluents will need to be disposed of properly, especially if the in situ soils are contaminated or potentially acid sulphate soils.

Even if the vibro-displacement dry method is used, when filling takes place, excess pore pressure generated due to the load imparted on the ground has to dissipate, which necessitates the expulsion of pore water. Even if the settlement is only 500 mm or so, the volume of water removed from the soil strata is 0.5 m^3 per square meter on the plan area of a site and cannot be ignored. Hence, drainage control is essential in a stone column site.

Containment and disposal of water becomes complex if the soil is contaminated or has acid sulphate potential because such water needs to be treated prior to disposal. In the latter case where acid sulphate soils are present, as mentioned in Chapter 9, the treatment could be simply spreading hydrated lime along surface drains where the water is allowed to run. The amount of hydrated lime needed should be assessed by an experienced engineer.

More attention should be given if contaminants are found in the soil as to containment and processing of water coming out.

The stone column system is environmentally positive since no external materials other than gravel and water are used in construction, i.e., the installation process does not use cement or steel used in some other rigid inclusion systems such as semi-rigid inclusions, for example. Therefore, the method reduces the carbon footprint significantly when a large area is improved.

Stone columns could accommodate recycled materials if necessary by assigning appropriate parameters for the materials selected in the design. This could be very useful in large developments where traffic movement associated with borrow materials and disposal could be significantly reduced.

With regards to noise generated, stone column machines emanate very low sound and therefore do not disturb the community during the installation process.

10.10 DESIGN AND CONSTRUCTION ISSUES

Some of the design and construction issues related to stone columns are given below.

- Depending on the spacing, significant heaving and lateral displacement could occur if the columns are too close. This is particularly important with the vibro-displacement method where soils are displaced laterally rather than being brought up to the ground surface.

While soil heaving is not a major issue *per se*, if heaving causes columns already constructed to lift up, the density of the stone columns could reduce and separation cracks could occur causing soft soil to fill up any space. In this case, a weak plane is introduced, which may not be acceptable from the point of view of lateral stability if that is an issue. In addition, the drainage path will be blocked causing time delays for pore water to dissipate. To avoid or reduce such issues, columns could be installed in a hit and miss basis.

- Lateral displacement of the surrounding soil during vibro-displacement could push any structures or underground utilities in close vicinity sideways. The lateral displacement on any adjacent piles could exert lateral pressure on the piles causing additional bending moment and shear force on the piles. Construction sequencing is therefore important in order to avoid any undesirable effects on the adjacent structures.

- One of the important design issues for stone columns is the modulus value of the gravel or crushed rock used. While laboratory investigations can be carried out to guide an assessment, it is not easy to assess the compaction achieved in the gravel after installation. Therefore, geotechnical testing and/or instrumentation become very valuable. A simple test such as a plate load test on selected columns could provide information to the designer to calibrate the assumed model. Monitoring of settlement is considered essential for stone column sites to inform the designer whether the design intent has been met and, if not, take appropriate remedial measures.

- Stone column installation machinery are heavy and therefore an assessment of bearing capacity should be carried out and, if necessary, a working platform created to ensure that bearing failure does not occur under the heavy load of the installation machinery.

10.11 DESIGN AND CONSTRUCTION OF DYNAMIC REPLACEMENT COLUMNS

While the design of dynamic replacement columns is similar to vibro-replacement and vibro-displacement stone columns, design parameters for the rock fill used in dynamic replacement are likely to be different as follows:

- Strength parameters of rockfill – $c' = 0$ kPa and $\phi' = 35$ deg.
- Young's modulus = 25 to 40 MPa (adopt the lower value if the rock fill is poor and subject to breakdown)
- Permeability = 0.1 to 1 m/day (or 10^{-4} to 10^{-3} cm/s)
- Smear effect – Either reduce the column size to account for smear or adopt a reduced permeability for the smear zone.

It should be noted that there is no limitation on adopting higher values for such properties if the rock fill is of very high quality.

In order to ensure successful execution of dynamic replacement, it is imperative that appropriate site observations and monitoring be carried out during and after the dynamic replacement process. Such work may include the following:

- Condition survey of adjacent infrastructure;
- Measurement of vibration and taking necessary action if measurements indicate unusual or undesirable effects;
- Drilling of selected dynamic replacement columns to confirm the depth of the columns;
- Installation of settlement plates and settlement points, and inclinometers to monitor movements; and
- The use of piezometers to monitor dissipation of excess pore water pressure generated during the pounding process.

10.12 ADVANTAGES AND DISADVANTAGES

Advantages and disadvantages associated with stone columns are presented in Table 10.1.

Table 10.1 Advantages and disadvantages of stone columns

Item	Advantage/disadvantage
Method	Advantage – Generally considered simple in design and construction.
Pore pressure dissipation	Advantage – Although increasing the soil stiffness is the central objective of a stone column, it has a secondary function of assisting excess pore pressure dissipation.
Fill material	Advantage – As fill material is gravel/crushed rock, it is relatively easy to obtain borrow materials locally. Even if the material is not fully competent, the design could be adjusted to accommodate lower stiffness value. Recycled materials could be a potential fill material if desired.
Water	Disadvantage – The volume of water needed for the process is very high for the wet process. Containment and disposal of water, especially if contaminated, could be expensive and difficult. The dry method is more advantageous in these instances if viable.

(Continued)

Table 10.1 (Continued)

Item	Advantage/disadvantage
Spoil	Advantage – Very little spoil is generated in the installation process, especially in the dry method where hardly any spoil occurs. Disadvantage – Especially in the wet process where there will be at least excess water that needs to be disposed of. In addition, especially with the dry process, when columns are installed, significant heaving could occur to accommodate the soil that is displaced and some soil may need to be removed. If the soils or water is contaminated or contain acid sulfate, containment and treatment prior to disposal could be expensive.
Sensitive soils	Disadvantage – Vibration of the probe could lead to significant drop in shear strength of sensitive clays, rendering the stone column option impractical.
Lateral movement	Disadvantage – Stone columns when installed too close to each other lead to high lateral displacement and associated earth pressures. This could lead to any adjacent infrastructure being affected. During installation, construction sequencing of the stone columns should be that the column installation order moving away from any infrastructure rather than towards it. Ideally, the stone columns should be installed prior to the construction of any new structures in the vicinity.
Vibration	Disadvantage – Due to vibration generated during the compaction process, in particular dynamic replacement, this ground improvement method may not be deemed suitable for areas where sensitive structures are present.
Time	Disadvantage – Stone columns are usually designed in combination with preloading. Hence, there is a delay from installation to site opening.
Complexity of analysis	Advantage – Analysis is generally considered not complex. Further, simple instrumentation such as settlement plates, inclinometers, etc. could be used to monitor the behaviors and take necessary remedial measures if adverse behaviors is observed.
Cost	Advantage – Less expensive compared to some other ground improvement techniques such as semi-rigid inclusions.
Preload height	Advantage – Generally, the height of preload could be reduced when stone columns are used if available waiting time for preloading is limited.
Design parameters	Disadvantage – Similar to wick drain analysis, there are many design parameters involved, which makes it difficult to back-analyze.
Edge effects	Disadvantage – The stone column configuration could create differential settlement issues adjoining a non-stone column area and therefore transition should be taken into account in the design.

(Continued)

Table 10.1 (Continued)

Item	Advantage/disadvantage
Stability	Advantage – The rate of strength gain in soft clay is increased which allows faster construction and/or less high strength geotextiles to manage stability.
Remedial measures if stone columns not properly installed or inadequate	Advantage – Additional stone columns could be installed if necessary, but casings may need to be used to ensure no remolding of the surrounding soil occurs. As discussed in Chapter 9, the same outcome cannot be achieved for wick drains.
Environment	Advantage – Reduces carbon footprint as no cement or steel is used to improve the ground.
Settlement	Disadvantage – Stone columns are simply semi-rigid soil columns, not structural members. As a result, stone columns can bulge or fail. While settlement of the treated ground can be greatly reduce, settlement can still be significant. The design must take the residual settlement into account.

10.13 CASE STUDIES

There are many case studies in geotechnical literature in journals, conferences and seminars, as well as in books. Two case studies are very briefly introduced below and references to the two case studies are provided so that readers could decide to peruse further if necessary.

CASE HISTORY 1   DYNAMIC REPLACEMENT COLUMNS

This case study refers to a project in Alexandria, Egypt, as documented in Wong and Lacazedieu (2004). The subsurface profile consists of very soft clay to a depth of 4 m to 9 m. The ground improvement adopted was a combination of dynamic replacement (DR) with wick drains.

Although DR was adopted as the main ground improvement technique, it was accepted that DR would not be able to fully penetrate the soft clay layer. The thickness of untreated clay thickness below DR columns was estimated to be 2.2 m. Therefore, the design approach was to have an additional measure of preloading the site with wick drains taken down to the base of the soft clay layer. The DR column diameter was 2.5 m and spacing was 5.5 m and 7 m with wick drains installed in a square grid with 1.1 m to 1.25 m spacing.

Settlements were underestimated by about 200 mm but time for consolidation compared well with prediction. Extensometer results suggested that the observed settlements within DR columns were higher than the design, which led to the conclusion that the Young's modulus adopted for DR was closer to 25 MPa rather than the design value, which was doubled.

CASE HISTORY 2   VIBRO-DISPLACEMENT COLUMNS AND DYNAMIC REPLACEMENT COLUMNS

Poon et al. (2011) described the design of a reinforced soil wing wall (RSW) over 6 m soft ground treated with DR columns, supplemented by stone columns when DR columns failed to penetrate the soft clay layer. Prior to the RSW construction, a post-DR investigation indicated that many of the DR columns were driven short of the design depth. Re-designs were implemented, which involved the installation of remedial full depth stone columns (SC).

The design configuration of the RSW is outlined in Figure 10.31. The DR and SC are 2.5 m and 1 m in diameter respectively at 5 m triangular spacing. To limit the differential settlements of the wall facing, a ground beam spanning over a row of remedial SC was provided to support the wall panels. Further, a dead-man anchor was adopted to tie back the ground beam into the platform in order to reduce the horizontal force on the front row of SC.

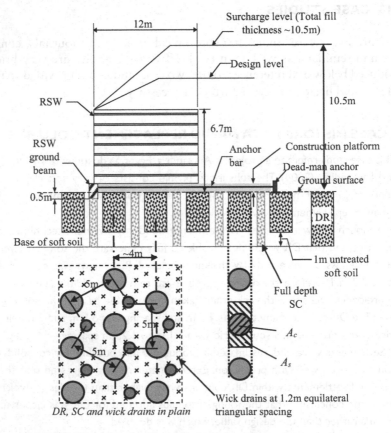

Figure 10.31 Reinforced soil wing wall design configuration.

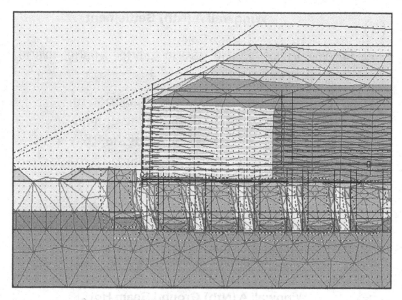

Figure 10.32 2D FEA result for RSW wing wall.

The FEA results have indicated that the lateral soil movements of the SC and adjacent soil beneath the wall facing have a direct influence on the overall performance of the wing wall design. For example, if the first row of SC supporting the wing wall ground beam is omitted, both horizontal and differential vertical movements may exceed the design criteria. Poon et al. (2011) demonstrated that the adopted SC configuration together with the dead-man anchor was able to greatly reduce the outward movement of the wall facing.

Figure 10.32 shows an exaggerated deformed mesh of the RSW Wing Wall A under short term fill load with surcharge. Figure 10.33a shows that the FE predictions for the vertical settlements are in good agreement with field measurements at the wall facing. The prediction for the horizontal movement (Figure 10.33b), is slightly conservative, but within the design limit of 150 mm.

10.14 SUMMARY AND GENERAL COMMENTS

Stone column technique is frequently used in soft ground treatment, slope stabilization and treatment of potential liquefaction in sand. This chapter introduces various types of stone column systems that are commonly used in the industry at present.

Stone columns should not be treated as structural members as they have limited bearing capacity with minimal bending and tensile capacities.

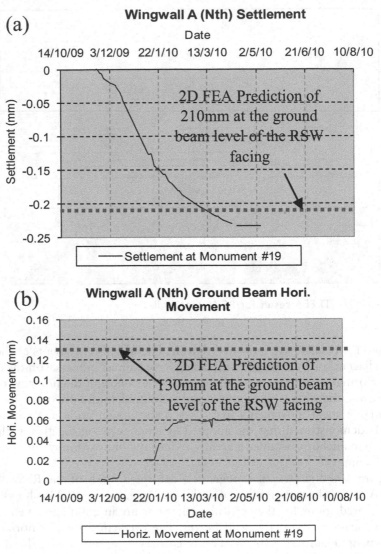

Figure 10.33 (a) measured and predicted settlement at wing wall facing, (b) measured and predicted settlement horizontal movement at ground beam level of wing wall facing.

The columns would exhibit bulging failure under excessive loading. Simple close form and empirical solutions have been outlined in this chapter to determine bearing capacity of a single stone column. While a single stone column is rarely installed in isolation in engineering practice, the bearing capacity of a single stone column provides a fundamental understanding of the failure mechanism that is essential in the design of a group of stone columns. Several additional concepts pertinent to ground treatment design using stone column group have also been outlined in this chapter, namely, the unit cell concept, the area replacement ratio and the stress concentration ratio. These design concepts and the theoretical bearing capacity of a single stone column can be used to describe the load transfer mechanism of stone columns. This chapter provides a simplified design procedure to determine the load distribution of stone columns accounting for the yielding of columns under vertical loading.

The reduction in primary settlement due to ground treated with stone column may be assessed using the well-known Priebe (1995) method. The Priebe method is based on a combination of elasticity and Rankine earth pressure theory, and it takes account of limited bulging of the columns by limiting the improvement factor. However, Priebe's improvement factor does not reduce with high applied load. The equilibrium method also does not appear to consider stone column yield unless the designer judiciously limits the stress ratio between the stone columns and the surrounding soil. Further, Priebe (1995) does not provide advice on the calculation of secondary (creep) settlement for stone column improved soil.

Assuming settlement is approximately proportional to the applied stress, the efficiency of stone columns to reduce settlement compared to that of the untreated ground may be expressed by an alternative settlement reduction ratio to that considered in the Priebe method. The new settlement reduction factor, μ_{soil}^*, can be defined by Eq. (10.39) as the ratio of the load being transferred to the soil surrounding the column after the consideration of stone column yield (σ_{soil}^*) to the applied load $\Delta\sigma$ on the soil. This chapter outlines that the μ_{soil}^* factor may be used to calculate the improved soil compressibility ratios for the assessment of primary and secondary consolidation settlements after the treatment with stone columns.

The different methods outlined in this chapter for the settlement prediction of stone column treated soft soils are using either simplistic theoretical method (e.g. Priebe 1995) or a composite material approach in which equivalent strength and deformation parameters are derived using semi-empirical correlation to represent the entire reinforced soil. While these approaches have been accepted as reasonable methods for vertical displacement prediction, they are less certain for the prediction of horizontal displacement. Recently, some researchers (e.g. Tan et al. 2008; Chan and Poon 2012) have started to simulate rows of stone columns using equivalent plane-strain strips in 2D FEA modelling. This approach is considered more appropriate to analyze stone columns subject to lateral loading. This chapter outlines the bases of this design approach and the comparison the predicted lateral behavior with measurements in the case studies presented at the end of this chapter.

Finally, this chapter also highlights the environmental considerations and the construction issues associated with stone column installation.

REFERENCES

Balaam, N.P. and Booker, J.R. (1981). "Analysis of rigid rafts supported by granular piles." *International Journal for Numerical and Analytical Method in Geomechanics,* 5, 379–403.

Balaam, N.P. and Poulos, H.G. (1982). "*The behavior of foundations supported by clay stabilized by stone columns.*" *Proc. 8th European Conference on Soil Mechanics and Foundation Engineering,* Helsinki.

Barksdale, R.D. and Bachus, R.C. (1983). "Design and construction of stone columns." *Federal Highway Administration Office of Engineering and Highway Operations, Volume I and II,* Report No. FHWA/RD-83/026, Georgia Inst. of Tech., Atlanta. School of Civil Engineering.; Federal Highway Administration, Washington, DC.

Bergado, D.T. and Lam, F.L. (1987). "Full scale load test of granular piles with different densities and different proportions of gravel and sand in the soft Bangkok clay." *Soils and Foundations Journal,* 27(1), 86–93.

Chan, K.F. and Poon, B. (2012). "*Designing stone columns using 2D FEA with equivalent strips.*" *International Conference on Ground Improvement and Ground Control (ICGI),* Wollongong, Vol. 1, 609 – 620.

Chan, K. and Poon, B. (2015). "Chapter 18 New analytical approach for predicting horizontal displacement of stone columns." In: Indraratna, B., Chu, J. and Rujikiatkamjorn, C. (eds), *Ground Improvement Case Histories – Embankments with Special Reference to Consolidation and Other Physical Methods.* Elsevier Ltd, Butterworth Heinemann, Oxford, UK.

FHWA (1983) U.S. Department of Transportation Federal Highway Administration (Dec,1983) – *Ground Modification Methods Reference Manual – Volume II.* Report No. FHWA/RD-83/026.

FHWA (2017) U.S. Department of Transportation Federal Highway Administration (April 2017) – *Design and Construction of Stone Columns, Vol 1.* Report No. FHWA-NHI-16-028. Authors V.R. Schaefer, R.R. Berg, J.G. Collin, B.R. Christopher, J.A. DeMaggio, G.M. Filz, D.A. Bruce and D. Ayala.

Frikha, W. and Bouassida, M. (2015). "Prediction of stone column ultimate bearing capacity using expansion cavity model." *Proc. Institution of Civil Engineers – Ground Improvement,* 168(2), 106–115.

Gibson, R.E. and Anderson, W.F. (1961). "In situ measurement of soil properties with the pressuremeter." *Civil Eng and Public Works Review,* 56(658), 615–618.

Han, J. and Ye, S.L. (1991). "*Field tests of soft clay stabilized by stone columns in coastal areas in China.*" *Proc. 4th Int. Conf. on Piling and Deep Foundations,* Balkema, Rotterdam, the Netherlands, 243–248.

Hughes, J.M.O. and Withers, N.J. (1974). "Reinforcing of soft cohesive soils with stone columns", *Ground Engineering,* 7(3), 42–49.

Kundu, P.K., Sharma, K.G. and Nanda, A., (1994). "*Analysis of stone column foundation for storage tank by FEM*", *Proc. Settlement '94, Vertical and Horizontal Deformations of Foundations and Embankments.* ASCE GSP 1 (40), 701–710.

Ladd, C.C. and Foott, R. (1974). "New design procedure for stability of soft clays." *Journal of the Geotechnical Engineering Division,* 100(GT7), 763–786.

Ladd, C.C., Foott, R., Ishihara, K., Schlosser, F. and Poulos, H.G. (1977). "*Stress-deformation and strength characteristics.*" *Proc. 9th ICSMFE,* Tokyo, 2, 421–494.

Madhav, M. R. and Nagpure, D.D. (1996) "*Design of granular piles for embankments on soft ground.*" *Proc. 12th Southeast Asian Geotechnical Conference*, Kuala Lumpur, Malaysia, 285–289.

Madhav, M.R. and Vitkar, P.P. (1978). "Strip footing on weak clay stabilized with a granular trench or pile." *Canadian Geotechnical Journal*, 15, 605–609.

Mesri, G. and Feng, T.W. (1991). "*Surcharging to reduce secondary settlements.*" *Proc. International Conference on Geotechnical Engineering for Coastal Development*, 1, Yokohama, Japan, 359–364.

Mitchell, J.K. (1981). "*Soil improvement state-of-the-art.*" *Proc.10th ICSMFE*, Stockholm, Vol. 4, 509–565.

Poon, B., Chan, K. and Kelly, R. (2011). "*Analysis and performance of a reinforced soil wing wall constructed over treated soft ground.*" *Proc. 14th Pan-American Conference on Soil Mechanics and Geotechnical Engineering*, Toronto, Canada, Paper 247.

Priebe, H.J. (1995). *The Design of Vibro Replacement.* Keller Grundbau GmbH Publication.

Stuedlein, A.W. and Holtz, R.D. (2013). "Displacement of spread footings on aggregate pier reinforced clay." *Journal of Geotechnical & Geoenvironmental Engineering*, 140, 36–45.

Tan, S.A., Tjahyono, S. and Oo, K.K. (2008). "Simplified plane-strain modelling of stone-column reinforced ground." *Journal of Geotechnical and Geoenvironmental Engineering*, 134(2), 185–194.

Wong, P.K. and Lacazedieu, M. (2004). "Dynamic replacement ground improvement – Field performance versus design predictions for the Alexandria City Centre Project in Egypt." *Advances in Geotechnical Engineering. The Skempton Conference*, 2, 1193–1204.

Chapter 11

Semi-rigid inclusions

Thayalan Nall

11.1 INTRODUCTION

Semi-rigid inclusions are installed in the soil for the purpose of enhancing the global modulus of an otherwise weak soil and is a widely accepted ground improvement technique across the world. According to FHWA (2017), quoting Reid and Buchanan (1984), the first documented highway application of a road supported on columns was in Europe in 1984, and in 1994 for a storage tank foundation in USA. However, Menard is credited with the development of Control Modulus Columns (CMC)™ and obtaining a patent in 1994 (Chu et al. 2009). For ease of reference, semi-rigid inclusion in this book is abbreviated to SRI.

SRI is basically a stiff column, usually concrete, that is used to improve the overall strength and stiffness of a weak layer. While it could be used for both sandy and clayey soils, in this chapter we will only discuss the improvement of clayey soils.

SRIs are used as a viable ground improvement technique in many civil engineering projects. They are most useful on highway projects where thick compressible clay layers have to support tall embankments and where designing a preload to cater for tight settlement criteria is difficult. This is commonly the case at bridge approaches and/or road widenings where large settlements cannot be tolerated. SRIs are also used in storage yards, port works and large factory buildings. A typical arrangement of SRIs supporting a highway embankment is shown in Figure 11.1.

Development of sites with weak ground conditions often poses challenges with respect to bearing failure, intolerable settlement that induce large lateral pressures/displacements and global/local instability. In civil engineering construction, often there are structures such as road/rail embankments, earth dams, breakwater structures and container terminals, etc. where structures can tolerate only small deformations but could accommodate flexible behavior. In such circumstances, shallow foundation solutions may not be viable and a rigid solution such as a deep foundation solution can be an overkill. For such intermediate situations, ground improvement with SRIs

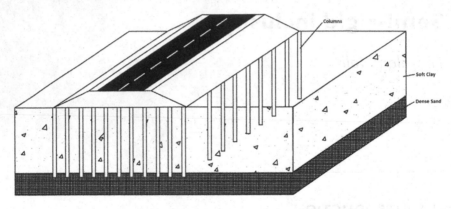

Figure 11.1 Typical SRI arrangement supporting a highway embankment.

could be an effective solution. SRIs are classified as semi-rigid inclusions/ reinforcement columns by ISSMGE (TC-17, working group D).

Often, the purpose and performance of SRIs are misunderstood by engineers and are expected/designed to meet the criteria of pile foundations. The ISSMGE working group describes

> Semi-Rigid Inclusion soil improvement technique is intended to improve the soil globally and to reduce its deformability using semi-rigid soil reinforcement columns. It does not aim to bypass the compressive ground or to build piles that will directly support the entire load from the structure. The objective is to reduce global and differential settlement by minimising the part of the loads actually supported by the soil and usually between 60 and 90%, is transmitted through the inclusions to the soil strata.

> (Plomteux 2020)

Further, the load from the structure is not directly transferred to the SRIs. Instead, the load transfer is developed through an intermediate load transfer platform (LTP) similar to a flexible raft, constructed between the heads of the columns and the structure to be supported. This LTP is commonly made of well compacted granular material with geotextile reinforcement in most instances. There may be circumstances, where high loads are expected, when a reinforced concrete raft may be required to spread the load. The load transfer concept of SRI compared to other conventional foundation systems is described in Figure 11.2.

Conventionally, the columns are designed to be socketed or founded within a stiffer/denser soil layer underlying the weak compressible clays. But there can be instances where the columns are entirely embedded within the weak layer itself. A classic example is a bridge approach where the issue of

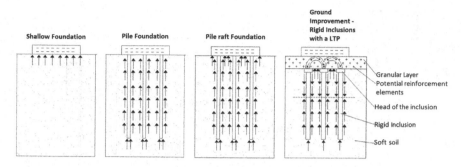

Figure 11.2 Load transfer concepts of foundation systems (adapted from Varaksin 2016).

differential settlement for a smooth transition can be managed by gradually shortening the lengths of SRIs, moving away from the bridge abutment.

The primary objective of this ground improvement technique is that the grid of installed SRIs will distribute the load within the weak soil mass and act as a composite column-soil, with an improved stiffness of the soil mass within the footprint of the development. To arrive at this, the size, spacing and column material properties are optimized depending on the loads and the subsoil characteristics of the project site.

11.2 LOAD TRANSFER MECHANISM

The load transfer mechanism in operation in the SRI supported foundation system can be rationalized in the following four phases:

o Phase 1: Consolidation – Design load reaching the weak compressible clay and triggering settlement;
o Phase 2: Mobilization of downdrag – As the compressible clay settles, load induced on the SRI from the adjacent soil;
o Phase 3: Resistance on SRI – The above two stages and portion of the design load carried directly by the SRI make it settle, resulting in the development of positive (upward) skin friction and base resistance in the SRI; and
o Phase 4: Equilibrium – The foundation system reaching a state of force equilibrium. The top of the SRI is still subjected to a downdrag force or negative skin friction acting downward. The bottom of the SRI has developed an upward skin friction and base resistance.

They are graphically presented in Figure 11.3.

Based on the load transfer mechanism described above and resulting deformation contours, we could infer that, within this composite system,

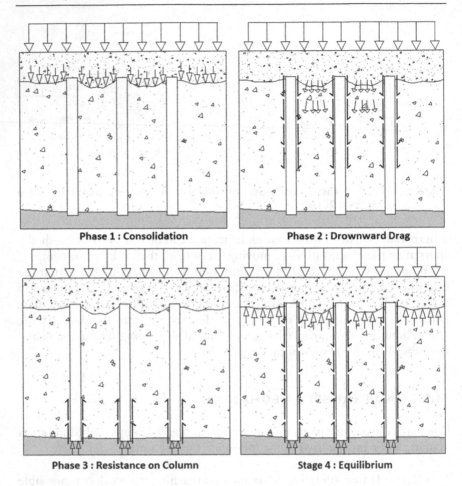

Figure 11.3 Load transfer mechanism.

there will be three neutral planes, otherwise known as planes of equal deformation, levels at which the displacements of both the SRI and surrounding soil are equal.

- o Upper neutral plane (NP-1) – above column head level – above this level settlement will be uniform;
- o Middle neutral plane (NP-2) – within the length of the semi-rigid inclusion – level between which the negative and positive skin friction are in operation; and
- o Lower neutral plane (NP-3) – located beneath the toe level of the column – settlement becomes uniform again.

This is described in Figure 11.4 with neutral planes shown in dash lines.

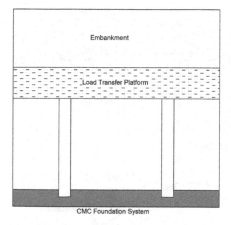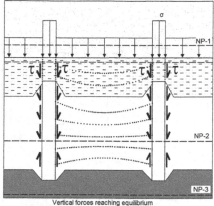

Figure 11.4 Inferred neutral planes.

The load transfer mechanism forms a general basis for the design and assessment of the SRI foundation system.

11.3 CRITICAL HEIGHT OF EMBANKMENT

Depending on the order of settlement the weak compressible clay layers are subjected to, the upper neutral plane (NP-1) (as shown in Figure 11.4) will advance into the embankment fill. If the embankment height above the LTP is smaller than the height of the upper neutral plane above the column head, the vertical deformations experienced by the foundation system could be partially reflected at the road surface as differential settlement. This is commonly known as the 'mushrooming' effect or 'egg carton' surface. This is demonstrated in Figure 11.5.

To effectively manage differential settlements reflecting at the embankment surface, the clear spacing between columns is designed in comparison with the embankment height. A number of standards and references provide guidelines for the critical embankment height, with respect to column spacing to avoid mushroom effects. Some of them are summarized in Table 11.1 where the following notations apply.

S_1 and S_2 = center to center spacing between SRI columns (see Figure 11.6)
d = equivalent diameter of column
$S = S_1 = S_2$ for square configuration of the SRI
$S - d$ = clear spacing between columns
$S^* = \sqrt{S_1^2 + S_2^2}$)
H_{crit} = critical embankment height to ensure no mushroom effect

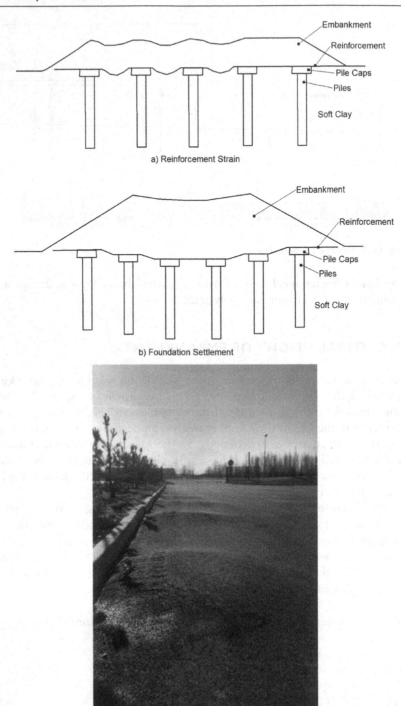

Figure 11.5 Differential settlement reflecting on finished surface.

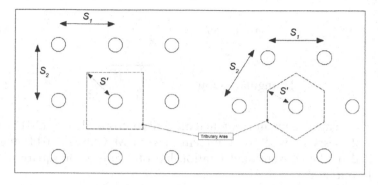

Figure 11.6 SRI layouts commonly considered in construction.

Table 11.1 Some published guidelines for critical height

Reference	Critical height of embankment	Comments
British Code, BS8006 (2010)	$H_{crit} = 0.7\ (S-d)$	
German Code, EBGEO (2010)	$H_{crit} = 0.7\ (S-d)$ $H_{crit} = 2.0\ (S-d)\#$	#when subjected to impact loading
Dutch Code, CUR226 (2016)	$H_{crit} = 0.66\ (S-d)$	
Rogbeck et al. (1998), Collin (2004)	$H_{crit} = (S-d)$ and > 0.9 m	
Kempfert et al. (2004)	$H_{crit} = 1.4\ (S^*-d)$ and > 0.9 m	
Hewlett and Randolph (1988)	$H_{crit} = 2.0\ (S-d)$	
French ASIRI guideline (2013)	$H_{crit} = 0.7\ (S-d)$	Based on Prandtl's failure mechanism
	$H_{crit} = \{(S/\pi)\ -(d/2)\}\ /\ \tan\phi'$	Based on punching shear failure mechanism
Swedish method (Carlsson 1987)	$H_{crit} = 1.87\ (S-d)$	
McGuire (2011)	$H_{crit} = 1.15\ s'+1.44\ d$	
Sloan (2011)	Greater of $H_{crit} = 1.5\ (S-d)$ or $H_{crit} = 1.15\ s'+1.44\ d$	

$$\text{Square layout} s' = \frac{\sqrt{S_1^2 + S_2^2}}{2} - \frac{d}{2} \qquad (11.1)$$

$$\text{Triangular layout} s' = \frac{\sqrt{2S_1^2 + S_2^2}}{3} - \frac{d}{2} \qquad (11.2)$$

a. The critical embankment height derived from Sloan's (2011) field-scale tests showed good agreement with McGuire's (2011) findings and of the conventional relationship of $1.5(S\text{-}d)$ for square column layout; and

b. S' is the radius of the tributary area of the column and determined using Eq. (11.1) and Eq. (11.2), respectively for square and triangular SRI layouts.

11.4 SIZE AND SPACING OF SEMI-RIGID REINFORCING ELEMENTS

In the selection of column type and column dimensions, the primary consideration is that the semi-rigid inclusion should perform as part of the composite system and improve the deformation modulus of the soil mass within the footprint.

11.4.1 Column size

The size, spacing and the material of the semi-rigid column are chosen based on the development of an optimal combination of support from the columns and the soil mass in order to limit settlements of the treated site within the allowable range, and/or to achieve a desired equivalent deformation modulus of the improved ground.

Commonly, precast columns, vibro-concrete columns, deep-soil mixing columns and other cement-based columns are considered as semi-rigid inclusions. Some of them are constructed using a displacement auger (i.e. no spoil to be removed), deep soil mixing tools or constructed within a pre-drilled hole. Although concrete may be the most popular material used in SRIs, columns made of other materials, such as timber columns and stone columns, etc. could also be used as support columns.

Conventionally, column diameter varies between 0.3 m and 0.6 m. Table 11.2 provides typical dimensions of columns as given in FHWA (2017). The design concept is equally applicable for dimensions outside the typical ranges; however, machine capacity and other practical construction limitations may require consideration.

Table 11.2 Typical column dimensions (adapted from FHWA 2017)

Column type	Typical column length (m)	Typical column diameter (m)
Timber pile	6 to 18	0.3 to 0.45
Steel H pile	9 to 30	0.25 to 0.35
Steel pipe pile	9 to 36	0.25 to 1.2
Precast concrete pile	9 to 15	0.25 to 0.6
Cast-in-place concrete shell (mandrel driven)	6 to 36	0.25 to 0.45
Shells driven without mandrel	6 to 23	0.3 to 0.9
Continuous flight auger pile	6 to 23	0.3 to 0.6
Deep soil mixed column	6 to 27	0.6 to 2.0
Aggregate column	3 to 9	0.6 to 1.2
Vibratory compacted column	6 to 27	0.45 to 0.6
Combined soil stabilization	3 to 9	0.1 to 0.2

11.4.2 Column spacing

For the composite column/surrounding soil to behave as a block, conventionally the center-to-center spacing of the columns is maintained between three and four times the column diameter ($S = 3d$ to $4d$). Some of the published guidelines on typical column spacing are given in Table 11.3.

The column spacing determines the area replacement ratio (A_r), which is defined as the ratio of the column cross sectional area to the tributary area of the column as shown in Figure 11.6. The variation of A_r with S/d (spacing/diameter) is shown in Figure 11.7.

Table 11.3 Typical SRI spacing

Reference	Typical column spacing
FHWA (2017)	1.2 m to 1.8 m
Kempfert et al. (2004)	$S*-d \leq 3$ m – for static loading $S*-d \leq 2.4$ m – under heavy live loading
Collin (2004)	$S-d \leq 3$ m
German Code, EBGEO (2010)	$S*-d \leq 2.5$ m
German Code, EBGEO (2011)	$d/S > 0.15$ ($S < 6.7d$)
Wong and Muttuvel (2011)	1.3 m to 2.0 m

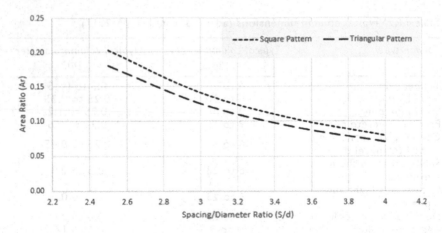

Figure 11.7 Area ratio versus spacing/diameter ratio.

Rogbeck et al. (1998) recommend that A_r should be a least 10% and this seems to be the lower limit reported by Han and Gabr (2002) based on case histories for geosynthetic-reinforced, column supported embankments.

For typical *S/d* values of 3 to 4, the area ratio value falls within 7% to 13% for the recommended column spacing as per Figure 11.7.

11.4.3 Equivalent parameters

At the initial stages of a design, it is worthwhile undertaking a preliminary assessment of the system displacement considering equivalent block parameters of the column treated site. This will help to arrive at appropriate column dimensions and material properties to achieve the intended outcomes.

The assessment of the equivalent strength of the soil mass improved by SRI can be estimated in the same manner as that for stone columns, proposed by Madhav and Nagpure (1996), and equations proposed for the estimation of equivalent strength properties are given below.

$$N_{eq} = A_r N_1 + (1 - A_r) N_2 \tag{11.3}$$

$$c_{eq} = \frac{c_1 A_r \sqrt{N_1} + c_2 (1 - A_r) \sqrt{N_2}}{A_r N_1 + (1 - A_r) N_2^{0.5}} \tag{11.4}$$

$$A_r = f_1 \left[\frac{d}{S}\right]^2 \tag{11.5}$$

$$\gamma_{eq} = A_r \gamma_1 + (1 - A_r) \gamma_2 \tag{11.6}$$

where: $N_{eq} = \tan^2(45 + \phi_{eq}/2)$; ϕ_{eq} = equivalent angle of friction
A_r = area ratio; $f_1 = 0.907$ and 0.7854 for triangular and square column patterns
d = column diameter; s = column spacing (center-to-center)
$N_1 = \tan^2(45 + \phi_1/2)$; $N_2 = \tan^2(45 + \phi_2/2)$
c_{eq} = equivalent cohesion; γ_{eq} = equivalent unit weight
c_1, ϕ_1, γ_1 = cohesion, friction angle and unit weight of column material
c_2, ϕ_2, γ_2 = cohesion, friction angle and unit weight of soil

A number of expressions have been published to assess the equivalent stiffness of soft ground reinforced by semi-rigid columns. They include:

o Balaam et al. (1976) – Provided design charts for fully and partially penetrating stone columns based on finite element studies;
o Shahu et al. (1998) – Carried out a parametric study of soft ground reinforced with granular piles and with a granular mat on top;
o Bergado and Lam (1987) – Compared the results of their parametric study with field measurements.

We have considered the equivalent stiffness relationship based on Young's modulus proposed by Watts et al. (2000) and Poulos (2002) for preliminary settlement assessment.

The expression equivalent modulus proposed by Watts et al. (2000) for assessing the settlement of the soil mass reinforced by stone columns supporting a strip footing, is given by Eq. (11.7):

$$E_{eq} = E_c \left[A_r + \frac{E_s}{E_c}(1 - A_r) \right] \tag{11.7}$$

E_c and E_s are Young's modulus of the column and the untreated soil, respectively.

A modified form of the above expression for stone columns supporting general fill embankments was proposed by Poulos (2002) for the assessment of equivalent modulus as given in Eq. (11.8).

$$E_{eq} = E_c \left[A_r^2 + \frac{E_s}{E_c}(1 - A_r^2) \right] \tag{11.8}$$

The above two expressions are compared in terms of Settlement Ratio (SR) and presented in Figure 11.8. The SR is defined as given in Eq. (11.9):

$$SR = \frac{S_T}{S_{UT}} \tag{11.9}$$

where S_T and S_{UT} are settlements of the treated and untreated ground, respectively.

Settlement being proportional to the Young's modulus, the settlement ratio, SR, can be rewritten as follows.

Adopting equivalent modulus as per Watts et al. (2000):

$$SR = \frac{E_s}{E_c}\left[A_r + \frac{E_s}{E_c}\left(1 - A_r\right)\right]^{-1} \tag{11.10}$$

Adopting equivalent modulus as per Poulos (2002):

$$SR = \frac{E_s}{E_c}\left[A_r^2 + \frac{E_s}{E_c}\left(1 - A_r^2\right)\right]^{-1} \tag{11.11}$$

The settlement ratios computed as per Eq. (11.10) and Eq. (11.11) are shown in Figure 11.8.

According to Wong (2004), based on a comparison of these two equations on settlement assessment and field measurements reported by Bergado and Lam (1987), Poulos' expression for equivalent stiffness shows better agreement with field measurements.

Wong further indicates that the equation proposed by Watts et al. (2000) is based on equal strains in the soil and columns, and this would only apply if a rigid raft (LTP) exists over the columns, or if the area replacement ratio is sufficiently large that the applied load will be taken mostly by the columns via arching action.

However, some column types can be subjected to bulging, particularly near the top of the soft soil profile where confining pressures are low, and could cause non-uniform strain conditions. Therefore, the use of the equal strain approach in such instances may under-estimate total and differential settlements.

On this basis, for the purpose of preliminary designs, the method proposed by Poulos (2002) appears to provide a relatively conservative design, which could then be refined by numerical methods during the detailed design stage.

It is also recognized that in certain favorable situations where the column bulging is limited, rigid LTP is present or a large area replacement ratio has been adopted, the equal strain method proposed by Watts et al. (2000) could provide reasonable results.

From the SR distribution (Figure 11.8), it may be inferred that the change in SR noticeably decreases for A_r values beyond 0.3, suggesting that improving area replacement ratios beyond 0.3 may not result in any further increase the performance of the SRI foundation system, hence become less economically attractive;

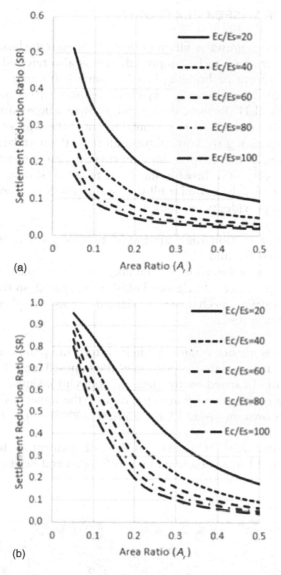

Figure 11.8 Range of area replacement ratios for recommended column spacing (a) *SR* vs. Area Ratio (Watts et al. 2000); (b) *SR* vs. Area Ratio (Poulos 2002).

It is important to note that, for simplicity an 'equal strain' method is being used, and this approach is not rational because it is not compatible with the concept of differential settlements that are necessary to mobilize stress concentrations in less deformable materials, leading to arching in the soil.

11.5 LOAD TRANSFER PLATFORM

The LTP is made of granular fill and installed between the heads of the columns and the structure to be supported. This is also referred to as a load distribution platform or bridging layer in some publications. To enhance load transfer to the columns, geosynthetic reinforcements are commonly employed within LTP. In some designs the geosynthetic reinforcement also plays the role of counteracting horizontal thrust at the sides of the embankment and of separating the embankment fill from the foundation soils (Lin 2000). The scope of this chapter does not cover design of geosynthetic reinforcement for horizontal thrust induced due to lateral sliding. Gravel is the designer's choice for the granular fill since it is easily compacted and for the high stiffness it provides when compacted.

- o Generally, a friction angle around 35 to 40 degrees is expected from the granular fill; and
- o Geosynthetic reinforcement (number of layers, their stiffness and strength) are selected in geotechnical design based on the load transfer/deformation mechanism considered across the elements of this foundation system.

There are several methods suggested in literature and practiced by geotechnical engineers to calculate the portion of the imposed load (from embankment or structure) carried by the geosynthetic reinforcement and the load carried by the column, and that supported by the upper soil layers. Each method has its own merits and limitations, though in some cases they vary significantly.

The load distribution at the base of the embankment at the interface is shown in Figure 11.9, as described by Van Eekelen and Bezuijen (2012).

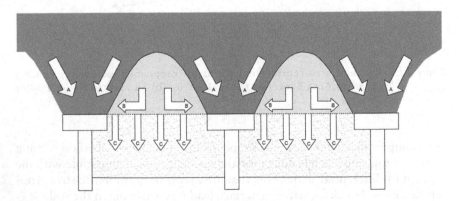

Figure 11.9 Load distribution at the base of the embankment.

Generally, it is accepted that the following three load components come into play with respect to load transfer from the embankment/structure onto the composite foundation system.

o Load component A – Carried by the vertical reinforcement (semi-rigid columns);
o Load component B – Carried by the horizontal reinforcement (geosynthetic layers); and
o Load component C – Carried by the soil beneath the geosynthetic reinforcement and between the columns.

Hence it is essential to assess each load component to undertake an effective design of LTP that would perform satisfactorily over the design life.

This process can be undertaken in the following three phases:

o Phase 1: Initial condition – Based on the column arrangement, identify the tributary area for columns and estimate the total load contribution (components A, B and C);
o Phase 2: Soil arching – Adopting an appropriate arching model and estimating load component A that will be transferred on column heads and non-arched load that will be shed within the area between column heads (components B + C); and
o Phase 3: Membrane action – Assessing the tensile load taken up by the geosynthetic reinforcement (component B) and separating the load supported by the subsoil (component C).

This process may be illustrated conceptually as shown in Figure 11.10.
There are over 20 models that exist for calculating the arching stresses. Van Eekelen et al. (2013) broadly categorized them as rigid, limit equilibrium, frictional, mechanical or empirical models. Similarly, there are different approaches that model the membrane action.
So, the primary difference between published design approaches is the way in which the load components are computed and implemented in the design.

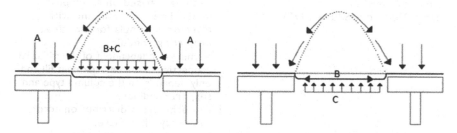

Figure 11.10 Soil arching and membrane action.

Terzaghi (1943) first described soil arching based on his trap door experiment and the vertical equilibrium of a soil element. He described the average vertical stress (σ_z) above the trap door by Eq. (11.12) and demonstrated that due to soil arching vertical stress acting above the trap door is much smaller compared to the geostatic vertical stress computed from the overburden pressure as γZ.

$$\sigma_z = \frac{S(\gamma - 2c'/S)}{2K\tan\phi'}\left[1 - e^{-2K\tan\phi'\frac{Z}{S}}\right] + q\,e^{-2K\tan\phi'\frac{Z}{S}} \qquad (11.12)$$

where:
S – Trap-door width;
Z – depth below surface;
c', ϕ' and γ – effective cohesion, friction angle and unit weight of the soil;
K – earth pressure coefficient, equal to 1.0 based on Terzaghi's experiment; and
q – surcharge at the soil surface
Some of the published methods and their method of computation of the load components are summarized in Table 11.4 and the concept of some of these methods is illustrated in Figure 11.11 to Figure 11.15.
Within the framework of the ASIRI project (2013):

- Centrifuge test results were compared with limiting vertical pressures on the column head (q_P) calculated from Prandtl's theory, and good agreement was found between them (Figure 11.16).
- Prandtl's approach was also investigated by performing finite element calculations for various uniformly distributed loads. The plastic points observed just before failure suggest the failure mechanism is similar to Prandtl's failure mechanism.

Table 11.4 Soil arching considered in published methods

Reference	Method employed to estimate load components
Jones et al. (1990) British Method (BS8006 1995)	• Arching is based on an empirical method developed in line with Martson's formula for a positive projecting conduit. • Arching is independent of the material properties of the embankment fill and only depends on the column type and support condition. • Uniform stress distribution across the geosynthetic layer. • No support from the soft soil between the columns.

Reference	Method employed to estimate load components
Hewlett and Randolph (1988)	• Arching is based on a limit equilibrium analysis of a shell in a hemispherical dome above the columns. • The thickness of the arching shell is $b/\sqrt{2}$ (b width of a square pile). • Above and below the arching shell, the stress is increasing linearly with depth, similar to geostatic condition.
Kempfert et al. (1997) German Method (EBGEO 2010)	• Modified version of Hewlett and Randolph (1988) and considers multi shells arching between the columns. • Also considers, soft soil providing support to geosynthetic reinforcement based on modulus of subgrade reaction.
Swedish Method (Rogbeck et al. 1998)	• A simplified approach: pyramids of 75° wall inclination. • Independent of type and strength of embankment fill.
Van Eekelen et al. (2015b) Dutch Method CUR 226 (2016)	• Considering concentric 2D and 3D arches and as a result relatively large load is exerted on the columns and the geosynthetic reinforcement between adjacent columns. The radius of arches indicating the load intensity and found good agreement with field measurements. • Two types of load distribution on geosynthetic layer is adopted on the basis of subsoil support.
French national project ASIRI (2013)	• For sufficiently thick embankment based on Prandtl's failure mechanism. • For relatively thin load transfer platforms, based on punching shear failure mechanism.
Carlsson (1987)	• Wedge of soil whose cross-sectional area under the arching soil is approximated by a wedge with an internal angle at the apex of the wedge equal to 30°. • This wedge load is carried by the geosynthetic layer and the rest is carried by the column.

(*Continued*)

Table 11.4 (*Continued*) Soil arching considered in published methods

Reference	Method employed to estimate load components
Bell et al. (1994)	• LTP is made of two layers of geosynthetic reinforcement. • Based on the work by Guido et al. (1987), the angle of load spread through the geosynthetic reinforced granular fill is considered at 45 degrees. • This wedge load is carried by the geosynthetic layers and the rest of the load is carried by the column.
Russell and Pierpoint (1997)	• Modified Terzaghi's arching analysis to account for the three-dimensional shape of the settling embankment fill between the columns. But no change in the earth pressure coefficient (K).
Svanø et al. (2000)	• Considers a pyramid type arching and enables to calculate the load carried by the column.
Russell et al. (2003)	• Similar to Russell and Pierpoint (1997), but with an earth pressure coefficient (K) of 0.5, resulting in an increased load on the geosynthetic reinforcement. • Also considers, soft soil providing support to geosynthetic reinforcement based on modulus of subgrade reaction.
Kempfert et al. (2004)	• Arching is based on lower bound plasticity theory of a hemispherical dome above the columns. • Also considers, soft soil providing support to geosynthetic reinforcement based on modulus of subgrade reaction.
Collin (2004)	• Made further refinement to the work by Guido et al. (1987) and Bell et al. (1994). • LTP with minimum three layers of geosynthetic reinforcement. • Soil load on geosynthetic as a pyramid inclined at 45 degrees.

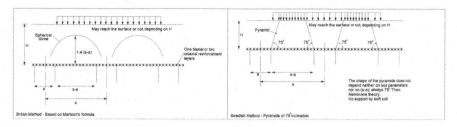

Figure 11.11 British and Swedish methods.

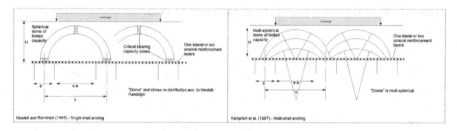

Figure 11.12 Single and multi-shell arching.

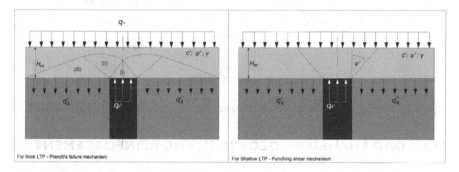

Figure 11.13 French approach ASIRI 2013.

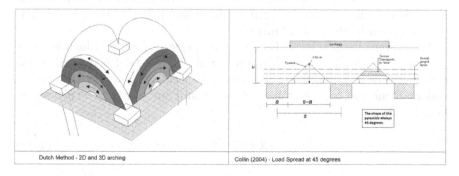

Figure 11.14 Dutch method and simplified Collins method.

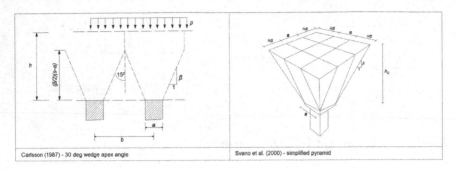

Figure 11.15 Simplified methods by Carlsson and Svanø.

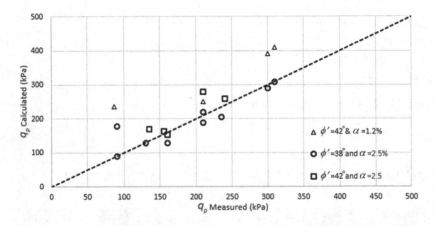

Figure 11.16 Comparison of vertical stress on the column head (ASIRI 2013).

11.6 LOAD SHARED BY GEOSYNTHETIC REINFORCEMENT

The portion of the load shed between the columns (load components B and C in Figures 11.9 and 11.10) has been assessed by researchers as an average vertical stress between the columns (σ_{avR}). The average stress is used in some methods to calculate the induced strains and tension in the geosynthetic reinforcement by the embankment load.

The average vertical stress is expressed in terms of stress reduction ratio, SRR, a parameter introduced by Low et al. (1994). SRR is defined as the ratio between σ_{avR} and average geostatic vertical stress at the base of the embankment due to the embankment fill and surcharge, as expressed in Eq. (11.13).

$$SRR = \frac{\sigma_{avR}}{\gamma H + q} \qquad (11.13)$$

where:

  σ_{avR} – average stress on the geosynthetic reinforcement or subgrade soil (if the LTP is not reinforced);
  γ – unit weight of the embankment fill;
  H – embankment height; and
  q – surcharge pressure at the surface of the embankment.

The derivations used for calculating the SRR in some published methods are summarized below.

Terzaghi (1943) – Modified by Russell and Pierpoint (1997)

$$SRR = \frac{\left(S^2 - d^2\right)}{4Hd\tan\phi'}\left\{1 - e^{\left[\frac{-4HdK\tan\phi'}{\left(S^2 - d^2\right)}\right]}\right\} \tag{11.14}$$

$$K = 1 - \sin\phi \tag{11.15}$$

Guido et al. (1987) – Modified by Bell et al. (1994) and Collin (2004)

$$SRR = \frac{(S-d)}{3\sqrt{2H}} \tag{11.16}$$

Hewlett and Randolph (1988)
For low embankments – critical arching is at the crown.

$$SRR = \left(1 - \frac{d}{S}\right)^{2(K-1)}\left(1 - \frac{2S(K-1)}{\sqrt{2}(2K-3)H}\right) + \frac{2(S-d)(K-1)}{\sqrt{2}(2K-3)H} \tag{11.17}$$

For high embankments – critical arching is at the column head.

$$SRR = \left\{\left(\frac{2K}{K+1}\right)\left[\left(1 - \frac{d}{S}\right)^{1-K} - \left(1 - \frac{d}{S}\right)\left(1 + \frac{dK}{S}\right) + \left(1 - \frac{d^2}{S^2}\right)\right]\right\}^{-1} \tag{11.18}$$

$$K = \frac{1 + \sin\phi'}{1 - \sin\phi'} \tag{11.19}$$

Low et al. (1994)

$$SRR = \frac{\left[\sigma_s - \left(\frac{tE_s}{D}\right)\right]}{\gamma H} \tag{11.20}$$

$$\sigma_s = \frac{\gamma(S-d)(K-1)}{2(K-2)} + \left(\frac{S-d}{S}\right)^{K-1}\left[\gamma H - \frac{\gamma S}{2}\left(1+\frac{1}{K-2}\right)\right] \quad (11.21)$$

This method considers contribution from the foundation soil:

E_s – elastic modulus of the foundation soil

t – maximum vertical displacement of the soil (midway between two adjacent column)

D – thickness of the foundation soil

K – earth pressure coefficient same as given in Eq. (11.19)

Kempfert et al. (2004)

$$SRR = \frac{1}{\gamma H}\left\{\lambda_1^\chi\left(\gamma + \frac{q}{H}\right)\left[\begin{array}{l} H\left(\lambda_1 + h_g^2\lambda_2\right)^{-\chi} \\ +h_g\left(\left(\lambda_1 + \frac{h_g^2\lambda_2}{4}\right)^{-\chi} - \left(\lambda_1 + h_g^2\lambda_2\right)^{-\chi}\right) \end{array}\right]\right\} \quad (11.22)$$

$$\lambda_1 = \frac{1}{8}(S_d - d)^2 \quad (11.23)$$

$$\lambda_2 = \frac{S_d^2 + 2dS_d - d^2}{2S_d^2} \quad (11.24)$$

$$\chi = \frac{(K-1)d}{\lambda_2 S_d} \quad (11.25)$$

$$h_g = \frac{S_d}{2} \quad \text{for} \quad H \geq \frac{S_d}{2} \quad (11.26)$$

$$h_g = H \quad \text{for} \quad H > \frac{S_d}{2} \quad (11.27)$$

q – surcharge load

S_d – diagonal spacing between columns; and

K – earth pressure coefficient same as given in Eq. (11.19)

BS8006 (2010)

For partial arching:

$$SRR = \frac{2S}{(S+d)(S^2-d^2)}\left[S^2 - d^2\left(\frac{P_c}{\gamma H}\right)\right] \quad (11.28)$$

For full arching:

$$SRR = \frac{2.8S}{(S+d)^2 H}\left[S^2 - d^2\left(\frac{P_c}{\gamma H}\right)\right] \tag{11.29}$$

$$\left(\frac{P_c}{\gamma H}\right) = \left[\frac{d\,C_c}{H}\right]^2 \tag{11.30}$$

$$C_c = 1.5\frac{H}{d} - 0.07 \quad \text{arching coefficient for friction piles} \tag{11.31}$$

ASIRI Guidelines (2013) for a granular LTP fill material

$$SRR = \frac{q_0 - H_r q_p}{1 - H_r}\left(\frac{1}{\gamma H_m + q_0}\right) \tag{11.32}$$

For Prandtl's failure mechanism, q_p is defined as:

$$q_p = \frac{N_q}{1 + H_r(N_q - 1)}q_0 \tag{11.33}$$

$$N_q = \tan^2\left[\frac{\pi}{4} + \frac{\phi'/f_\phi}{2}\right]e^{\pi\tan\left(\frac{\phi'}{f_\phi}\right)} \tag{11.34}$$

For punching shear failure mechanism q_p is defined by Eq. (11.35) and Eq. (11.36) for $H_m < H_{cri}$ and $H_m > H_{cri}$ respectively.

$$q_p = \frac{H_m}{3}\left[\frac{R_c^2}{0.5d^2} + 1 + \frac{R_c}{0.5d}\right]\frac{\gamma}{f_\gamma} + \frac{R_c^2}{0.5d^2}q_0 \tag{11.35}$$

$$q_p = \left\{\frac{H_m}{3}\left[\frac{R_c^2}{0.5d^2} + 1 + \frac{R_c}{0.5d}\right] + (H_m - H_{cri})\frac{S^2}{d^2}\right\}\frac{\gamma}{f_\gamma} + \frac{R_c^2}{0.5d^2}q_0 \tag{11.36}$$

$$R_c = 0.5d + H_m \tan\left(\frac{\phi'}{f_\phi}\right) \tag{11.37}$$

q_0 – vertical stress on top of the LTP;
q_p – limiting vertical stress on the column head;
H_m – LTP thickness;

H_r – area replacement ratio;
ϕ' – friction angle of LTP material;
γ – unit weight of LTP material;
f_ϕ – partial material factor for ϕ;
f_γ – partial material factor for γ;
S – column spacing; and
d – column diameter.

Ariyarathne and Liyanapathirana (2015) compared the *SRR* estimated by the above methods with the results of finite element 2D and 3D modelling performed on a column supported embankment. They investigated the influence of three parameters, viz., column size, column spacing, and embankment height, on the estimated *SRR* and the estimated *SRR* values are summarized in Figure 11.17 to Figure 11.20.

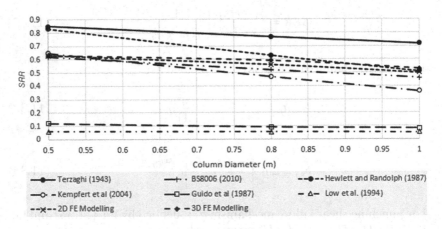

Figure 11.17 Variation of SRR with column diameter.

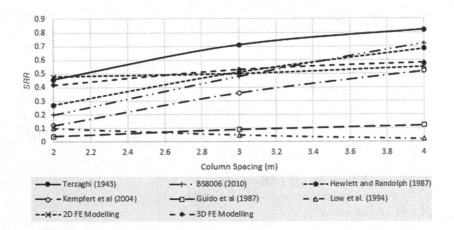

Figure 11.18 Variation of SRR with column spacing.

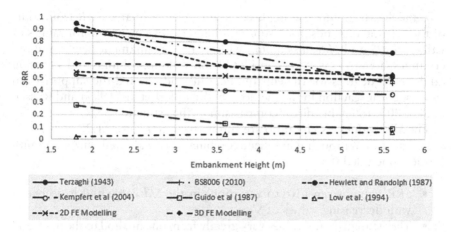

Figure 11.19 Variation of SRR with embankment height.

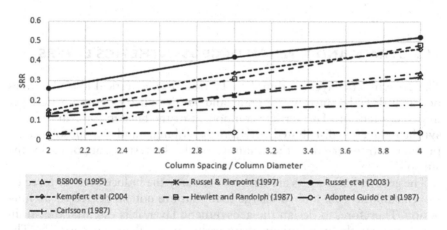

Figure 11.20 Variation of SRR for H/S = 4.

In the assessment by Ariyarathne and Liyanapathirana (2015), for the method by Low et al. (1994) which requires the elastic modulus of the foundation soil, a value equal to the weighted average of the elastic modulus of the soil layers was adopted.

By comparing the outcomes with the results of numerical modelling, they conclude:

- The Low et al. (1994) method highly underpredicts the stress reduction ratios and shows an inverse variation compared to the other design methods and numerical results; and
- Among the other methods, the Guido et al. (1987) method significantly underpredicts the stress reduction ratio.

Filz and Smith (2006) also made comparison on estimated SRR from some of the existing methods, for a square column arrangement and for ranges of values of d/S and H/S (d –column width, H – embankment height and S – column spacing). This study considered non-yielding columns and a friction angle of 35 degrees for the embankment fill and ignored the support from subgrade soil reaction in the methods by Kempfert et al. (1997) and Russell et al. (2003) to enable direct comparison of the estimated SRR.

The SRR values Vs H/S ratio are summarized in Figure 11.20.

For the foundation embankment combinations evaluated, Filz and Smith (2006) concluded that:

- SRR value is sensitive to variations in the S/d value and it decreases with decreasing values of S/d;
- The estimated SRR values vary greatly from one method to the next; and
- The adapted Guido method generally gives very low values of SRR compared to the other methods and it is almost one eighth of that estimated from the BS8006 Method.

11.7 STRAIN AND TENSION IN GEOSYNTHETICS LAYERS

The geosynthetic layers within the LTP pick up the vertical load that does not arch directly onto the columns, and as a result the geosynthetic layer undergoes tension. The vertical component of the tension force in the geosynthetic layer is distributed between the columns and the subsoil, i.e., a portion is transferred to the columns, while the rest is supported by the subsoil.

The geosynthetic layers are designed to carry the induced tensile force by the embankment fill and surcharge load that are not supported by arching action. Therefore, to design the geosynthetic layer, it is essential the strain developed on the geosynthetic reinforcement is calculated correctly. The induced tensile force is then calculated by multiplying the computed strain by the stiffness of the geosynthetic layer. Finally, to ensure the foundation system to perform satisfactorily, the developed tensile force should be limited to the long-term tensile strength of the geosynthetic layer.

The computation of tension load on the geosynthetic layer depends on:

- The soil arching model adopted since it determines the portion of the embankment fill and traffic load shared between the columns (stiffer member of the foundation), and the geosynthetic and the subsoils between the adjacent columns; and
- Distribution of the vertical load across the geosynthetic layer and the load deflection behavior (commonly known as membrane behavior) of the geosynthetic.

There are several methods published and practiced in the design of geosynthetic reinforcement by engineers. Based on the above two factors, the outcomes could be considerably different (Van Eekelen et al. 2015a).

There are a number of analytical approach-based design guidelines commonly adopted in practice. Numerical approaches are not common as some believe they tend to underpredict strain in geosynthetic layers (Farag 2008). Within Europe alone there are six different guidelines usually cited:

- Dutch CUR 226 (2010, updated in 2015);
- German EBGEO (2010);
- British Standard BS8006 (1995, updated in 2010);
- Finnish Liikennevirasto (2012);
- Nordic Handbook on Reinforced Soils and Fills (NGG 2002); and
- French ASIRI (2013).

BS8006 and EBGEO calculation models assume the strain in the geosynthetic to occur mainly in the tensile strips that lie between adjacent piles.

Commonly, there are three modes of load distribution on the geosynthetic reinforcement considered in the estimation of mobilized strain as shown in Figure 11.21.

- German EBGEO (2010) and Dutch CUR226 (2010) consider a triangular distribution;
- British Standard BS8006 (2010) considers that the load is distributed uniformly across the geosynthetic strip between the columns; and
- Revised Dutch CUR 226 (2015) considers an inverse triangular distribution.

The load distribution dictates the deflection shape of the geosynthetic reinforcement. By solving the cable equation, it may be demonstrated that an equally distributed load gives a parabolic second order deformation pattern, while an inverse triangular load gives a third order power law function.

Measurements made on a series of model tests (Van Eekelen et al. 2012) indicate that the deformation pattern of the geosynthetic layer follows a power law of third order or higher. Based on this outcome it is believed the load on the geosynthetic layer is of an inverse triangle and the Dutch CUR 226 (2015) was updated based on this evidence.

When comparing triangular and inverse triangular load distributions of the geosynthetic layers, the inverse triangular distributions result in 30% lesser maximum vertical displacement, hence inducing about 25% lesser strain (thus tensile load). From the design perspective, adopting an inverse triangular distribution enables a more economic design but may come at a cost of increased risk.

Another area that is investigated in geosynthetic design is ways of improving arching so that the vertical load component reaching the geosynthetic

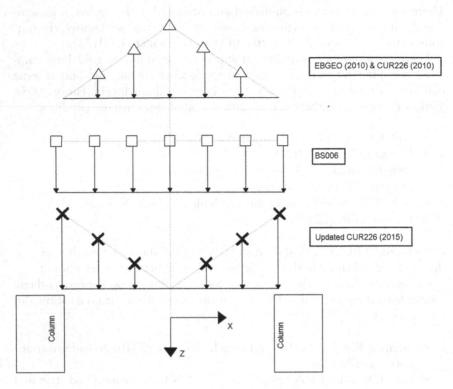

Figure 11.21 Load distribution on geosynthetic reinforcement.

layer can be reduced. Based on outcomes of model tests (Van Eekelen et al. 2012), the following are recommended:

- Constructing the LTP with material having higher angle of shearing resistance improves arching performance. The ASIRI project (2013) also confirms this outcome.
- Adopting geotextiles or geogrids of similar stiffness properties as the reinforcing layer shows no difference in performance. However, when rockfill is used as LTP material, sometimes particles getting trapped within the geogrid cell tend to enhance the stiffness of the geogrid and results in improved arching.
- Approaches of EBGEO (2010), CUR226 (2010) and BS8006 (2010) are based on a single geosynthetic layer at the base of the embankment. But in practice, commonly the required strength and stiffness is divided into two layers (easy procurement of medium-strength range geosynthetic layers compared to high strength) with granular material in between. The two-layer system showed improved arching.

- Strain in the geosynthetic layer is mainly limited to the tensile strips that lie between adjacent piles as adopted by BS8006 (2010) and EBGEO.(2010)

11.7.1 Estimation of strain in the geosynthetic layer

In this section, the method of estimation of induced strain in the geosynthetic layer adopted in the following methods is summarized:

- German EBGEO (2010);
- Dutch CUR 226 (2010);
- Dutch CUR 226 (updated in 2015); and
- British Standard BS8006 (1995).

11.7.1.1 German EBGEO (2010) and Dutch CUR 226 (2010)

Both adopt the model developed by Zaeske (2001).

As the first step, based on a 3D soil arching model, the load components B and C (refer to Figure 11.9) not shared by the columns are estimated. The model adopted only depends on the geometry of column supported embankment and the friction angle of the fill.

For the load displacement behavior of the geosynthetic layer, a triangular load distribution is considered as shown in Figure 11.22.

The subsoil support is modelled as an elastic spring with a constant modulus of subgrade reaction, K_s. The upward support reaction, F_{ux}, is expressed by Eq. (11.38).

$$F_{Ux} = K_s Z(x) \tag{11.38}$$

where $Z(x)$ is the vertical deflection of the geosynthetic reinforcement strip.

The vertical deflection due to the triangular loading $q(x)$ is expressed by differential equation, Eq. (11.39).

$$\frac{d^2Z}{dx^2} = \frac{q(x) - Ks\,Z(x)}{T_H} \tag{11.39}$$

where T_H is the horizontal component of the tensile force in the geosynthetic layer.

Solving the above differential equation, the computed average strain (ε) in the geosynthetic layer is provided in a series of graphs as given in Figure 11.23. The tensile force (T) on the geosynthetic layer is determined by Eq. (11.40).

$$T = J\varepsilon \tag{11.40}$$

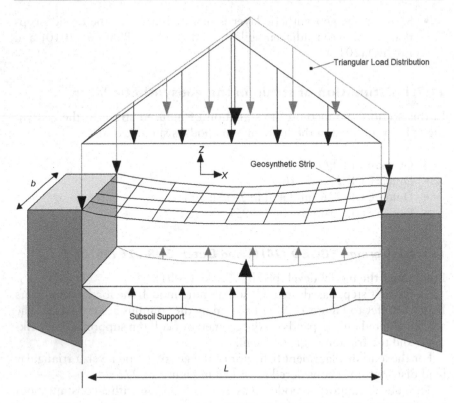

Figure 11.22 Load distribution adopted in EBGE (2010) and CUR 226 (2010).

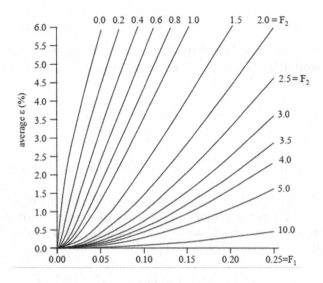

Figure 11.23 The average strain in the geosynthetic layer.

Note that F_1 and F_2 in Figure 11.23 are defined in Eq. (11.41) and Eq. (11.42); symbol J in Eq. (11.40) is given below.

$$F_1 = \frac{0.5(B+C)/b}{J} \qquad (11.41)$$

$$F_2 = \frac{K_s L^2}{J} \qquad (11.42)$$

J – is the stiffness of the geosynthetic layer, commonly determined from the isochronic curves relevant to the specific product.

11.7.1.2 Revised Dutch CUR 226 (2015)

In the light of the findings from model tests, the revised version of the CUR 226 made the following two changes.
First change:

> The contribution of subsoil to be increased from the area directly below the section of the strips to the entire area of the subsoils below the geosynthetic layer ($L_x \times L_y$). hence the upward support F_{ux} is expressed by the Eq. (11.43).

$$F_{ux} = \frac{A_{xy}}{[L \times b]} K_s Z(x) \qquad (11.43)$$

Only incorporating the above change, the design chart given in Figure 11.21 can be used adopting a revised F_2 as given in Eq. (11.44).

$$F_2 = \frac{K_s A_{xy} L}{J \times b} \qquad (11.44)$$

With respect to soil support, good engineering judgement is essential. If the subsoil support is expected to disappear over time, it may be safe to evaluate on the basis of no subsoil support. For such scenarios, the above suggested revision has no influence.
Second change:

> The second change is the distribution of the vertical load on the geosynthetic layer between adjacent piles. Based on the vertical deflection of the geosynthetic layer measured in the model tests, the distribution of the load is considered to be an inverse triangle (Van Eekelen et al. 2012). The load distribution considered is shown in Figure 11.24.

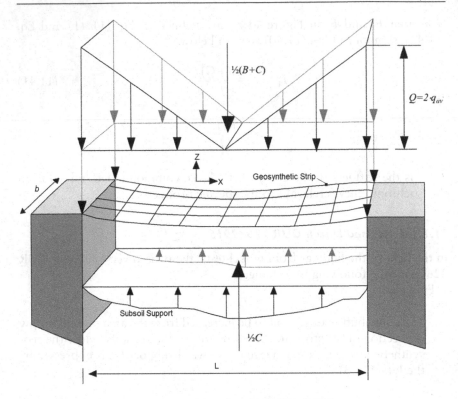

Figure 11.24 Load distribution adopted in CUR 226 (2015).

Considering the equilibrium of a section of the reinforcement as shown in Figure 11.25, the induced tensile force and the strain on the geosynthetic layer are computed using Eqs. (11.45) to (11.49).

Resolving both vertically and horizontally, the relationship between the load and deflection of the geosynthetic layer can be expressed as given below:

$$\frac{dT_v}{dx} = T_H \frac{d^2z}{dx^2} \tag{11.45}$$

Solving the differential equation, a series of design graphs have been published to estimate the strain (ε) in the geosynthetic layer. They are presented in Figure 11.26.

$$F_2 = \frac{A_{Lxy}.L_{xy}}{[J.a_{eq}]} K_s \tag{11.46}$$

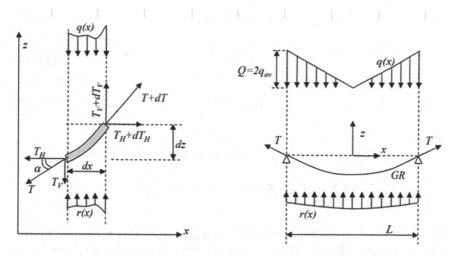

Figure 11.25 Deformation and load on geosynthetic layer.

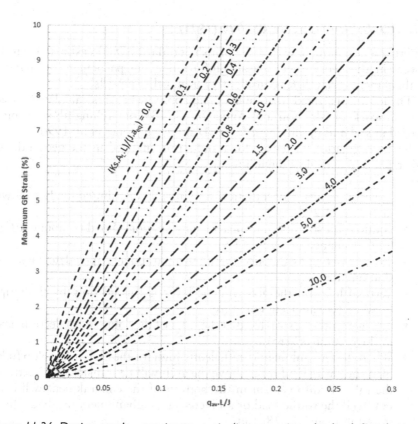

Figure 11.26 Design graph – maximum strain (inverse triangular load distribution)

$$a_{eq} = 0.5d\sqrt{\pi} \qquad\qquad (11.47)$$

$$Q = \frac{B+C}{A_s} \qquad\qquad (11.48)$$

$$q_{av} = 0.5Q \qquad\qquad (11.49)$$

where:

A_s – area of the geosynthetic strip between adjacent piles B, C – refer to Figure 11.9;

A_{Lxy} – tributary area of the geosynthetic layer; and

L_{xy} – diagonal length.

EBGEO (2010) and CUR 226 (2010) designs are based on average strains. However, the CUR 226 (2015) design is based on the calculated maximum strains at $x = L/2$.

11.7.1.3 British Standard BS8006 (1995)

There is a common view that the British Standard BS8006 estimates tensile forces in the geosynthetic basal reinforcement, in particular for shallow embankments, that differ considerably from other models.

The load transferred to columns (load component A) is calculated based on the work of Marston and Anderson (1913) for arching above a pipe buried in soil and modifying it for a 3D situation (Jones et al. 1990).

In the estimation of tensile load and strain induced in the geosynthetic layer, BS8006 considers the following:

- The load on the geosynthetic is a uniformly distributed load as shown in Figure 11.27;
- Foundation soil does not provide any support (Load component C = 0); C is defined in Figure 11.9;
- The above two assumptions result in a parabola-shaped deformed tension membrane;
- Vertical load on the geosynthetic layer is carried only by the strips between two adjacent columns;
- The geosynthetic strip fixed at the sides of the columns (deflection next to the column is zero);
- When the embankment is sufficiently high for the arch to develop fully, then the entire load from above the arch will be transferred directly to the piles. Reinforcement in the bottom of the embankment will thus not 'feel' the traffic load or an increasing embankment height as illustrated in Figure 11.28; and
- The arch height is estimated to be equal to the diagonal distance between the pile caps, i.e., 1.4(S-a).

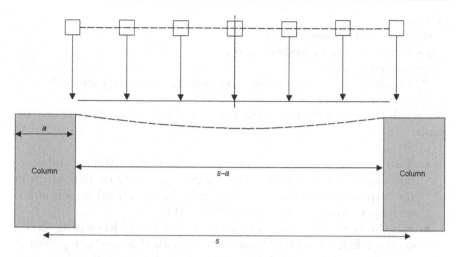

Figure 11.27 Load distribution adopted in BS 8006.

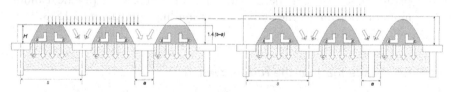

Figure 11.28 Partial and full arching adopted in BS 8006.

With no support from the subsoil (C) being considered, thereby considering vertical equilibrium the load component B is calculated, thus the uniformly distributed line load W_T is estimated.

For an embankment supported by square columns installed in a square pattern, W_T is given by the following equations. Note, they do not contain partial load factors, since SRI are commonly designed for serviceability conditions, partial load factors are not considered in the equations below.

For full arching by Eq. (11.50):

$$W_T = \frac{1.4S\gamma(S-a)}{S^2 - a^2}\left\{S^2 - a^2 \frac{P_c'}{\sigma_v'}\right\} \qquad (11.50)$$

For partial arching by the Eq. (11.51)

$$W_T = \frac{S(\gamma H + q)}{S^2 - a^2}\left\{S^2 - a^2 \frac{P_c'}{\sigma_v'}\right\} \qquad (11.51)$$

Where:

H – embankment height

q – surcharge load on embankment;

S – column spacing;

P'_c – stress on the column estimated as per the arching theory; and

a – column diameter;

On review of this method by researchers, the following observations are made:

- With the view to change this plane-strain concept to 3D case and make this approach more robust, BS8006 considers load on both strips in the perpendicular directions. Consequently, the vertical equilibrium is no longer satisfied (Van Eekelen et al. 2011).
- Van Duijnen et al. (2010), Haring et al. (2008) and Jenck et al. (2005) compare BS8006 with 2D tests and conclude that load component A is strongly overestimated when H increases beyond full arching. This assumption may lead to relatively low calculated tensile forces. It is believed that to reduce this risk, BS8006 recommends geosynthetic reinforcement should be designed to carry at least a practical minimum proportion of the embankment loading equivalent to 15%.

11.7.1.4 Tensile load on the reinforcement

Solving the differential equation for the tension membrane, the resulting tensile force T in kN in the geosynthetic strip is derived by Eq. (11.52).

$$T = \frac{W_T (S - d)}{2} \sqrt{1 + \frac{1}{6\varepsilon}} \text{kN} \tag{11.52}$$

Dividing T by the width of the strip 'a' gives the tensile force T_{rp} in the strip in kN/m:

$$T_{rp} = \frac{W_T (S - d)}{2a} \sqrt{1 + \frac{1}{6\varepsilon}} \tag{11.53}$$

Here, the allowable strain ε is an input parameter to calculate the tensile force T_{rp}. The apparent stiffness of the geosynthetic reinforcement can therefore be calculated using Eq. (11.54).

$$J = \frac{T_{rp}}{\varepsilon} \tag{11.54}$$

11.7.1.5 Other aspects covered in BS8006

In the shoulder regions of the embankment, the geosynthetic layer will be subjected to additional tensile load imposed by the outward thrust load of the embankment. This additional load is computed separately in BS8006. These details are not discussed herein.

11.7.1.6 Proposed modifications

For shallow embankments in particular, BS8006 estimates a relatively large tensile load for geosynthetic (basal) reinforcement in comparison with other design methods. Van Eekelen et al. (2011), reviewed the BS8006 approach and proposed the following modifications to the estimation of the distributed line load W_T.

For partial arching:

$$W_T = S(\gamma H + q)X \quad \text{to} \quad W_T = 0.5(\gamma H + q)(S + a)X \tag{11.55}$$

For full arching:

$$W_T = 1.4S\gamma(S - a)X \quad \text{to} \quad W_T = 0.7\gamma(S^2 - a^2)X \tag{11.56}$$

where

$$X = \frac{\left[S^2 - a^2\left(\dfrac{P_c'}{\gamma H + q}\right)\right]}{S^2 - a^2} \tag{11.57}$$

11.7.1.7 Comparison with finite element modelling

A comparison study was performed by Van Eekelen and Jansen (2008) where the outcomes from Modified BS8006 (2010), BS8006 (1995), and EBGEO (2010) were compared with the results of finite element calculations using PLAXIS. Their conclusions are summarized in Figures 11.29 and 11.30.

Full arching develops at $H = 1.89$ m; the tensile forces of both BS8006 (1995) and the Modified BS8006 (2010) decrease when the embankment height reaches this height as traffic load is no longer accounted in the geosynthetic load estimation. The tensile forces remain constant with increasing embankment height, while the PLAXIS analysis and EBGEO (2010) approach indicate an increase in the tensile force with increasing height in the embankment.

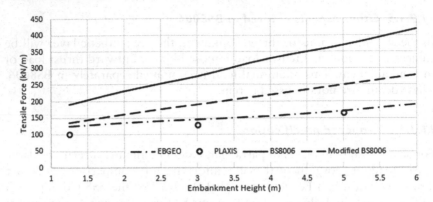

Figure 11.29 Comparison of tensile force – partial arching.

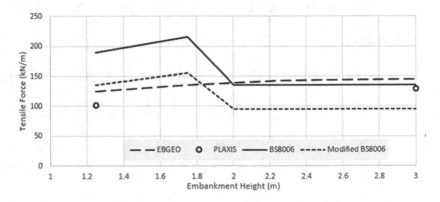

Figure 11.30 Comparison of tensile force – transitioning from partial to full arching.

11.7.1.8 Comparison with field test results

The 'Kyoto Road' shallow road embankment constructed over 9 m of soft clay was supported on 13 m long timber piles in Giessenburg, the Netherlands. This was monitored for around three and a half years. The monitored results were compared with those estimated by the Modified BS8006, BS8006, and EBGEO methods as reported by van Eekelen et al. (2011) and are shown in Figure 11.31.

The estimated load component B without the subsoil support seems significantly higher than that measured on site. Further, the estimated modulus of subgrade reaction based on subsoil conditions was around 500 kN/m³; but a reasonably close agreement between monitored results and EBGEO (2010) was found to be with a value of 1000 kN/m³.

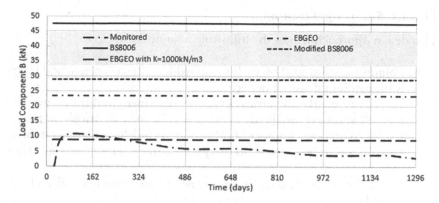

Figure 11.31 Comparison of load component B – From field monitoring.

11.8 GEOTECHNICAL DESIGN OF COLUMNS

The semi-rigid elements are designed to carry the load transferred by the system, without undergoing excessive settlement and structural failure.

Commonly, there are two approaches followed in practice in designing SRIs. They are as rigid inclusions or structural elements, and semi-rigid inclusions or geotechnical elements.

In both approaches, the columns are designed to have geotechnical capacity under the serviceability limit state, considering the total load acting within the tributary area (refer to Figure 11.6).

The diameter of the tributary area (d_T) depends on the spacing and configuration of columns and when columns installed in a square pattern d_T can be calculated using Eq. (11.58).

$$d_T = 1.128S \tag{11.58}$$

The total applied load on the tributary area (F_T) is estimated using Eq. (11.59).

$$F_T = A_T(\gamma h + q) \tag{11.59}$$

where
  S (m) – spacing between columns;
  A_T (m²) – tributary area;
  h (m) – height of the embankment;
  γ (kN/m³) – unit weight of the embankment fill; and
  q (kPa) – surcharge on the embankment.

11.8.1.1 SRI as rigid inclusions

This design approach makes the following assumptions:

- Columns are founded on rock or embedded into competent soil layers (stiff, medium dense or better);
- They carry nearly all vertical loads (embankment and traffic in highway, fill and structure loads in other developments). British Standard, BS8006 is among a number of other design guidelines that recommend the columns to be designed to carry the entire embankment load. They include:
  o Load portion A, directly on to the column through soil arching; and
  o Load portions B + C indirectly through membrane action of the geosynthetic reinforcement.

Each approach has its own merits and demerits, and they include:

- The total and post construction settlement will be considerably small and as a result settlement and lateral movement would not impact on adjacent structures (bridge abutment piles, retaining structures, buildings, etc.) through induced loads in subsoil layers.
- Columns attract greater load, including bending moments and shear forces in situations where non-uniform loading or ground conditions exist. In such situations reinforcement requirements need to be evaluated and may become necessary to be designed as structural elements, i.e., similar to piles. This means, all other traditional structural design requirements such as load factors, strength reduction factors, concrete strength and durability will kick in and make this solution economically very unattractive.

11.8.1.2 SRI as semi-rigid inclusions

Design standards Dutch CUR 226 (2016) and German EBGEO (2011) recommend designing the column supported developments as a whole to share a portion of the load with the surrounding soil. This is primary implemented by designing the geosynthetic reinforcement considering subsoil support as described in the previous section or without any geosynthetic reinforcements.

Further measures such as adopting relatively compressible columns types (constructed using grout, concrete or a combination of cementitious materials including waste products such as fly ash and slag) and allowing yielding of the column toe, make the design economically more attractive (Wong and Muttuvel 2012a).

11.8.1.3 Floating columns

Columns not founded or embedded into competent soil layers are also sometime referred to as floating columns. The column toes will be subject to yielding upon loading. Muttuvel and Wong (2012) recommend a force-balance approach to assess the portion of the embankment load transferred to the soil and thereby to assess the settlement behavior of column-soil composite foundation system. This approach is summarized below, and it is more applicable for columns constructed without a geosynthetic reinforced LTP, since it considers a good portion of the load is carried by the adjacent ground. The idealized stress distributions are shown in Figure 11.32.

In the bottom positive zone, the column moves past the soil thereby mobilizing positive skin resistance of the soil. The positive zone can be divided into a number of sublayers with varying shear strength and overconsolidation ratio (see Figure 11.33).

Adopting the force balance approach and considering equilibrium of forces in the vertical directions the following expressions can be derived.

$$F_t = F_{st} + F_{pt} = F_{s0} + F_{p0} \tag{11.60}$$

$$F_{s0} = F_{s1} + L_1 \pi d_c f_{s1} \tag{11.61}$$

$$F_{s0} + F_{p0} = F_{s1} + F_{p1} \tag{11.62}$$

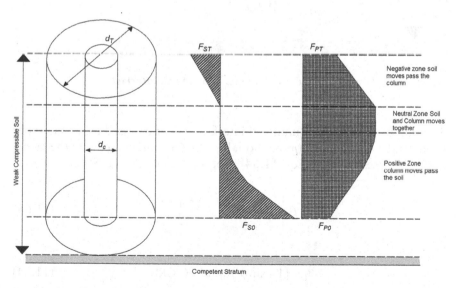

Figure 11.32 Idealized stress distribution between column and surrounding soil.

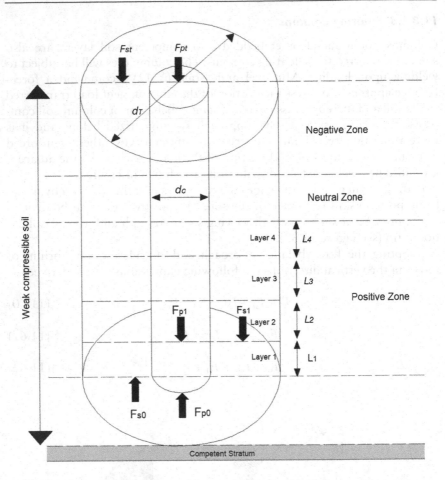

Figure 11.33 Forces on column elements and subsoil interface.

The skin friction f_{s1} at column-soil interface can be assessed using a number of methods. For the beta (β) method, Eq. (11.63) can be used and the factor β_1 can be assessed using Eq. (11.64) proposed by Poulos (1989).

$$f_{s1} = \beta_1 \left[\frac{\left(\sigma'_{v0} + \sigma'_{v1} \right)}{2} + \frac{\left(F_{S0} + F_{S1} \right)}{2A_s} \right] \qquad (11.63)$$

$$\beta_1 = \left(1 - \sin \phi' \right) \tan \phi' \times OCR^{0.5} \qquad (11.64)$$

where
  F_{s0} (kN) – load increase in the soil annulus area at the column toe level;

F_{p0} (kN) – end bearing resistance of the column;

F_{p1} (kN) – load acting on the column at the top of sublayer 1;

F_{s1} (kN) – load increase in the soil annulus area at the top of sublayer 1;

f_{s1} (kPa) – average skin friction at the column-soil interface within sublayer 1;

L_1 (m) – thickness of sublayer;

σ'_{vo} (kPa) – initial effective vertical effective stress at toe level;

σ'_{v1} (kPa) – the initial effective vertical stress at the top of sublayer 1;

A_s (m²) – area of soil annulus within sublayer 1;

ϕ' (degrees) – drained friction angle of sublayer 1; and

OCR – final over consolidation ratio of sublayer 1.

Upon solving for the forces (F_{S1} and F_{P1}) at the interface of layers 1 and 2, the same procedure can be repeated for layer 2 and so on for other layers within the positive zone. Through this process, the stress distribution in the soil within the positive zone can be established.

In the upper negative zone, the soil will take up a portion of the applied load (F_{ST}).

- The load F_{ST} can be estimated based on the arching theory adopted and the column configuration considered.
- F_{ST} will diminish with depth based on the available skin friction in each soil layer.
- Dividing up this zone into a number of sublayers and adopting the force-balance approach described above, the stress distribution in the negative zone can also be established.

Knowing the vertical stress distribution, the primary consolidation settlement for both layers can be assessed using the one-dimensional (1D) consolidation theory.

Further, based on the final stress distribution within and below the treated zone, the OCR for each subsoil layer can be assessed. Adopting an appropriate creep strain rate, the creep settlement can be estimated.

11.8.1.4 Non-yielding columns with geosynthetic reinforced LTP

Commonly, column-supported embankments are designed with a geosynthetic reinforced LTP. Further, the columns are founded in competent material such as soil having a consistency/density stiff/medium dense. Although the intent of the design is to transfer a major portion of the load within the tributary area directly onto the columns and indirectly through the geosynthetic layer, an understanding of the ground reaction mechanism and the construction sequence is essential to reduce potential risks associated with the geotechnical design of columns.

Often, where ground treatment with semi-rigid inclusions are proposed, the upper soil layers are weak and consists of clayey soils. Invariably, a working platform will be required to facilitate the installation of the columns. This is commonly around 1.0 m to 1.5 m in thickness and placed directly on the weak subsoils. The LTP is constructed therefore to be placed on top of this layer. Hence the working platform will become an additional permanent load on the weak subsoil.

A ground reaction curve describes the relationship between the development of arching stresses and subsoil settlements, commonly presented with respect to stress reduction ratio and relative displacement (defined as trap door displacement/trap door width × 100%). A typical ground reaction curve (to describe more generally the development of arching in a soil mass) (Iglesia et al. 2011) is given in Figure 11.34.

Adopting a total vertical stress at the top of the LTP equal to 100 kPa, the change in vertical stress between the columns due to arching and that carried by the geosynthetic reinforcement and the subgrade can be inferred. This is described in Figure 11.35.

Based on the above, consideration shall be given to the following:

- The additional load from the working platform on the subgrade will be eventually transferred to the column as part of the negative skin friction component. For example, a 1.5 m thick working platform over a tributary area diameter of 2.5 m could impose an additional load around 150 kN on the column. This could trigger potential yielding of the column and perhaps leading to unforeseen settlements.
- During construction and design life, the subsoils between the columns are subjected to additional vertical stress, both intentionally and unintentionally. As a result, they undergo both primary and creep settlement.

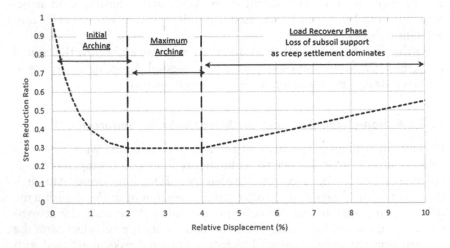

Figure 11.34 Characteristic ground reaction curve (adapted from Iglesia et al. 2011).

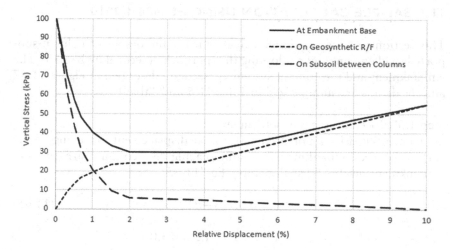

Figure 11.35 Load shared by the geosynthetic layer and the subsoils (inferred).

Often, the upper region of the weak compressible clays will be pushed passed their pre-consolidation pressure. Consequently, the subsoil support to the geosynthetic layer will reduce with time and leading to larger displacements in the reinforcement layer and development of further shear bands within the embankment fill. This will eventually push the ground reaction mechanism into the load recovery phase (Figure 11.34). During this phase, load on the geosynthetic reinforcement could increase beyond that estimated based on maximum arching. Therefore, it is prudent to design the geosynthetic reinforcement capacity taking the above into account (Figure 11.36).

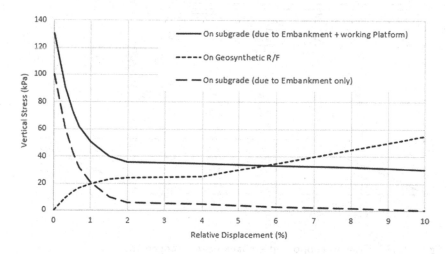

Figure 11.36 Potential increase in the load shared by and the subsoil (inferred).

11.9 SAMPLE CALCULATION USING BS8006-1 2010

This section provides a sample calculation for the design of column-supported road embankment, adopting the approach described by BS8006. The embankment and the column system assessed are described in Figure 11.37 and the details are summarized in Table 11.5 and Table 11.6.

Step 1: Reduced differential settlement
To reduce the risk of potential differential displacement occurring at the surface of the embankment, column spacing (S), column size (d) and embankment height (H) shall comply Eq. (11.65).

$$H \geq 0.7(S - d) \tag{11.65}$$

$$0.7 \times (2.0 - 0.4) = 1.12\,\text{m} < 5.0\,\text{m}$$

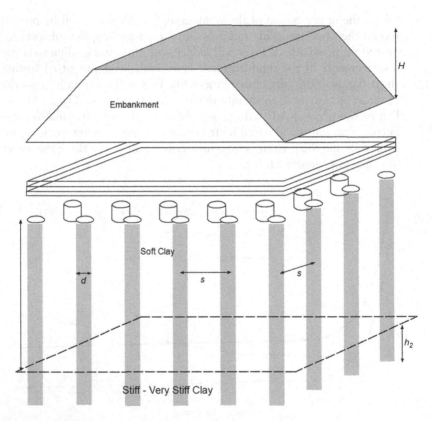

Figure 11.37 Column supported embankment arrangement.

Table 11.5 Typical column dimensions

Description	Parameter
Embankment height – H (m)	5
Surcharge load on embankment – W_S (kPa)	20
Embankment batter gradient	1V:2H
Embankment fill friction angle – ϕ (deg)	30
Embankment fill density – γ_f (kN/m³)	20
Column size – d (m)	0.4
Column shape	Circular
Column arrangement	Square
Column spacing – $S=5d$ (m)	2.0
Soft soil thickness – h_1 (m)	6

Table 11.6 Ground model

Description	γ_s (kN/m³)	c_u (kPa)	α	f_s (kPa)	f_b (kPa)
Soft Clay	16.5	20	1.0	20	180
Stiff to very stiff Clay	18.0	100	0.55	55	900

γ_s – bulk unit weight; c_u – Undrained shear strength; α – adhesion factor; f_s – skin friction; and f_b – end bearing

Step 2: Tributary area (A_T)
For square column arrangement, diameter of the tributary area

$$d_T = \frac{2S}{\sqrt{\pi}} = 1.128\,S \tag{11.66}$$

$$d_T = 2.0 \times 1.128 = 2.256\,m$$

$$A_T = \pi \times 2.256^2 / 4 = 4.0\,m^2$$

Step 3: Total load on the pile (Q_P)

$$Q_P = \left[\gamma_f H + W_S \right] A_T \tag{11.67}$$

$$Q_P = \left[20 \times 6 + 20 \right] \times 4 = 560\,kN$$

Step 4: Estimation of soil resistance for column length
By equating the total load on the column to axial capacity of the column, the depth of embedment into the competent layer is estimated.

$$Q_R = Q_s + Q_b \tag{11.68}$$

$$Q_S = \pi d \sum f_{si} h_i \tag{11.69}$$

$$Q_b = \pi d^2 / 4 f_b \tag{11.70}$$

$$Q_s = \pi \times 0.4 \times \{20 \times 6 + 55 \times h_2\} = 151 + 69 h_2$$

$$Q_b = \pi \times 0.4^2 / 4 \times 900 = 113 \text{kN}$$

Note: f_{si} – skin friction in individual layers
f_b – Toe resistance

Minimum depth of embedment (h_2) into stiff to very stiff soils could be calculated by equating Q_P with Q_R

Minimum depth of embedment into stiff to very stiff clay layer (h_2) = 3.1 m

Step 5: Column head Diameter

In the assessment of vertical load shedding, BS8006 considers a square cap on the column of size 'a'. Where circular caps are to be used, an effective pile cap width of a_{equ} is employed in the calculation of load shedding.

In our case, where no cap is used, the column head becomes the cap ($a = d$) and it being circular in shape, the effective cap width is estimated employing Eq. (11.71).

$$a_{equ} = \sqrt{\frac{\pi D^2}{4}} \tag{11.71}$$

$$a_{equ} = \left(\pi \times 0.4 \times 0.4 / 4\right)^{0.5} = 0.3544 \text{m}$$

Step 6: Arching coefficient and stress ratio

Portion of the load transferred directly to the column is estimated based on the arching concept by Jones et al. (1990).

The ratio of the vertical stress exerted on top of the column (P'_c) to the average vertical stress at the base of the embankment (σ'_v) is given by Eq. (11.72).

$$\frac{P'_c}{\sigma'_v} = \left[\frac{C_c a_{equ}}{H}\right]^2 \tag{11.72}$$

where C_c is the arching coefficient and it is given by Eq. (11.73) and Eq. (11.74) for end bearing and frictional columns.

$$C_c = \frac{1.95H}{d} - 0.18 \tag{11.73}$$

$$C_c = \frac{1.5H}{d} - 0.07 \qquad (11.74)$$

A larger portion of the load capacity being mobilized from the stiff to very stiff clay layer underlying the soft clay, the columns under study can be categorized as end bearing columns.

$$C_c = [1.95 \times 6 / 0.4] - 0.18 = 21.33$$

$$P'_c / \sigma'_v = \{(21.33 \times 0.3544) / 5\}^2 = 3.753$$

Step 7: Distributed load on the geosynthetic layer
The distributed load W_T carried by the reinforcement between adjacent pile caps is determined from Eq. (11.75) and Eq. (11.76):
For $H > 1.4(S-a)$

$$W_T = \frac{1.4Sf_{fs}\gamma(S - a_{equ})}{(S^2 - a_{equ}^2)}\left[S^2 - a_{equ}^2\left(\frac{P'_c}{\sigma'_v}\right)\right] \qquad (11.75)$$

For $0.7(S-a) \leq H \leq 1.4(S-a)$

$$W_T = \frac{(f_{fs}\gamma H + f_q W_s)S}{(S^2 - a_{equ}^2)}\left[S^2 - a_{equ}^2\left(\frac{P'_c}{\sigma'_v}\right)\right] \qquad (11.76)$$

BS8006 recommends, irrespective of the arching theory, pile layout and embankment geometry, the geosynthetic reinforcement should be designed to carry at least a practical minimum proportion of the embankment loading equivalent to 15%. The minimum distributed load W_{Tmin} is given by Eq. (11.77).

$$W_{Tmin} = 0.15S[f_{fs}\gamma H + f_q W_s] \qquad (11.77)$$

Load factors recommended by BS8006 are given in Table 11.7.

Table 11.7 Load factors

Description	ULS	SLS
Load factor for soil mass (f_{fs})	1.3	1.0
Load factor for external live load (f_q)	1.3	1.0

1.4$(S-a)$ = 2.24m Therefore $H > 1.4(S-a)$

Hence as per Eq. (11.75)

W_{TULS} = 109.2 kN/m and W_{TSLS} = 83.9 kN/m

W_{Tmin} = 46.80 kN/m as per Eq. (11.77)

Design distributed load on the geosynthetic layer under ULS and SLS, are respectively:

W_{TULS} = 109.2 kN/m and W_{TSLS} = 83.9 kN/m

Step 8: Tension in the geosynthetic layer

The tensile load (T_{rp}) on the geosynthetic layer is estimated using Eq. (11.78).

$$T_{rp} = \frac{W_T (S-a)}{2a} \sqrt{1 + \frac{1}{6\varepsilon}} \qquad (11.78)$$

The above equation has two unknowns T_{rp} and the strain (ε). Hence, T_{rp} is estimated by taking into account the maximum allowable strain in the reinforcement and understanding the load/strain characteristics of the geosynthetic reinforcement at different load levels.

Therefore, to undertake this assessment, we need to obtain from the geosynthetic manufacturer the load/strain characteristic. This is generally known as isochronous curves as shown in Figure 11.38 or alternately this information could be obtained in tabular form.

Further, other strength reduction factors associated with the performance of the geosynthetic layer also require consideration.

The factors commonly considered, and the typical values adopted are given in Table 11.8.

The tensile capacity of geosynthetic reinforcement is referred to using the following terms:

a. Characteristic strength (T_{ch}) – This is the ultimate initial strength at rupture strain (ε_R). For the isochronous curves given above, the rupture strain is 11%.

b. Creep rupture strength (T_{CR-ULS}) – This is estimated by factoring down T_{Ch} by the material creep factor f_C.

Table 11.8 Factors associated with geosynthetic performance

Description of factors	Factor
For installation damage – f_D	1.10
For weathering – f_W	1.00
For environmental effects – f_E	1.06
For creep effect – f_C	1.43
For data extrapolation – f_S	1.00
For limiting strain to 6% – f_R	0.52

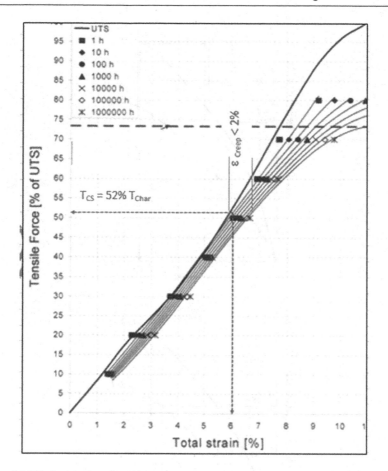

Figure 11.38 A sample of isochronous curves.

c. Tensile strength at limiting strain $(T_{CS\text{-}SLS})$ – This is estimated by factoring down T_{Ch} for a limiting strain (ε_L). $T_{CS\text{-}SLS} = T_{CH} \times f_R$

d. Design tensile strength (T_D) – This is estimated by factoring down $T_{CR\text{-}ULS}$ or $T_{CS\text{-}SLS}$ by all the other material factors $(f_m = f_D \times f_W \times f_E \times f_s)$

In Figure 11.39, the approach to estimating the relevant tensile capacities are graphically described.

For the sample calculations the following have been adopted:

- Rupture strain – ULS (ε_R) = 10%;
- Limiting strain – SLS (ε_L) = 6%;
- Maximum creep strain – SLS (ε_{Creep}) = 2%;
- Total allowable strain – SLS $(\varepsilon_{T\,max})$ = $\varepsilon_L + \varepsilon_{Creep}$ = 6%+2% = 8%.

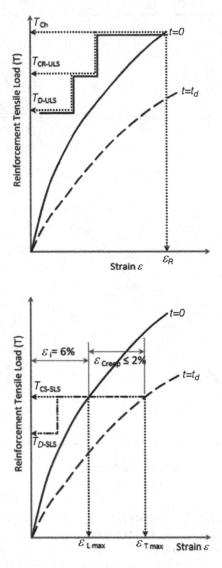

Figure 11.39 Estimating tensile capacity (a) Reinforcement tensile load vs. Strain; (b) Reinforcement tensile load vs. Total strain including creep.

The design tensile load on the geosynthetic layer estimated using Eq. (11.78):

$$T_{D-ULS} = \left\{109.1 \times (2.0 - 0.3544)\right\} \times \left\{1 + \left[1/(6 \times 10 \times 0.01)\right]\right\} = 413.7 \, kN/m$$

$$T_{D-SLS} = \left\{83.0 \times (2.0 - 0.3544)\right\} \times \left\{1 + \left[1/(6 \times 6 \times 0.01)\right]\right\} = 378.7 \, kN/m$$

$$f_m = f_D \times f_W \times f_E \times f_s \qquad (11.79)$$

$$f_m = 1.10 \times 1.0 \times 1.06 \times 1.0 = 1.166$$

Under ULS loading

$$T_{CR-ULS} = f_m \times T_{D-ULS} \qquad (11.80)$$

$$T_{CH-ULS} = f_c \times T_{CH-ULS} \qquad (11.81)$$

$$T_{CR-ULS} = 1.166 \times 413.7 = 482.3 \, \text{kN/m}$$

$$T_{CH-ULS} - 482.3 \times 1.43 = 689.7 \, \text{kN/m}$$

Under SLS loading

$$T_{CS-SLS} = f_m \times T_{D-SLS} \qquad (11.82)$$

$$T_{CH-SLS} = \frac{T_{CS-SLS}}{f_R} \qquad (11.83)$$

$$T_{CR-SLS} = 1.166 \times 378.7 = 441.6 \, \text{kN/m}$$

$$T_{CH-SLS} = 441.6 / 0.52 = 849.2 \, \text{kN/m}$$

Step 9: Outcomes

In this example the geosynthetic tensile capacity is determined by the SLS loading conditions.

The characteristic tensile capacity required is 850 kN/m. This does not include additional tensile loading due to horizontal thrust resulting from potential lateral sliding of embankment.

Suitable single or multi bi-axial layers of reinforcement could be adopted.

11.10 CASE STUDIES

CASE HISTORY 11.1 WONG AND MUTTUVEL (2012B)

Economic Design of Semi-rigid inclusions for Ground Improvement – ANZ 2012 Conference Proceedings

The case history refers to the Pacific Highway upgrade in Australia where the northern approach to Macleay bridge is underlain by alluvial soft to firm clays with interbedded loose sand to 6.5 m to 10.5 m depth above extremely weathered rock. SRIs were used in this project to speed up bridge construction works but limit the lateral movement to 30 mm in 100 years. The design was carried out using Plaxis 2D.

Wong and Muttuvel (2012b) describe the traditional SRI design where it is designed more like a piled embankment, i.e., the columns carrying the full design load with the soft soil not designed to carry any load. They specifically discuss the LTP and its usefulness and iterate the need for the LTP to be designed to ensure the full load is transferred to the columns. It is stressed that an inadequate LTP will give rise to the mushroom effect in low embankments. Where high embankments occur, LTP may be redundant because soil arching will help to dissipate differential settlement through the embankment to the surface.

Axisymmetric finite element analysis was used to assess the total and differential settlement of an embankment on a gravel working platform, without an LTP. A typical example of a 10 m thick clay with 1 m working platform were adopted. The SRI diameter was fixed at 0.5 m placed 2 m square spaced grid installed through the soft clay of 10 m thick and embedded 0.5 m in stiff clay below.

Figure 11.40 shows the total and differential settlement at the top of embankment for varying embankment heights. The results clearly indicate that the differential settlement decreases with increasing embankment height whereas the total settlement increases. After the embankment height reaches about 3.5 m, the maximum differential settlement is not significant and suggests that a LTP is not required, in this example when the embankment height is greater than about 3.5m.

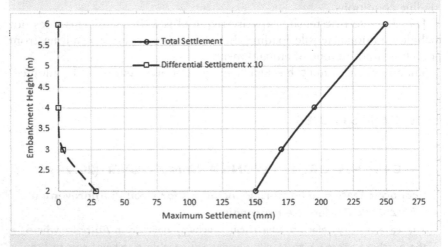

Figure 11.40 Embankment height versus estimated settlement.

On completion of SRI installation, the approach embankment was constructed in three months. Monitoring results indicated the settlements ceased
one month after completion of construction. The lateral deformation measured
was 22 mm and it was assessed that the total lateral movement at the end of
100 years would be less than the 30 mm stipulated.

CASE HISTORY 11.2 GUE AND TAN (2005)

This case history refers to a project constructed more than three decades
ago and is very relevant to our subject. The project refers to a highway constructed in the northern state of Malaysia in the 1980s. In the area of soft soils,
the embankment has been constructed with spun piles and pile caps with the
design based on transferring the embankment load to the pile caps solely via
arching in the embankment fill. However, a few years later, mushroom effects
(see Figure 11.41) were clearly visible with large differential settlements visible
on the pavement leading to frequent maintenance to fill in depressions on the
surface. A forensic examination had been carried out in 2001 to 2002 and the
case history is based on their findings.

The site is underlain by soft marine deposits of the Quaternary age, comprising very soft to soft silty clay and clayey/sandy silt with intermittent sand
layers. The embankment fill was mainly sandy or gravelly clay/silt. Soil investigations included boreholes, piezocones, Mackintosh probes, hand augers and vane
shear tests to assess the subsoil characteristics. Undisturbed block sampling
was carried out in the embankment fill. The fill materials indicate low strength
results of effective parameters of $c' = 2$ kPa and 25 degrees for friction angle.

Once the geotechnical characteristics were assessed and geotechnical models set up based on field and laboratory investigations, numerical analysis was
used to investigate the issues related to the mushroom effects. Plaxis 2D and
3D were the software used for the analysis to assess the possible causes of
failure. Piles were assumed to be installed in the stiff layers below.

The numerical results have indicated the differential settlements expected at
the pavement surface to be 64 mm to 156 mm with angular distortions up to
4%. The analysis concluded that the arching mechanism was not effective due to
poor fill materials and large pile spacing relative to the height of embankment.

In conclusion, remedial measures were carried out based on numerical analysis and consisted of placing a wearing course and a reinforced concrete raft
(250 to 300 mm thick) after excavation to accommodate them.

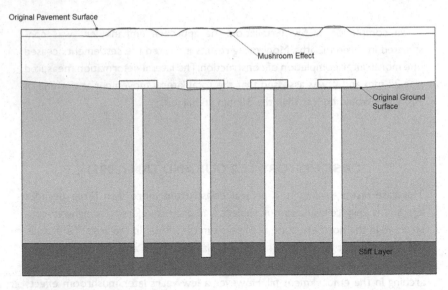

Figure 11.41 Mushroom effects.

11.11 PILED EMBANKMENTS

The design of a piled embankment is similar in many ways to the assessment of SRI supported embankments. In fact, the majority of the key European design standards and guidelines referred to in this chapter are directed towards the design of piled embankments. They include:

- EBGEO (German Recommendations for Geosynthetic Reinforced Earth Structures) – Geosynthetic reinforced pile-supported embankment;
- BS8006 (British Standards) – Design of piled embankments with basal reinforcement; and
- CUR226 (Dutch Design Guideline) – Basal reinforced piled embankments.

The design of semi-rigid inclusions is to be viewed as a hybrid approach of the piled embankment, and they are purposely designed for a safe and cost-effective outcome.

The ASIRI National Project (2012 French design guideline) outlines that shallow foundations are always preferred, as they will be the most cost-effective solution, granted they meet the design requirements of the structure with respect to stability and settlement. If both or either one of the design requirements is not met, then the conventional alternative will be the deep foundation, which is the pile solution. This design guideline views SRI foundation system as an intermediate solution that meets the requirements those shallow foundations fail to comply and at the same time not overdesigned as a pile foundation.

In the context of embankment design, whether the embankment is supported on rigid element (pile) or semi-rigid element (SRI), entirely rests on the design compliance. Piles are designed in compliance with the piling code, while SRIs are designed to meet the requirements of the specific structure. The following are some of the key differences:

Embankment being a semi-flexible structure they could perform satisfactorily even after undergoing a large order of total settlement (generally between 50 mm and 200 mm depending on location). Further, such semi-rigid structures can tolerate some order of differential settlement. Hence it will be an overkill if the SRIs are designed as piles to limit their settlements to a percentage of the element's diameter.

Piles are designed as load-bearing elements; hence they are designed with an appropriate geotechnical reduction factor, which varies between 0.45 and 0.7 in the Australian context. However, the settlement reduction columns are commonly designed with a geotechnical reduction factor of 1.0 since the semi-rigid elements behave as a composite element along with the surrounding soils.

Piles are designed with the required minimum steel reinforcements to withstand estimated compressive, tensile and shear stresses. Further, based on durability requirements adequate cover and concrete strength are incorporated. However, in SRI designs, to support embankments none of the above requirements are considered. A central reinforcement bar holds the settlement column intact to prevent cracking due to shear loads or tension. Only those located along the toe region of the embankment (commonly two to three outer rows) are reinforced to avail shear capacity, if additional support is required to provide stability to the embankment.

In summary, piled embankments are designed to a more stringent requirement and to evaluate some of the key pile performance indicators, numerical modelling may become necessary.

11.12 SUMMARY

This chapter deals with semi-rigid inclusions (SRIs), a widely accepted ground improvement technique to improve soft soil sites. The method revolves around the introduction of a stiff column, usually concrete, into the ground to primarily to perform as a settlement reduction element by improving the overall stiffness of the weak soil mass. The introduction of the column also improves the overall strength of the soil mass.

The chapter describes the load transfer mechanism in the design and construction of SRIs. A critical height as proposed by several authors is defined and guidelines are provided to assess it. The selection of the column type, column dimensions and spacing are discussed in detail. The key steps involved in the design are outlined and a set of sample calculations of a column supported embankment is given to provide further direction on the design.

One of the key concepts in this ground improvement technique is the contribution of the LTP. Design guidelines of LTP including parameter selection is described. The LTP is reinforced with a geosynthetic layer and the load transfer mechanism is the key in the derivation of the load shared by the geosynthetic reinforcement and critical in the design of LTP. Several methods have been proposed by various authors to arrive at the loads shared by the geotextile reinforcement and such methods are covered in this chapter.

REFERENCES

Ariyarathne, P. and Liyanapathirana, D.S. (2015). 'Review of existing design methods for geosynthetic-reinforced pile-supported embankments.' *Soils and Foundations*, 55(1), 17–34.

ASIRI (2013). *Recommendations for the Design, Construction and Control of Rigid Inclusion Ground Improvements*. Presses des Ponts, Paris, 383p.

Balaam, N.P., Booker, J.R. and Poulos, H.G. (1976). '*Analysis of granular pile behaviour using finite elements.*' *Proceedings International Conference Finite Elements in Engineering*, Adelaide, Australia, 1–13.

Bell, A.L., Jenner, C., Maddison, J.D. and Vignoles, J. (1994). '*Embankment support using geogrids with vibro concrete columns.*' *Proceedings, 5th International Conference on Geotextiles, Geomembranes, and Related Products*, Singapore, 335–338.

Bergado, D.T. and Lam, F.L. (1987). 'Full scale load test of granular piles with different densities and different proportions of gravel and sand in the soft Bangkok clay.' *Soils and Foundations*, 27(1), 86–93.

British Standards BS8006 (1995). *Code of Practice for Strengthened/Reinforced Soils and Other Fills*. BSI, London, UK.

British Standards BS8006-1 (2010). *Code of Practice for Strengthened/Reinforced Soils and Other Fills*. BSI, London, UK.

Carlsson, B. (1987). *Reinforced Soil, Principles for Calculation*. Terratema AB, Linköping, Sweden.

Chu, J., Varaksin, S., Klotz, U. and Mengé, P. (2009). '*State of the art report: Construction processes meeting ground improvement.*' *Proceedings 17th International Conference on Soil Mechanics & Geotechnical Engineering*, TC17, Alexandria, Egypt.

Collin, J. G. (2004). '*Column supported embankment design considerations.*' *Proceedings 52nd Annual Geotechnical Engineering Conference*, University of Minnesota.

CUR 226 (2010). Ontwerprichtlijn paalmatrassystemen (Design Guideline Piled Embankments), ISBN 978-90-376-0518-1 (in Dutch).

CUR 226 (2015). Ontwerprichtlijn paalmatrassystemen (Design Guideline Piled Embankments), updated version (in Dutch).

CUR226 (2016). CUR 226 – *Design Guideline Basal Reinforced Piled Embankments*. Revision of the Design Guideline CUR226.

EBGEO (2010). 'Recommendations for design and analysis of earth structures using geosynthetic reinforcements – EBGEO.' *Translation of the 2nd German edition*. German Geotechnical Society. Ernst & Sohn, Germany.

EBGEO. (2011). 'Recommendations for design and analysis of earth structures using geosynthetic reinforcements – EBGEO.' Ernst & Sohn – A Wiley company. DGGT – Deutsche Gesellschaft für Geotechniek e. V. – German Geotechnical Society.

Farag, G. (2008). 'Lateral spreading in basal reinforced embankments supported by pile-like elements,' PhD thesis. Universität Kassel. Institut für Geotechnik und Geohydraulik. Kassel Univ. Press.

FHWA (2017). *Ground Modification Methods, Reference Manual – Volume II*, U.S. Department of Transportation Federal Highway Administration, Washington, DC.

Filz, G. M. and Smith, M. E. (2006). *Design of Bridging Layers in Geosynthetic-Reinforced, Column-Supported Embankments*. Virginia Transportation Research Council, Charlottesville, Virginia, 46p.

Finnish Transport Agency (2012). *Georeinforced Earth Structures 2012*. Liikenneviraston ohjeita nro/2012.

Gue, S.S. and Tan, Y.C. (2005). *'Failure in ground improvement works in soft ground.'* In: Indraratna, B. and Chu, J. *Ground Improvement – Case Histories*. Elsevier.

Guido, V.A., Kneuppel, J.D. and Sweeney, M.A. (1987). *'Plate loading test on geogrid reinforced earth slabs.'* *Proceedings Geosynthetics '87 Conference*, New Orleans, 216–225.

Han, J. and Gabr, M.A. (2002). 'Numerical analysis of geosynthetic-reinforced and pile supported earth platforms over soft soil.' *Journal of Geotechnical and Geoenvironmental Engineering*, 128(1), 44–53.

Haring, W., Profittlich, M. and Hangen, H. (2008). *'Reconstruction of the national road N210 Bergambacht to Krimpen a.d. IJssel, NL: design approach, construction experiences and measurement results.'* *Proceedings 4th European Geosynthetics Conference*, Edinburgh, UK.

Hewlett, W.J. and Randolph, M.F. (1988). 'Analysis of piled embankments.' *Ground Engineering*, 21(3), 12–18.

Iglesia, G.R., Einstein, H.H. and Whitman, R.V. (2011). 'Validation of centrifuge model scaling for soil systems via trapdoor tests.' *Journal of Geotechnical and Geoenvironmental Engineering*, 137(11): 1075–1089. doi:10.1061/(ASCE) GT.1943-5606.0000517.

Jenck, O., Dias, D. and Kastner, R. (2005). 'Soft ground improvement by vertical rigid piles two dimensional physical modelling and comparison with current design models.' *Soils and Foundations*, 45(6), 15–30.

Jones, C.J.F.P., Lawson, C.R. and Ayres, D.J. (1990). *Geotextile Reinforced Piled Embankments, in Geotextiles, Geomembranes and Related Products*, Den Hoedt (ed.), Balkema, Rotterdam, ISBN 90 61911192, 155–160.

Kempfert, H.G., Gobel, C., Alexiew, D. and Heitz, C. (2004). *'German recommendations for reinforced embankments on pile-similar elements.'* *Proceedings of the EuroGeo3*, Munich DGGT, 279–284.

Kempfert, H.-G., Stadel, M. and Zaeske, D. (1997). 'Berechnung von geokunststoffbewehrten Tragschichten über Pfahlelementen.' *Bautechnik*, 75(12), 818–825.

Lin, K.Q. (2000). 'Behaviour of DCM columns under highway embankment at Bridge approaches.' Thesis for the Degree of Doctor of Philosophy, Nanyang Technical University, Singapore.

Low, B.K., Tang, S.K., and Chao, V. (1994). 'Arching in piled embankments.' *Journal of Geotechnical Engineering, ASCE*, 120(11), 1917–1938.

Madhav M.R. and Nagpure, D.D. (1996) '*Design of granular piles for embankments on soft ground.*' *Proceedings 12th SE Asian Geotechnical Conference*, Kuala Lumpur, 1: 285–290.

Marston, A. and Anderson, A.O. (1913). 'The theory of loads on pipes in ditches and tests of cement and clay drain tile and sewer pipe.' *Bulletin No. 31*, Engineering Experiment Station.

McGuire, M. P. (2011). 'Critical height and surface deformation of column-supported embankments.' PhD thesis. Department of Civil Engineering, Virginia Polytechnic Institute and State University, Virginia, USA.

Muttuvel, T. and Wong, P. (2012). '*Design of semi-rigid inclusions for bridge approach embankments.*' *Proceedings 2nd International Conference on Ground Improvement & Ground Control*. Wollongong, Australia.

NGG (2002). *Nordic Handbook – Reinforced Soils and Fills*. Nordic Geotechnical Society, Stockholm, Sweden.

Plomteux, C. (2020). 'Semi-rigid inclusions.' www.bbri.be/homepage/download.cfm?lang=en&dtype=services&doc=WGD_3_Semi_rigid_Inclusions.pdf. Last visited 11/17/2020.

Poulos, H.G. (1989). 'Pile behaviour-theory and application.' *Geotechnique*, 39(3), 365–415.

Poulos, H.G. (2002) Personnel communication.

Reid, W.M. and Buchanan, N.W. (1984). '*Bridge approach support piling.*' *Piling and Ground Treatment*, Thomas Telford Ltd, London, UK.

Rogbeck, Y., Gustavsson, S., Sodergren, I. and Lindquist, D. (1998). '*Reinforced piled embankments in Sweden – design aspects.*' *Proceedings, Sixth International Conference on Geosynthetics*, Vol 2, Atlanta, USA. 755–762.

Russell, D. and Pierpoint, N. (1997). 'An assessment of design methods for piled embankments.' *Ground Engineering*, 30(10), 39–44.

Russell, D., Naughton, P.J., and Kempton, G. (2003). '*A new design procedure for piled embankments,*' *Proceedings 56th Canadian Geotechnical Conference and 2003 NAGS Conference*, Vol. 1, Winnipeg, MB, 858–865.

Shahu J.T., Madhav M.R., Hayashi S. and Miura N. (1998) '*Parametric study of soft ground reinforced with granular pile-mat system.*' *Proceedings Thirteenth Southeast Asian Geotechnical Conference*, Taipei, Taiwan, ROC, 415–420.

Sloan, J.A. (2011). 'Column-supported embankments: full-scale tests and design recommendations.' PhD thesis, Department of Civil Engineering, Virginia Polytechnic Institute and State University, Virginia, USA.

Svanø, G., Ilstad, T., Eiksund, G. and Want, A. (2000). '*Alternative calculation principle for design of piled embankments with base reinforcement.*' *Proceedings of the 4th GIGS in Helsinki*.

Terzaghi, K. (1943). *Theoretical Soil Mechanics*. John Wiley and Sons, New York.

Van Eekelen, S.J.M. and Bezuijen, A. (2012). '*Basal reinforced piled embankments in the Netherlands, field studies and laboratory tests.*' *Proceedings ISSMGE - TC 211 International Symposium on Ground Improvement IS-GI*, Brussels.

Van Eekelen, S.J.M., Bezuijen, A., Lodder, H.J. and van Tol, A.F. (2012). 'Model experiments on piled embankments Part I.' *Geotextiles and Geomembranes*, 32, 69–81.

Van Eekelen, S.J.M., Bezuijen, A. and Van Tol, A.F. (2013). 'An analytical model for arching in piled embankments.' *Geotextiles and Geomembranes*, 39, 78–102.

Van Eekelen, S.J.M., Bezuijen, A. and van Tol, A.F., (2011). 'Analysis and modification of the British Standard BS8006 for the design of piled embankments.' *Geotextiles and Geomembranes*, 29, 345–359.

Van Eekelen, S., Bezuijen, A. and Van Tol, A. (2015a). 'Validation of analytical models for the design of basal reinforced piled embankments.' *Geotextiles and Geomembranes*, 43(1), 56–81.

Van Eekelen, S.J.M., Bezuijen, A. and van Tol, A.F. (2015b). '*Axial pile forces in piled embankments, field measurements.*' *Proceedings of XVI ECSMGE*, Edinburgh.

Van Eekelen, S.J.M. and Jansen, H. (2008). Op weg naar een Nederlandse ontwerprichtlijn voor paalmatrassen 1, Verslag van een casestudie, GeoKunst nr. 3, 2008 (in Dutch).

Van Duijnen, P.G., van Eekelen, S.J.M. and van der Stoel, A.E.C. (2010). '*Monitoring of a railway piled embankment.*' *Proceedings of 9 ICG, Brazil,* 1461–1464.

Varaksin, S. (2016). 'Ground Improvement vs. Pile Foundations?' ISSMGE – ETC 3 *International Symposium on Design of Piles in Europe. Leuven, Belgium.* www.wtcb. be/homepage/download.cfm?lang=en&dtype=services&doc=Keynote_TC211_ at_ETC3_Leuven_Symposium.pdf. Last visited 17/11/2020.

Watts, K.S., Johnson, D., Wood L.A. and Saadi A. (2000). 'An instrumented trial of vibro ground treatment supporting strip foundations in a variable fill.' *Geotechnique* 50(6), 699–708.

Wong, P.K. and Muttuvel, T. (2011). 'Design of embankments supported on semi-rigid inclusions.' *International Journal of Geotechnical Engineering*, 6(2), 207–213.

Wong, P.K. and Muttuvel, T. (2012a). '*Economic design of semi-rigid inclusions for ground improvements.*' *Proceedings 11th International Conference ANZ,* Melbourne, Australia.

Wong, P.K. and Muttuvel, T. (2012b). 'Design of embankments supported on semi-rigid inclusions.' *International Journal of Geotechnical Engineering*, 6(2), 207–213. doi:10.3328/IJGE.2012.06.02.207-213

Wong, P.K. (2004). 'Ground improvement case studies chemical lime piles and dynamic replacement.' *Australian Geomechanics Journal*, 39(2), 47–60.

Zaeske, D. (2001). Zur Wirkungsweise von unbewehrten und bewehrten mineralischen Tragschichten über pfahlartigen Gründungselementen. Schriftenreihe Geotechnik, Uni Kassel, Heft 10, February 2001 (in German).

Chapter 12

Lightweight fill

J. Ameratunga

12.1 INTRODUCTION

As far as soft soil engineering is concerned, lightweight fills are used mostly in highway embankments to reduce the stress imparted on weak soil below subgrade to reduce settlement and improve stability. Sometimes, it is forgotten that the stress imparted by embankment fill is significant. As an example, say you have to add 2 m of fill to elevate the existing ground surface to the final design level. If the unit weight of the engineered fill is 20 kN/m³, this means the stress from the fill is 40 kPa. This is equivalent to the stress expected from a four-story building if constructed on a raft foundation, allowing 10 kPa per floor. Hence, if the subsurface profile consists of a deep soft clay, the expected settlement could be very significant in addition to stability issues. Lightweight fills allow one to manage the settlement and stability where deep foundations or other types of ground improvement are too expensive or impractical.

There are some types of natural lightweight soil that could be used as fill material when they have a low density. According to Kikkawa et al. (2011), the maximum dry unit weight of pumice in Japan and New Zealand is less than 7 kN/m³. Similarly, there are lightweight materials as a by-product of processing of minerals, etc., a good example being fly ash. Most of these materials have a bulk unit weight of just over 10 kN/m³. This means a significant reduction from a unit weight expected of a normal compacted fill having a typical unit weight of 18 to 20 kN/m³. Such a reduction in the unit weight results in lower stresses imparted onto the ground. This could have a significant effect on a soft clay design with regards to settlement. This is demonstrated in Example 12.1.

EXAMPLE 12.1

As depicted in Figure 12.1, the site is underlain by 1 m of sand at the ground level, followed by 4 m thick slightly overconsolidated clay with the ground water table at the base of the sand layer. The overconsolidation ratio (OCR) of

---

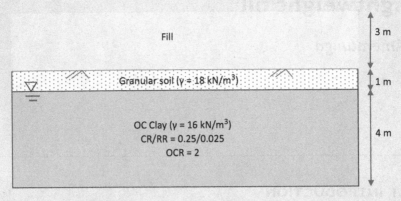

OC – Over Consolidated
OCR – Over Consolidation Ratio
CR – Compression Ratio
RR – Recompression Ratio

Figure 12.1 Example 12.1.

the clay layer has been assessed from laboratory tests to be 2. Other relevant parameters are as shown in Figure 12.1.

The development needs filling of 3 m above the existing ground surface. Ignoring the settlement compensation needed, calculate the primary consolidation settlement of the clay layer,

a) If the fill has a unit weight of 20 kN/m³; and
b) If lightweight fill is used having a unit weight of 12 kN/m³.

To keep the calculations simple, treat the clay as one layer without sub-dividing into multiple layers, i.e., stress changes to be calculated at the center of the layer.

Answer:

Initial stress at the center of the clay layer (σ'_{vo}) = 18 × 1 + (16–9.81) × 2 = 30.4 kPa

As OCR = 2, the maximum past pressure (σ'_{p}) = 30.4 × 2 = 60.8 kPa

a) **3 m regular fill**

Stress due to 3 m of fill = $\Delta\sigma$ = 3 × 20 = 60 kPa

Hence, the final stress at the end of primary consolidation would be:

$$\sigma'_v = \sigma'_{vo} + \Delta\sigma = 30.4 + 60 = 90.4 \text{ kPa}$$

This means the OC (overconsolidated) clay becomes NC (normally consolidated) at the maximum past pressure, i.e., at a stress of 60.8 kPa.

Hence, the total primary consolidation, based on Eq. 3.10, could be written as follows:

$$= RR \times H \times \log(\sigma'_p/\sigma'_{vo}) + CR \times H \times \log(\sigma'_v/\sigma'_p)$$
$$= 0.025 \times 4 \times 10^3 \times \log(60.8/30.4) + 0.25 \times 4 \times 10^3 \times \log(90.4/60.8)$$
$$= 30.1 + 172.3 \text{ mm}$$
$$\approx 202 \text{ mm}$$

b) **3 m lightweight fill**

Stress due to 3 m of lightweight fill $= \Delta\sigma = 3 \times 12 = 36$ kPa

Hence, the final stress at the end of primary consolidation would be:

$$\sigma'_v = \sigma'_{vo} + \Delta\sigma = 30.4 + 36 = 66.4 \text{ kPa}$$

This means the OC soil becomes NC at a stress of 60.8 kPa.

Hence, the total primary consolidation, based on Eq. 3.10, could be written as follows:

$$= CR \times H \times \log(\sigma'_p/\sigma'_{vo}) + CR \times H \times \log(\sigma'_v/\sigma'_p)$$
$$= 0.025 \times 4 \times 10^3 \times \log(60.8/30.4) + 0.25 \times 4 \times 10^3 \times \log(66.4/60.8)$$
$$= 30.1 + 38.3$$
$$\approx 68 \text{ mm}$$

i.e., in summary:

The primary consolidation settlement is 202 mm if regular fill is used but drops to 68 mm if lightweight fill of a unit weight 12 kN/m³ is used, which is a reduction of more than 65%.

In these instances, where traditional natural lightweight materials are used, the design process is no different to the conventional geotechnical design and is fairly straightforward. Depending on the material type, in addition to settlement, bearing capacity and stability, effects related to chemical and durability as well as environmental issues need to be addressed.

There are, however, other types of lightweight materials that are artificially made or non-traditional fill materials. They vary in density as summarized in Table 12.1. What is important is that the unit weight of some of these materials is significantly low, but compressibility is variable. Of the various materials considered, polystyrene stands out because its unit weight is at least one order less than other lightweight materials and about two orders of magnitude less than traditional earth fill.

Of the fill materials in Table 12.1, polystyrene is becoming more and more popular as more experience is gained and confidence is building up with each successful implementation, and the availability of case histories to the industry. Further, more use of the material leads to gradual reduction of cost

Table 12.1 Densities of various lightweight fill materials (after Christopher et al., 2006)

Fill type	Range in density (kg/m³)	Range in unit weight (kN/m³)
Polystyrene	12–32	0.1–0.3
Foamed concrete	320–970	3.1–9.5
Wood fiber	550–960	5.4–9.4
Shredded tires	600–900	5.9–8.8
Expanded shale and clay	600–1040	5.9–10.2
Fly ash	1120–1440	11.0–14.1
Boiler slag	1000–1750	9.8–17.2
Air-cooled slag	1100–1500	10.8–14.7

as supply volumes increase, making polystyrene more attractive to designers and constructors. It is also the lightest, generally one order less in weight than other lightweight materials, which means the lowest impact on soft soils at a site. Further, compared to other fill materials, material variability is very low because it is a manufactured product. Therefore, in this book, only polystyrene will be discussed as a potential lightweight fill.

There are two types of polystyrene, based on the manufacturing process, viz., expanded polystyrene (EPS) and extruded polystyrene (XPS), Generally, XPS is manufactured as planks or panels while EPS is formed as blocks. As it is much easier to deal with large blocks in earthworks related construction, only EPS rather than XPS is used in road infrastructure projects or building projects as a ground improvement technique. Hence, our discussion on polystyrene blocks in this book is limited to EPS.

12.2 EXPANDED POLYSTYRENE (EPS)

As previously mentioned, EPS weighs significantly less than traditional fill. As settlement due to development loads and instability due to lateral forces are related to the high unit weight of materials, obviously the use of EPS is attractive to the designers of soft soil projects (see Figure 12.2). There are other advantages as well with respect to EPS compared to standard ground improvement techniques. Some of the main advantages are presented in Table 12.2.

These are further discussed in Section 12.2.5.

Any method of ground improvement has negatives and it applies to EPS as well. Disadvantages are listed and discussed in Section 12.2.5.

12.2.1 EPS properties

EPS used in ground improvement are manufactured as blocks. They can be manufactured in different sizes or cut into different sizes. Block sizes vary

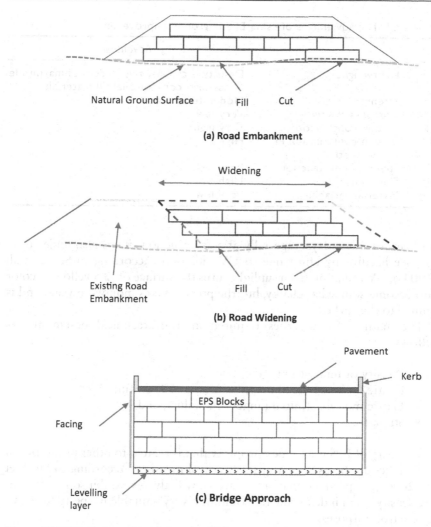

Figure 12.2 Typical projects adopting EPS.

but traditionally the least dimension is of the order of 0.5 m and the longest dimension 2 m to 3 m but could be as high as 5 m. Typical dimensions for civil engineering applications are 0.5 × 1.0 × 2.5 to 3.0 m weighing 25 to 30 kg (Aabøe et al. 2018). Generally, two people could handle a block on their own without further assistance from human or machine.

EPS is generally durable and has properties that are significantly better than other types of geosynthetics. According to Stark et al. (2004a), EPS is non-biodegradable and does not dissolve, deteriorate or chemically change when placed in the ground and under the water table. It is also understood that EPS does not attract animals and other living organisms. Exposure to

Table 12.2 The advantages of using EPS in ground improvement

		Advantage
i)	Lightweight	Density is at least two orders of magnitude less than conventional fill materials
ii)	Strength	Moderate to high
iii)	Compressibility	Very low
iv)	Ease in construction	Very high
v)	Ease in construction in wet weather	High
vi)	Speed of construction	Very high
vii)	Durability	High
viii)	Material variability	Very low

sunlight for a few days, especially at a construction site, is also understood to have hardly any effect due to UV radiation. According to Stark et al. (2004a), UV radiation from sunlight turns the surface of EPS yellow in color and become somewhat chalky, but the process takes months to years and is limited to the surface.

The main EPS properties required in a geotechnical design are as follows:

- Density or unit weight;
- Elastic limit stress/compressive stress at 1% strain;
- Elastic modulus (Initial tangent modulus); and
- Strength.

The density of EPS is an index property that is related to other properties of EPS and could be used as a quality control measure. According to Stark et al. (2004a), as production of any material will always exhibit some variability, density as an index property could be a very valuable property for quality control purposes.

Compressive stress is the main parameter used in the geotechnical design. To avoid long-term creep, it is generally accepted that the stress should be limited to the Elastic limit stress, which is taken as the compressive stress at 1% strain. This is further discussed later.

ASTM D6817 provides information on EPS material and typical physical properties are presented in Table 12.3. EPS is flammable and therefore ASTM D6817 stipulates that EPS should have a minimum Limiting Oxygen Index of 24 (volume) by incorporating sufficient flame retardants.

Table 12.3 indicates that a wide range of EPS materials are available. Higher density materials provide higher compressive strengths. It is also understood that the cost of EPS increases with material density. Value engineering could be carried out to assess the minimum extent of EPS for a particular design. Several types of material could be adopted in the same

Table 12.3 Range of EPS physical properties (minimum) according to ASTM D6817

Property	EPS12	EPS15	EPS19	EPS22	EPS29	EPS39	EPS46
ρ (kg/m³)	11.2	14.4	18.4	21.6	28.8	38.4	45.7
σ_{c1} (kPa)	15	25	40	50	75	103	128
σ_{c5} (kPa)	35	55	90	115	170	241	300
σ_{c10} (kPa)	40	70	110	135	200	276	345
f_y (kPa)	69	172	207	240	345	414	517

(ρ = density; σ_{c1}, σ_{c5} and σ_{c10} = compressive stress at 1%, 5% and 10% strain; f_y = flexural strength

cross section based on stability and other requirements. For example, higher density (i.e., higher strength) EPS blocks could be placed at the upper layers so that it could withstand imposed loads without exceeding stress and deformation criteria, with the rest of the embankment (i.e., lower layers) constructed using a lower density EPS.

It is generally accepted in the industry that the stress-strain behavior of EPS blocks is linear elastic to a compressive strain of about 1% (Stark et al. 2004a) and this is the compressive stress usually selected to limit long-term deformation under applied loads. This is demonstrated in Figure 12.3 where a typical stress–strain curve is plotted for EPS. The behavior indicates a linear elastic portion initially, then gradually yielding under the imposed stress. The stress at which the yielding commences is about 1 to 1.5% (Stark et al. 2004a) as evident from Figure 12.3. Considering the significant increase in

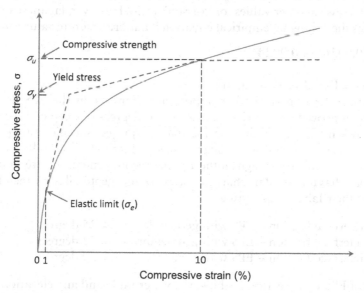

Figure 12.3 Stress strain curve of EPS (after Stark et al. 2004a).

strain beyond 1% for small changes to applied load as depicted in the figure, designs are usually carried out to ensure that the strain does not exceed 1%, i.e., stresses kept within Elastic limit stress. This is very important to reduce long-term creep. This stress is known as Elastic limit stress (σ_e). The slope of the line in the elastic region in the stress-strain plot is the Elastic modulus or the Initial tangent modulus, which is also a key design parameter. Beyond the Elastic limit, the stress-strain curve shows the yielding behavior, and after about 10% strain the stress increases linearly with strain. The ultimate compressive stress, σ_u, is considered the compressive strength at 10% strain. Extension of the initial linear portion (elastic region) meets the extension of the linear portion beyond 10% strain at the yield stress, σ_y.

Although it is generally advocated to limit strain to the Elastic limit, i.e., 1%, Penndot (2018) suggests that a more relaxed strain of 5% could be adopted for short-term loading such as due to stresses from construction vehicles.

According to Bartlett et al. (2012), the modulus value obtained from laboratory tests depend on the size of the sample tested. They cite the example where the modulus value from laboratory tests indicated the strength of a 60 cm cube to be double the value when 5 cm cubes were tested. Hence, the modulus value of a large block used in embankment construction may be much greater and the design is likely to be conservative especially in relation to settlement prediction.

Another parameter that is not considered significant in EPS ground improvement design is the Poisson's ratio (ν). According to Stark et al. (2004a), within the elastic range, ν is relatively small and of the order of 0.1 and could be assumed to be zero for practical design purposes. Elragi (2000) provides several other values for Poisson's ratio. However, Japanese research suggests the following empirical equation if a more accurate value is needed.

$$\nu = 0.0056\,\rho + 0.0024 \tag{12.1}$$

where ρ = EPS density in kg/m^3.

Interface friction is another important parameter in the EPS design of ground improvement. A friction angle of 30 degrees is accepted for an EPS/ EPS block interface. While iterating that site/project specific testing should be carried out, the following suggestions of Stark et al. (2004b) could be used for preliminary designs although the recommendations are limited to materials having similar characteristics to the geotextiles/geomembranes used in their laboratory testing.

- Interface friction – EPS with geotextile   = 25 degrees
- Interface friction – EPS with geomembrane = 52 degrees
- Interface friction – EPS with sand   = 30 degrees

As most EPS construction is above the water table and any elevation of the water table is temporary (such as after a flood), water absorption is not a significant issue unless the EPS blocks have been used in a 'dam' situation. Further, most projects using EPS include a HDPE liner for other reasons, such as to protect against hydrocarbon spills, which anyway acts as a barrier to water absorption. Even if water absorption does occur, mechanical properties are not expected to change (Stark et al. (2004a). However, the density of EPS will increase, and it is safe to assume a unit weight of 1 kN/m³ as proposed by Stark et al. (2004a).

12.2.2 Design considerations

The design of EPS as a ground improvement is not complicated but it needs to address the following:

- Settlement;
- Stability;
- Buoyancy;
- Bearing capacity; and
- Protection of EPS.

Settlement, stability and bearing capacity are checked in any conventional geotechnical design. The additional design considerations are EPS protection and buoyancy.

In the case of EPS, pavement design is also very important but is not covered in this book. However, it should be noted that, EPS has no influence on the pavement design if there is sufficient fill cover above the EPS blocks. According to Christopher et al. (2006) and re-iterated in SHRP2 (2014), the minimum gravel cover required to ensure that the pavement design is independent of EPS is 1 m. If the cover is less than 1 m, a detailed assessment is needed. The reader is referred to Stark et al. (2004a and 2004b) for relevant design procedures for flexible and rigid pavement designs.

EPS as ground improvement is mostly used in embankments and slopes. In soft clay engineering, EPS has become more popular over the last two decades. For those who wish to research further, the following documents provide further information and geotechnical guidance on EPS as a ground improvement material:

- Christopher et al. (2006) Geotechnical Aspects of Pavements
- FHWA (2017). Ground Modification Methods – Volume I
- Stark et al. (2004a). Geofoam Applications in the Design and Construction of Highway Embankments, NCHRP Web Document 65.
- PennDot (2018). Geotechnical Engineering Manual, Publication 293. Pennsylvania Department of Transportation.
- SHRP2 (2014). Geotech Tools: Geo-construction Information & Technology Selection Guidance for Geotechnical, Structural, & Pavement Engineers.

Table 12.4 Factors of safety for different design aspects

	Minimum factor of safety
Bearing capacity	3.0
Global stability	1.5
Seismic stability and overturning	1.2
Buoyancy/uplift	1.2
Sliding and overturning due to water	1.2
Sliding and overturning due to wind	1.2
Load bearing	1.2

The factors of safety that apply for EPS design according to Stark et al. (2004b), are presented in Table 12.4.

12.2.2.1 Settlement

EPS, as a ground improvement technique, is mostly used for settlement control, especially in major highway projects. In addition to reducing the total settlement, EPS is most useful to control differential settlement and grade changes. A classic example is the necessity to widen a highway, constructed on soft clays, which has been in existence for several years. If the highway is to be widened using conventional earth fill, one can expect significant settlement under the widening as the weak clay below will consolidate as a result of being the subject of high stresses. In the meantime, the existing embankment will hardly settle or settle at a significantly less rate as creep, because no additional stresses have been imposed (except at the widening interface).

Settlement in ground improvement projects involving EPS can be assessed using standard techniques already covered in previous chapters. Settlement due to fill above and below EPS blocks plus traffic loads can be assessed based on methods discussed in geotechnical text books referred to in Chapter 3.

Settlement due to EPS blocks itself has several components:

1. Subsurface settlement, immediate
   Immediate settlement due to imposed load by EPS blocks is usually neglected.
2. Subsurface settlement, consolidation
   Consolidation (primary and secondary) settlement of the underlying soils due to imposed EPS loads is also usually neglected. However, where the EPS block height is significant and the settlement criteria are tight, settlement can be calculated using conventional methods found in geotechnical textbooks referred to in Chapter 3.

3. Compression and creep of EPS
   Although it can be easily assessed based on available laboratory data, compression and creep of EPS are usually neglected because the magnitude is small relative to subsurface settlements as long as the stress is limited to elastic stress at 1% strain.

In soft clay projects, EPS is usually used to reduce settlement due to the application of development loads. As the EPS density is of the order of 1/100th of concrete or fill material, subsurface settlement due to the load exerted by EPS alone is very small. Yet, the load from the fill above the EPS blocks and the pavement structure could trigger some settlement. If this additional settlement is over the limit imposed by project criteria, excavation of some of the soil at the base and replacement with EPS could be an option to ensure the net load at the base is small or zero.

When EPS is used, the designer must give proper attention to the settlements and grade changes especially at the interfaces with other earthworks. For example, if EPS is adopted at a bridge approach, unless a gradual change in the usage of EPS blocks is adopted, abrupt changes to the settlement profile could occur leading to undesired outcomes. A similar situation arises when an existing road is to be widened. Therefore, the design should allow for a gradual change of the EPS block formation rather than an abrupt change from EPS to conventional earth fill.

12.2.2.2 Stability

In a highway embankment scenario on soft soils, stability is not a critical issue if the embankment is built up mainly using EPS blocks, because the applied stress is very low. Yet, one needs to assess the stability as part of the design process, which can be done using standard software. Analysis could appear to be complicated as the shear strength to be nominated for the EPS body has to take into account, both the EPS body and the interfaces.

One complication is the fact that critical failure surfaces due to imposed loads may pass through EPS blocks as well as thorough interfaces. To overcome the necessity to assign shear strength parameters to EPS body in the computations, one of the options discussed by Stark et al. (2004a) is ignoring the EPS embankment and place the loads (i.e., weight of EPS, soil cover, pavement, and any other external loads) directly at the base level of the EPS blocks assuming no dispersion of the load through the block. One disadvantage of this method is that it cannot be used in a pseudo-static analysis because the seismic force has to be applied at the center of gravity of the sliding mass (PennDot 2018).

Another method proposed by Stark et al. (2004a) is to simplify the stability analysis by assuming the contribution from both these elements (i.e., through EPS blocks and interfaces) together. Based on the assumption that 75% of a critical surface will pass though joints between blocks and 25%

passing through intact EPS, Stark et al. (2004a) suggests that stability analysis be carried out by ignoring the interfaces and treating the blocks as a single material having a cohesion equivalent to 1/4th the cohesion of EPS and 3/4th the friction angle at the interface. EPS cohesion is assumed to be half the compressive strength at 1% strain and the friction angle at interface is usually 30 degrees (Penndot 2018; SHRP2 2014). An example adapted from PennDot (2018) is presented below (Example 12.2).

EXAMPLE 12.2 ADAPTED FROM PENNDOT (2018)

An EPS embankment is constructed using Type EPS 22. According to EPS information presented in Table 12.3, the compressive stress at 1% strain is 50 kPa. Find suitable equivalent strength parameters to model EPS in slope stability analysis.

Answer:

Unconfined compressive strength at 1% strain of EPS = 50 kPa

Therefore, shear strength/cohesion of EPS = 50/2 = 25 kPa as shear strength is half of the unconfined compressive strength.

25% of shear strength = 25/100 × 25 = 6.25 kPa

Interface friction angle = 30 degrees (assumed based on Stark et al. 2004a)

75% of interface friction angle = 75/100 × 30 = 22.5 deg.

Therefore, the relevant parameters are cohesion = 6.25 kPa and friction angle = 22.5 deg.

Seismic stability could also be assessed using standard techniques such as carrying out a pseudo-static slope stability analysis (Stark et al. 2004b), and the effect of pseudo-static horizontal forces on the overturning of tall and narrow vertical embankments (such as bridge approaches) should be considered although it is a temporary condition.

There are no rigid rules on the slope angles to be adopted and the cover over the EPS blocks. According to PennDot (2018), sides of an embankment constructed using EPS should have a slope not steeper than 2H:1V to ensure cover material can provide adequate veneer stability for the slopes unless detailed analysis is carried out to verify and/or include structural support/facing.

12.2.2.3 Buoyancy

The buoyancy is a critical issue for EPS design because of EPS lightness, density being of the order of 1% of earth fill. Therefore, buoyancy should always be checked for any construction with EPS. It is important to neglect

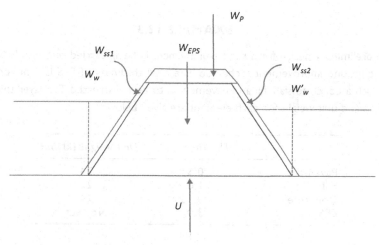

U   = Uplift load acting on the base of the EPS body
W_{EPS} = Weight of EPS-block embankment
W_p = Weight of Pavement Surcharge (pavement and fill above ESP)
W_{ss1} and W_{ss2} = Weight of soil cover on side slopes
W_w and W'_w = Vertical component of weight of water on the embankment face

Figure 12.4 Forces acting on EPS block body due to buoyancy.

any live load (such as traffic load) as a counterbalance load in the analysis because such loads are not permanent and may not be present, especially during a rain event when the buoyancy becomes a problem.

Stark et al. (2004b) provide a detailed description on the assessment of factor of safety and the vertical forces acting on the EPS body are summarized in Figure 12.4. For preliminary design purposes, the factor of safety can be simplified as the ratio of load on top of the EPS body, which includes the load exerted by the fill and the pavement, and the buoyant force. The load at the top of the EPS body can be calculated by the multiplication of unit weight, thickness and the width of the EPS block at the top. Buoyant force is calculated by the uplift water pressure multiplied by the width of the EPS body at the base.

A factor of safety of 1.2 is generally accepted by the industry to resist uplift forces. It is obviously necessary to adopt a credible flood event in calculations, generally to be on the conservative side, as the consequences could be the damage to the whole embankment due to uplift, leading for a rebuilt. If a significantly high return period is adopted, the author considers that the expected factor of safety could also be reduced accordingly.

A higher factor of safety such as 1.5 should be adopted if the buoyancy is not a short-term issue but could be prevalent long term such as in a dam situation. Where buoyancy is an issue, the design may need to include anchors to overcome uplift forces.

EXAMPLE 12.3

A preliminary design for a road embankment is being carried out with EPS as the ground improvement technique. The total thickness of EPS is 3 m above which a concrete slab, fill and pavement is to be constructed. The layer thickness and unit weight for each element are given below:

	Thickness (m)	Unit weight (kN/m³)
Pavement	0.5	24
Fill	1.0	22
Concrete	0.15	24
EPS	3.0	Neglect

As shown in Figure 12.5, the top width of the pavement is 10 m with side slopes of 1V:2H. If the flood design water level 1.4 m, calculate the preliminary factor of safety against buoyancy.

Answer:

As it is a preliminary design, ignore the weight of EPS, the downward water pressure on the side slopes and the weight of cover on the side slopes.

Cross sectional area of the pavement $= 0.5 \times (10 + 12) \times 0.5 = 5.5$ m²

Cross sectional area of the fill layer $= 0.5 \times (12 + 16) \times 1.0 = 14.0$ m²

Cross sectional area of the concrete slab $= 0.5 \times (16 + 16.6) \times 0.15 = 2.45$ m²

Weight of the pavement $= 5.5 \times 24 = 132$ kN/m length

Weight of the fill $= 14 \times 22 = 308$ kN/m length

Weight of the concrete slab $= 2.45 \times 24 = 59$ kN/m length

Hence, total weight or the downward pressure $= 132 + 308 + 59 = 499$ kN/m length

Width of the base of EPS $= 10 + (0.5 + 1.0 + 0.15 + 3) \times 2 \times 2 = 28.6$ m

Design water level $= 1.4$ m

Hence, buoyant force acting at the base of EPS $= 28.6 \times (1.4 \times 9.81) = 393$ kN/m length

Hence, simplified Factor of Safety $= 499/393 = 1.27 > 1.2$

If the load imparted on the subgrade cannot achieve the nominated factor of safety, it may be necessary to provide additional support such as anchors or geotextiles. To fix the anchors within the embankment, it would be necessary to place a concrete slab above the EPS blocks.

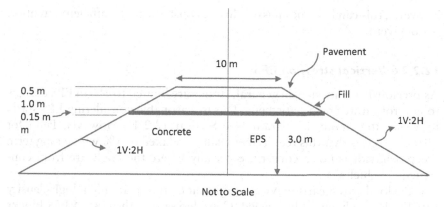

Figure 12.5 Example 12.3.

12.2.2.4 Bearing capacity

The external loads to be considered to assess the bearing capacity are limited to fill load, pavement load and live load. The weight of EPS blocks could be neglected as it is not significant. Bearing capacity can be assessed using conventional techniques that have been described in many textbooks and not repeated here.

12.2.2.5 Sliding and overturning

Sliding and overturning should be checked against unbalanced forces in an embankment. This is most relevant when unbalanced water pressure occurs after a flood event or if the embankment has been used to contain water.

Sliding could be an issue even during construction unless sufficient load is placed at the top to increase available resistance to external lateral forces such as unbalanced water pressure or due to pressure exerted by backfilling and compaction. Sliding is resisted by the interface friction between EPS blocks and either geotextiles used or subgrade soils. Interface friction angles have been discussed in Section 12.2.1. As discussed by Stark et al. (2004b), although the friction angle at the interface is high and almost close to the subgrade friction angle, resistance (stress times tan of friction angle) offered at the interface is small because the dead weight of the EPS embankment is small. Where the resistance is inadequate, mechanical inter-block connections could be used (Stark et al. 2004a).

Overturning is a potential failure issue, mostly for vertical tall embankments, and can also be assessed using basic principles.

Sliding and overturning against wind is not considered critical for normal applications but could be designed based on basic geotechnical principles especially if the embankment has to resist extreme events such as hurricanes.

However, this can become an issue during construction if sufficient attention is not given.

12.2.2.6 Vertical stress on EPS

As previously mentioned, in general, strains are limited to 1% in EPS designs to control future creep settlement. This is reasonable considering EPS starts to yield after about 1% strain (see Section 12.2.1). However, PennDot (2018) suggests that this requirement can be relaxed to 5% where long term creep behavior is not a concern, especially where the loads are from construction vehicles.

If the loads from earthmoving equipment or traffic are high, high-density EPS blocks are likely to be needed. Considering that the cost of EPS blocks increase with density, an option would be to use higher-strength blocks at the upper levels but change to lower density EPS blocks below for the bulk of the fill. Additional effort such as the use of concrete slabs to redistribute loads may also be needed if the EPS strains are high.

EXAMPLE 12.4   (ADAPTED FROM PENNDOT 2018)

Assuming a combined dual tire contact area of 500 mm × 125 mm and an axle load of 100 kN, assess whether EPS22 (see Table 12.3) could be used during construction, if the capping fill height over the EPS blocks is only 150 mm. Assume the bulk density of the capping material to be 20 kN/m^3.

Answer:

Axle load = 100 kN

Contact area = 0.5 × 0.125 m

Dual Tire pressure = [100/(0.5 × 0.125)]/2 = 800 kPa

Assuming a 1H:2V distribution through the cover material,

Pressure at the top of EPS due to tire pressure

$$= \frac{800 \times (0.5 \times 0.125)}{\{(0.5 + 0.150)(0.125 + 0.150)\}} = 280 \text{ kPa}$$

Dead load on EPS due to 150 mm capping = 0.15 × 20 = 3 kPa

Hence, total load pressure = 280 + 3 = 283 kPa

The total pressure of 283 kPa exceeds the compressive stress, both at 1% and 5% strain.

Hence EPS 22 cannot be used for the project and a higher strength EPS has to be considered.

12.2.2.7 Protection of EPS

EPS material is flammable. ASTM D6817 stipulates that sufficient retardants be included to ensure a minimum oxygen demand of 24. As EPS is covered by soil and the pavement, the risk of EPS catching fire is low when in operation. However, it is certainly possible during construction, especially in bushfire regions.

As previously mentioned, EPS material should be protected from hydrocarbons, fire, insects, etc. Where the risk of hydrocarbon spill is high, the blocks could be encapsulated in a suitable geomembrane. The geomembrane could be wrapped around by a separation geotextile as a protection. The cover provided on the sides and the pavement also helps to resist effects of hydrocarbon spillages.

EPS should also not be allowed to be damaged due to construction vehicles. Such machinery should not be allowed directly on top of EPS blocks. A thin concrete slab could be placed at the top to avoid such issues. Such a slab, in addition to being a safety feature against possible damage due to machinery, redistributes traffic loads that would be useful to ensure the elastic stress limit of the EPS is not exceeded. It also helps to reduce the risk of damage due to hydrocarbon spills.

12.2.3 Environmental considerations

There are no significant environmental issues that prop up due to the use of EPS. However, as EPS is flammable, although the material may contain a fire retardant, care should be taken to avoid contact with fire that could be triggered by simple visual indiscretions such as a cigarette butt. This is most important during the construction phase when EPS is exposed to the environment. The risk reduces significantly once the cover materials are placed. Stark et al. (2004a) state that the soil cover above the EPS blocks should be 300 mm to 500 mm with 400 mm thickness on the sides, to reduce effects of fire and unspecified hazards in addition to the assistance it provides to the growth of vegetation.

12.2.4 Construction issues

Some of the construction issues have already been discussed but repeated in here for completeness.

Subgrade preparation	The subgrade must be prepared to provide a smooth base, and this may require excavation or filling. It is advisable to place a thin layer, say 150 mm thickness, of bedding sand to make the surface suitable for EPS blocks to be placed.

Placement of blocks | In the longitudinal direction, the top surface of EPS blocks needs to be parallel to the pavement surface to facilitate construction and performance (Stark et al. 2004a). This means the thickness of the fill layer below the first layer of EPS blocks should be sloped using variable thickness of fill. Stark et al. (2004a) also suggest that the upper surface of the EPS blocks should be kept horizontal in the cross section and to allow the pavement thickness to be varied to obtain the desired final pavement surface.

According to PennDot (2018), EPS blocks and pavement sections perform better if the top layer of blocks are aligned perpendicular to the longitudinal axis of a road alignment.

The placement of blocks should be done manually.

Block arrangement | The blocks should be placed to ensure they work as interlocked blocks. That means blocks should be staggered to avoid continuous joints (see Figure 12.6), although interface strength is high. Liu and Negussey (2018) carried out experimental and numerical assessments and concluded that intercepted vertical joints were better than continuous vertical joints.

According to PennDot (2018), vertical joints should be offset by a minimum of 2 feet (~0.6 m).

According to Mijena (2018), based on NPRA (2015) guidelines, small gaps less than 3 cm could be tolerated, gaps between 3 and 5 cm should be filled with sand, and gaps exceeding 5 cm should not be allowed.

Wind effects | As EPS blocks are light, they need only minor forces to be dislodged. Therefore, the blocks should be anchored in some way or temporary loads placed such as sand bags to ensure EPS blocks are not shifted during construction due to high winds.

Compaction machinery | Loads imparted by compaction machinery could force the EPS to exceed elastic stress limits. To avoid this, a concrete slab, say 100 mm thick, could be placed on top of the EPS bocks to disperse stresses. As previously mentioned, no construction machinery should be allowed directly on the EPS blocks without at least a thin cover.

Cutting | It is best that blocks are manufactured/cut to the desired sizes in the factory so that cutting is not necessary in the field. Quite often, this is not possible and field cutting cannot be avoided. Cutting should be carried out to produce clean/smooth surfaces. PennDot (2018) recommends that only hot wire cutting should be allowed in the field.

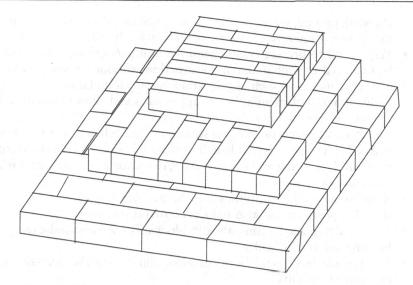

Figure 12.6 EPS block arrangement (adapted from Stark et al. 2004b).

12.2.5 Advantages and disadvantages

EPS provides the following advantages:

- Ease of construction.
- Very light and therefore could be manually handled.
- Construction not impeded significantly due to adverse weather such as heavy rain.
- Considered durable and inert.
- Not found attractive to insects and rodents.
- Water absorption very low.
- Footprint of an embankment could be narrowed by adopting steeper or vertical side slopes where space is limited to ensure the footprint is within the right-of-way.
- Could be recycled. Hence, useful where temporary road construction occurs, especially when services may be impacted.
- Speed of construction very high compared to any other type of ground improvement.
- Immediate handover of the site possible compared to other ground improvement techniques such as preloading.
- Significant reduction of preload heights; in some cases, no preload needed.
- Significant reduction in settlement, which also reduces the volume of fill necessary to compensate.

- On weak ground, staged construction not needed. Geotechnical stability enhanced and risk of failure significantly reduced.
- Very useful to protect critical/sensitive utilities/services, which could be impacted by embankment construction without resorting to very expensive measures such as relocation or deep foundations.
- Reduction of settlement at bridge approaches and allow gradual grade changes as expected in a design.
- Reduction of effects on existing structures (e.g., when an existing road and a bridge is to be widened) and avoid differential settlements between the existing and the new. Also, reduction of negative skin friction on existing piles.
- Relocation of critical utilities may be avoided which helps the construction program and also reduces construction costs.
- Soil could be placed against the EPS blocks. This allows landscaping to be done without difficulty.
- The outside face could be shotcreted, eliminating the necessity for expensive face units.

Some of the disadvantages include:

- EPS could be relatively expensive compared to some other ground improvement methods.
- Construction complicated where road furniture exists.
- EPS is buoyant. Although the design could be carried out to ensure no buoyancy due to high rainfall/flood events, it could be an issue during construction.
- It is very light and that means blocks need to be secured against high windy conditions during construction.
- EPS is combustible if subjected to high temperatures. However, according to Stark et al. (2004a), chemicals could be incorporated during the manufacturing process to make EPS flame retardant and self-extinguishing.
- EPS could be a fire hazard if left open to the environment unless suitably protected. Serious attention should be given for supervising construction and have site specific safety procedures.
- Liable to damage if exposed to some hydrocarbons. The design needs to take this into account and provide appropriate protection that could include cover such as fill, geotextiles and/or concrete slabs carefully designed to reduce this risk.
- Although EPS could be vulnerable to insect damage, no known instances are found in the literature. EPS is not an insect-attracting material.

CASE HISTORY 12.1 (WHITE ET AL. 2007)

Port of Brisbane Motorway Stage 1 is the main entry to the Port of Brisbane. It generally runs along gently sloping undulating terrain, but the land is low lying on the east. The subsurface profile in the latter area consists of recent alluvium, viz., soft to firm compressible clay extending up to 30 m depth. The compressible alluvium was only marginally overconsolidated and therefore expected settlements under an embankment of 5 m high were high.

One of the bridges, viz., Lindum Road Bridge, was added to the project towards the end of the main project and therefore construction time was limited. As the embankments were high it was not possible to adopt routine ground improvement methods such as preloading with wick drains.

Preliminary analyses had suggested that, if no ground improvement was considered, the maximum height of embankment to achieve the required factor of safety is only 3 m without staging, and the expected primary consolidation settlement is about 1.2 m over 10 years, which is much greater than the specification requirements. After investigating several other methods, based on cost and time considerations, it was decided to use EPS for the bridge approaches.

The EPS of a nominal density of 24 kg/m^3 was used, wrapped in a 0.5 mm thick textured HDPE membrane. Buoyancy was checked and the 1.3 m of fill and pavement above the EPS blocks provided adequate factor of safety against buoyancy. The HDPE membrane was covered with a layer of separation geotextile to protect against puncture. Above this, it is the usual practice to construct a thin concrete slab, but it posed significant time constraints due to curing and strength gain requirements. An alternative approach using cementitiously modified materials was proposed, which was adopted by the contractor. This approach used lightly cemented base course type fill material to create a working platform over the weak EPS materials. The material was end dumped and spread not allowing any machinery to directly operate on top of the EPS body or the geotextile and geomembrane. Compaction was carried out using a light roller, 1 tonne in mass, and without any vibration.

A layer of cement modified subbase quality material was placed to expedite construction but prevent damage to the EPS. Cement modification was light to achieve an unconfined compressive strength of 2 MPa at seven days by using an agent 60% cement and 40% fly ash. This layer was compacted using a 1 tonne roller and without vibration for at least eight passes. Select fill material and pavement materials were not placed until seven days later.

The construction was completed in 2001 and it is understood that the approaches are functioning well.

12.3 SUMMARY

In this chapter, the main thrust has been to introduce the use of EPS as a ground improvement technique. The characteristics of EPS are discussed in detail and design guidance has been provided. While the design itself is not complicated, especially with simple procedures highlighted, construction needs careful considerations. Useful hints and guidance have been provided for construction, some of the factors that need to be considered even during the design phase.

REFERENCES

Aabøe, R., Bartlett, S.F., Duškov, M., Frydenlund, T.E., Mandal, J.N., Negussey, D., Özer, A.T., Tsukamoto, H. and Vaslestad, J. (2018). '*Geofoam blocks in civil engineering applications.*' *Proceedings 5th International Conference on Geofoam Blocks in Construction Applications EPS 2018*, Springer. D. Arellano, A. Özer, S. F. Bartlett and J. Vaslestad Editors. Part I, 3–38.

ASTM D6817-17 (2017). *Standard Specification for Rigid Cellular Polystyrene Geofoam.*

Bartlett, S.F., Lawton, E.C., Farnsworth, C.B. and Newman, M.P. (2012). 'Design and evaluation of expanded polystyrene geofoam embankments for the I-15 reconstruction project, Salt Lake City, Utah.' *Report No. UT-1X.XX*, Department of Civil and Environmental Engineering, University of Utah.

Christopher, R.B., Schwartz, C. and Boudreau, R. (2006). 'Geotechnical aspects of pavements.' *Report No. FHWA-NHI-05-037*, US Department of Transportation, Federal Highway Administration, Washington, DC.

Elragi, A.F. (2000). 'Selected engineering properties and applications of EPS geofoam.' PhD thesis. State University of New York. College of Environmental Science and Forestry, Syracuse, NY.

FHWA (2017) *Ground Modification Methods Reference Manual – Volume I. Publication No. FHWA-NHI-16-027*. Federal Highway Administration, USDOT, Washington, DC.

Kikkawa, N., Orense, R.P. and Pender, M.J. (2011). '*Mechanical behaviour of loose and heavily compacted pumice sand.*' *Proceedings, 14th Asian Regional Conference on Soil Mechanics and Geotechnical Engineering*, Paper 214, 6pp.

Liu, C. and Negussey, D. (2018). '*Effects of installation of different density geofoam and continuous vertical gaps.*' *5th International Conference on Geofoam Blocks in Construction Applications Proceedings of EPS 2018*, Springer. D. Arellano, A. Özer, S.F. Bartlett and J. Vaslestad Editors. Part III, 219–229.

Mijena, E.H. (2018). '*EPS geofoam used in E16 Sandvika–Wøyen.*' *Proceedings 5th International Conference on Geofoam Blocks in Construction Applications EPS 2018*, Springer. D. Arellano, A. Özer, S.F. Bartlett and J. Vaslestad Editors. Part II, 57–70.

Norwegian Public Roads Administration (NPRA) (2015). *Prosesskode for Enterprise E02 Rud-Vøyenenga*. Norwegian Public Roads Administration (NPRA).

PennDot (2018). 'Geotechnical engineering manual.' *Publication 293*. Pennsylvania Department of Transportation.

Stark, T.D., Arellano, D., Horvath, J.S. and Leshchinsky, D. (2004a). 'Geofoam applications in the design and construction of highway embankments. *NCHRP Web Document 65*, Transportation Research Board, Washington, DC, http://trb.org/publications/nchrp/nchrp_w65.pdf. Accessed May 20 2019.

Stark, T.D., Arellano, D., Horvath, J.S. and Leshchinsky, D. (2004b). 'Guideline and recommended standard for geofoam applications in highway embankments.' *NCHRP Report 529 (Project 24-11)*, National Cooperative Highway Research Program, Transportation Research Board, Washington, DC.

SHRP2 (2014). *Geotech Tools: Geo-construction Information & Technology Selection Guidance for Geotechnical, Structural, & Pavement Engineers.* http://geotech-tools.org/

White, W., Shipway, I., Ameratunga, J. and Olds, R. (2007). '*Design and construction of closed cell expanded polystyrene foam embankment with a heavy duty pavement.*' *Proceedings 10th ANZ Conference*, Brisbane, Australia.

Chapter 13

Deep soil mixing

T. Muttuvel, S. Iyathurai and J. Ameratunga

13.1 INTRODUCTION

In recent years, the construction of infrastructure over compressible foundations such as soft ground has become an inevitable choice because of the lack of sites underlain by competent soil. To develop such land, various ground improvement methods need to be applied to improve the site in order to meet the project specific technical requirements depending on a number of factors such as ground conditions, associated cost and construction time. One of the ground improvement methods used by the industry is deep soil mixing (DSM) using cementitious or other binders, using dry and wet methods, to stabilize the soils to improve the strength and stiffness of the ground. This chapter will discuss some of the important aspects of the deep soil mixing.

The stabilization by DSM is conducted using a series of columns in square or triangular grids typically to control settlement, and in panels/grids to improve stability. The ground treatment could be full depth treatment or partial depth treatment depending on the specific requirements.

There are two types of mixing used in this process:

- Dry method;
- Wet method.

In the dry soil mixing method, dry binder such as lime, cement, lime-cement blend, etc., is mixed with existing soil by mechanical means, and in the wet method, slurry binder, primarily cement based, is used. During this process, clay mineralogy will be altered, and strength gain will occur as the curing process continues. The dry method is frequently used in soft clays and silts having very high moisture content (60 to 200%). The wet method is also used on soft clays and silts or fine sand where the moisture content is low (<60%).

In the DSM operation, the mixing tool is gradually advanced into the ground by continuous rotation. Once it reaches the design depth, the rotation of the mixing tool is reversed, and the binder is fed into the top of

Kelly bar via a tube as the binder pod is pressurized with air. The mixing tool is gradually withdrawn, with the rate of withdrawal controlled by the rate of binder flow, to achieve the target design binder content.

Column diameters vary from site to site and sometimes within the same site depending on the design requirements. Typically, diameter ranges from about 0.6 m to 1.5 m (FHWA 2013). The depth of improvement is generally limited to about 24 m but according to the Cement Deep Mixing Association (CDMA 2015), the maximum depth can be as high as 45 m.

According to EuroSoilStab (2002), the introduction of binders results in significant changes to chemical and physical properties of the soils with the pH value quickly rising to 11–12. Depending on the binder type, some reactions could occur in weeks but some may take months. EuroSoilStab (2002) cautiously advise that the undrained shear strength of the stabilized soil in DSM is within the range of 50 kPa to 150 kPa for the dry method of stabilization although laboratory test results could provide ten times these values because such strength increases hardly occur in situ.

The wet methods are suitable for any soil and typically used for structural support

  i. To improve bearing capacity;
 ii. For seismic stabilization; and
iii. Settlement control and/or shear strength improvement under embankments.

The dry methods are mainly suitable for cohesive soil with high moisture content and typically used for soil stabilization/reinforcement and settlement reduction.

One of the wet methods is called cutter soil mixing (CSM). CSM is a variation of DSM named after the cutter that is employed. The cutter consists of two sets of cutting wheels (see Figure 13.1) that are subjected to rotation, i.e., similar to excavating narrow trenches, leading to narrow panels when soil is mixed with the binder. The width of the panels generally ranges from 2 m to 3 m and the thickness of 0.5 m to 1 m. CSM is very useful when a wall type or a lattice/grid type DSM is warranted. According to Fiorotto et al. (2005), the rectangular panel, in comparison to a series of contiguous or secant columns, offers several advantages:

 • Structurally, if the properties of a rectangular shaped panel is compared to those of secant or contiguous columns whose diameter is equivalent to the width of the panel, the panel has a more efficient shape and the areas of treated soil in compression and tension are larger and the lever arm of the rectangular section is larger. This means a higher moment of resistance.
 • Column diameters need to be larger than panel thickness to produce a section of equivalent width.

Figure 13.1 CSM cutter (Ameratunga et al. 2009).

This means less soil needs to be treated when CSM panels are used and also less binder, and there is a saving in total energy expenditure.

According to FHWA (2013), the original concept of DSM originated in USA in the 1950s, i.e., more than six decades back, but most effort into DSM over the last four decades has been made by the Japanese and Scandinavians. Initial applications of DSM were in the areas where large developments took place such as major roads and highways, as well as ports and surrounds. Nowadays, the DSM technique is used in a variety of situations, sometimes as small as a warehouse site, commercial buildings, residences, tanks, etc.

The column configuration can be changed to suit the type of development and ground conditions. Such configurations include the following (see Figure 13.2):

1. Single columns;
2. Walls/Panels;
3. Blocks;
4. Lattice (Grid); and
5. Any combination of more than one single column with/without overlapping.

A comparison among various column patterns is presented in Table 13.1 and an appropriate pattern can be selected depending on the purpose.

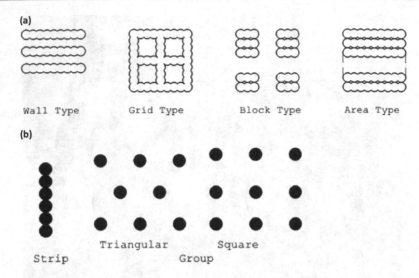

Figure 13.2 (a) Wet mixing patterns (BS EN 14679:2005 (E)); (b) Dry mixing
patterns.

Table 13.1 Comparison of DSM arrangement types (adapted from Kitazume
and Terashi 2013)

Feature	Comparison of features
Block	Large block provides higher resistance to external forces, hence highly stable. Volume of improvement greater, hence higher cost. Longer construction time as columns are overlapped.
Wall/panel	Where all improved walls are linked firmly, high stability is obtained. Volume of improvement is smaller than block type, hence lower cost. Requires precise operation of overlapping of long and short units. Requires consideration of unimproved soil between walls. Improvement size affected by internal stability.
Grid/lattice	More stable than wall type but less stable than block type. Cost range is between block type and wall type. Installation sequences are complicated because grid/lattice shape must be formed. Requires design on three-dimensional internal forces.
Group (either triangular or square pattern)	Where lateral forces are small, high stability is obtained. Installation requires short period, and volume of improvement is small. Low cost. Overlapping operation is not required. Requires design on overall stability and on internal stresses of columns. Generally, not adopted for stability control unless spacing is reduced significantly. For stability purposes, wall/panel types could be used.

Column strengths vary depending on the soil and binder type and, according to Kitazume (2017), column strength achieved could vary from 200 kPa to 2000 kPa in the field for wet method.

13.2 DESIGN ASPECTS

The design procedure outlined here is mainly applicable to ground treated with dry soil mixing columns. However, the general design philosophy can also be appropriately applied to the wet mixing technique.

DSM design requires two important assessments:

- Settlement of the embankment, long term (serviceability limit state) – based on the equal strain assumption that the distribution of load between columns and unimproved soils result in the same compression of columns and unimproved soils.
- Stability of the embankment supported by DSM columns, panels/grids (ultimate limit state) – failure of column itself and overall failure through the columns and untreated ground.

To carry out the above assessments, the strength and stiffness properties of the DSM columns and interaction between columns and surrounding soil need to be well understood. Considering the DSM columns are semi-rigid, the heterogeneous nature of treated material due to mixing process and potential excessive deformation once it reaches its ultimate capacity, the application of excessive loads on the columns may lead to undesirable settlement and stability issues. In order to address this matter, emphasis should be given to two main aspects (a) Settlements due to overstressing of columns, and (b) Stability, as discussed below.

a) Overstressing of columns and its impact on settlement:

DSM columns and the surrounding soils should move together as a treated block. This should be achieved through adopting appropriate spacing between columns and length of columns if they are floating. The load-sharing mechanism between soil and columns needs to be studied carefully to ensure the load transferred to columns is kept within specified limits. The applied load is initially distributed over soil and DSM columns based on the stiffness of the soil and DSM columns, and subsequently part of the load acting on the soil is transferred to the columns during consolidation. If the column is overstressed/yielded, then part of the load carried by the column will be transferred to the soil due to the degradation of column stiffness. Therefore, the consolidation process may take a longer time for yielded condition compared to non-yielded condition. Column yielding typically starts from the top part of the column considering its proximity to the drainage boundary and low confining stress. This may progress through depth with time. As such, yielding is a time-dependent process, leading to excessive deformation and uncertainty in the prediction of treated ground. Therefore, it is very important to ensure the columns are kept in non-yielded conditions.

The maximum design undrained shear strength of DSM columns should be limited to 150 kPa (SGF 1997; EuroSoilStab 2002). In addition, EuroSoilStab (2002) recommends that the maximum stress transferred to columns should be limited to 0.60 to 0.95 of the ultimate strength of columns to reduce the risk of creeping of DSM columns. The selection of appropriate factor is influenced by types of soil and homogeneity of DSM material. For example, the limiting factors (i.e., creep factor) are 0.65, 0.75 and 0.85 for peat and organic soil, clay and clayey silt soil, respectively.

b) **Stability of embankment on treated ground:**

The effectiveness of DSM columns to stabilize an embankment is governed by the mechanisms by which they fail. The columns located under the passive side offer little if any resistance to failure as they will be under tension. Therefore, resistance from columns in the active zone and direct shear zone can be considered in the design. The selection of appropriate strength parameters for these zones will be discussed in the next sections. If the factor of safety of untreated ground is greater than one (1), then the contribution from columns can be considered in the stability assessments, together with stability berms, to achieve a minimum factor of safety required for short term stability such that the DSM columns don't undergo unnecessary stresses, leading to long-term deformation.

If the factor of safety is less than one (1) for existing conditions (i.e., without any ground improvement) and there is no space to install any stability berms, using single columns to stabilize an embankment is ineffective, and it is recommended that panels or grids as shown in Figure 13.3 should be adopted. The extent of these treatments could be limited to active and direct shear zones to produce an economical design. It is important to extend the DSM columns such that the factor of safety for a slip circle going under the treated block meets the minimum factor of safety requirements and any substantial failure through the DSM treated block should meet the minimum factor of safety.

In general, the following calculation checks are recommended as part of the design:

- Factor of safety (FoS) without column stabilization;
- FoS after column installation during fill placement/load application to aid with construction sequence;

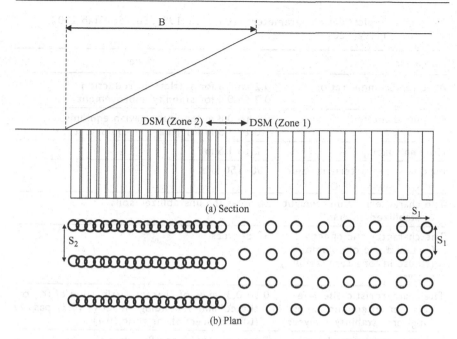

Figure 13.3 Schematic sketch of DSM beneath embankment footprint.

- FoS during the serviceability stage;
- Column yielding; and
- Long-term settlement.

13.2.1 Typical design parameters

Some typical design parameters are presented in Table 13.2.

It should be noted that during installation of columns, mixing in some zones may be substandard and the strength may therefore be lower. In order to capture such behavior in design, the following limitations apply:

- A disturbed zone in the unstabilized soil below each column – the extent of disturbed zone depends on the mixing tool and the column size. Normally, a 0.5 m thick soil layer located below the tip of the column is considered disturbed and reconsolidation could occur in the long term (ref: SGF 1997).
- Reduced strength over the top meter length of the column – in general, the pressure feed of binder should be terminated at about 0.5–1.0 m below ground level to prevent the binder being blown back along the shaft. Depending on the platform thickness, the strength of the crust layer disturbed in such a way may be lower than that of the original crust material.

Table 13.2 Typical design parameters (refer SGF 1997; EuroSoilStab 2002; FHWA 2013)

Parameter	Value
Area replacement ratio	0.2 to 0.3 for settlement reduction 0.2 to 0.4 for stability improvement
Column diameter	0.5 to 0.8 m for Scandinavian equipment 1.0 to 1.5 m for Japanese equipment
Column spacing	1.0 to 1.5 m
Maximum characteristic value of c_u of the column	100–150 kPa
The characteristic unit weight of stabilized column	Same as the unstabilized soils
The characteristic effective friction angle of column (to use in long-term stability analyses)	30 degrees
The characteristic effective cohesion of column (to use in long-term stability analyses)	0 to 0.3 times of characteristic value of (c_u) of the column depending on active (0.3), passive (0) and direct shear zone (0.1)
The characteristic value of Young's modulus	Lime column: 50 to 100 times c_u Organic soil: 50 times c_u Silty clays: 100 times c_u Columns with other binders such as Lime Cement: 50 to 150 times c_u
Drainage/permeability of columns	Columns are usually considered as drains because the permeability of columns is higher than that of the original soil. For example: The permeability of lime stabilized organic soil may be assumed to be approximately 1000 times that of unstabilized clay. The permeability of soil stabilized with other binders (e.g., lime/cement) can be assumed to be 200–600 times that of non-stabilized soil.

13.2.2 Design procedure

As discussed previously, in general, single columns are adopted within the crest of the embankment to control settlement, and panels/grids are adopted under embankment batter to improve stability of the embankment. Spacing between DSM columns under the crest can be selected such that DSM treated ground can be considered as an equivalent mass and meets bearing capacity and settlement requirements. The equivalent mass will then be treated as over consolidated material with equivalent recompression parameters as described in the following sections:

Figure 13.3 shows the zone under the footprint of the embankment improved using DSM.

13.2.3 Settlement assessment

Settlements within the stabilized soil block are influenced by the following factors:

- The ratio of the compression modulus of the columns to that of unstabilized soil;
- The proportion of the stabilized surface occupied by columns, i.e., area replacement ratio;
- The consolidation properties of the soil;
- The bearing capacity of the columns; and
- The time of load application in relation to column installation.

Zone 1 shown in Figure 13.3 represents the area where DSM columns shall be installed as single columns in a grid pattern to carry the maximum embankment load and to limit post-construction settlement to acceptable limits.

Spacing between DSM columns shall be assessed to meet the bearing capacity requirements for a given embankment height. As recommended in Swedish Geotechnical Society Report 4:95E (SGF 1997), the ultimate bearing capacity/strength of a column ($\sigma_{ult, col}$) is given by Eq. (13.1).

$$\sigma_{ult,col} = 2c_{col} + 3\sigma_h \qquad (13.1)$$

where c_{col} (kPa) is the undrained shear strength of the column and σ_h (kPa) is the initial horizontal stress from surrounding soil. A portion of vertical load transferred to soil can be considered in the prediction of the ultimate bearing capacity of the column (typically 0.5 times the vertical load). The maximum axial load taken by the column (design bearing capacity of the column) should be limited to a certain ratio of $\sigma_{ult, col}$. The ratio, known as creep factor μ, depends on the type of soil and some typical values for various soils are summarized below:

- $\mu = 0.85$ for clayey silt soil;
- $\mu = 0.75$ for clay; and
- $\mu = 0.65$ for peat and organic soil.

The load (σ_{col}) taken by the column has been predicted assuming that strain compatibility exists between soil and DSM columns, i.e., the settlement at the column, soil in between the columns and the improved blocks are equal. The load (σ_{col}) taken by the column is calculated by Eq. (13.2).

$$\sigma_{col} = \frac{M_{col}}{M_{eq}} \Delta\sigma \leq \mu\sigma_{ult,col} \qquad (13.2)$$

where M_{col} (kPa) is the constrained modulus of the column; M_{eq}(kPa) is the constrained modulus of the equivalent block and $\Delta\sigma$(kPa) is the applied vertical load. The equivalent constrained modulus of the DSM block can be calculated using Eq. (13.3).

$$M_{eq} = a_r M_{col} + (1 - a_r) M_{soil} \qquad (13.3)$$

where M_{soil} (kPa) is the constrained modulus of soil and a_r is the replacement ratio, which can be calculated using Eq. (13.4).

$$a_r = \frac{r^2}{R_{eq}^2} \qquad (13.4)$$

where r (m) is the radius of the DSM column and R_{eq} (m) is equivalent radius of influence zone/unit cell for each DSM column, which depends on the spacing S and layout pattern.

The equivalent undrained shear strength of DSM and soil equivalent block c_{eq} (kPa) is calculated using Eq. (13.5).

$$c_{eq} = a_r c_{col} + (1 - a_r) c_{soil} \qquad (13.5)$$

where c_{soil} (kPa) is the undrained shear strength of the soil. The equivalent overconsolidation ratio (OCR_{block}) for DSM treated ground, prior to any load applied, can be calculated using Eq. (13.6).

$$OCR_{block} = \left[\frac{c_{eq}}{c_{soil}} \right]^a OCR_{soil} \qquad (13.6)$$

where OCR_{soil} is the initial overconsolidation ratio of each soil layer; a is an empirical constant and can be taken as $a = 1.05$ (inferred from information provided in Ladd and DeGroot 2004). The compression ratio (CR) for each layer can be assessed based on the constrained modulus of the equivalent block (M_{eq}) for the stress range applied and a percentage of the calculated CR can be used as creep strain rate $(C_{ae})_{NC}$ for normally consolidated DSM block. The primary settlement of the treated ground can be assessed using the following equation.

$$S_p = \Sigma \frac{\Delta h . \Delta\sigma}{M_{eq}} \qquad (13.7)$$

where:
Δh = sub-layer thickness

The creep strain rate $(C_{ae})_{OC}$ for overconsolidated DSM block can be calculated using Eq. (5.33) in Chapter 5. The creep settlement (S_{creep}) of each layer treated with DSM can be calculated using Eq. (3.24b) in Chapter 3. The recommencement time for the secondary compression can be calculated using Eq. (8.1) in Chapter 8.

The consolidation behavior of DSM treated soil can be predicted assuming the DSM column acts as a vertical drain as recommended in Swedish Report 4:95E 1 (SGF 1997). A brief method of assessing the degree of consolidation is summarized below:

$$U = 1 - \exp\left[\frac{-2c_h t}{R_{eq}^2 f(n)}\right]$$ (13.8)

where:

U = degree of consolidation

c_h = coefficient of consolidation in untreated clay in the horizontal direction

t = period of consolidation

$$f(n) = \frac{n^2}{n^2-1}\left[\ln(n) - 0.75 + \frac{1}{n^2}\left(1 - \frac{1}{4n^2}\right)\right] + \left[\frac{n^2-1}{n^2}\frac{1}{r^2}\frac{k_{clay}}{k_{col}}L_D^2\right]$$ (13.9)

where:

$$n = \frac{R_{eq}}{r}$$

L_D = drainage path length, i.e., column length with drainage upwards only or half column length with drainage both upwards and downwards

k_{clay} = permeability of unstabilized clay

k_{col} = permeability of DSM column

The above rate of consolidation is approximate only and applies only for the stabilized block. The rate of consolidation below the stabilized zone should be performed in the normal way.

13.2.4 Stability assessment

Zone 2 shown in Figure 13.3 represents the area beneath the batter of the embankment, where DSM columns should be installed in panels/walls, grids or blocks to provide the necessary stiffness and shear resistance against lateral deformation and instability. The effectiveness of treatment on stabilizing the ground is minimal when it is under tensile loads. Therefore, the width of the treated zone should be selected based on the extent of columns experiencing compression and direct shear zone as discussed previously.

13.2.5 Numerical modelling – Useful tips

DSM columns can be modelled using commercially available computer programs to understand the load distribution between columns and soils. Columns can be modelled in two-dimensional or three-dimensional computer programs such as PLAXIS 2D and 3D using the Mohr-Coulomb soil model. An idealized equivalent strip with equivalent strength, permeability and stiffness parameters should be adopted in two-dimensional modelling. Soil between columns should be modelled as appropriate. For example, if the existing soil is soft, soft soil model can be used.

13.2.6 Case studies

CASE HISTORY 13.1 (AMERATUNGA ET AL. 2009)

A large storage tank was to be built on a site where the subsoil conditions were reasonably competent but not competent enough to be founded on shallow foundations to meet the tight differential settlement criteria specified. Due to time constraints, the client initially proposed 15 m long driven piles but later agreed to a DSM solution with cutter soil mix (CSM) columns.

The design was carried out using the steps discussed in Ameratunga et al. (2009). Two-dimensional finite element analysis was the main tool used to assess differential settlement. The soil and rock materials were modelled as Mohr-Coulomb material and the CSM columns as a linear elastic material. The analysis was carried out using an axisymmetric model (i.e., 2D), which was set up for the circular tank structure with a uniform radial cross section and loading scheme around the central axis, where the deformation and stress state are assumed to be identical in any radial direction. As the CSM columns were not radially placed (see Figure 13.4), this layout was transformed to axisymmetric using an equivalent area approach. The tank CSM foundation was divided into nine zones radially (see Figure 13.4) and the CSM columns and caps were modelled as equivalent 'doughnuts' of stiff material in the axisymmetric analysis. The width of each 'doughnut' was assessed using a simplified approach, i.e., averaging the area of CSM at a particular radius.

An innovative change was the CSM 'pile cap' that was introduced to facilitate the distribution of the vertical load from the structure. The 'pile cap' was pre-dug at each column location to contain this overflow of binder (see Figure 13.5) and allow no waste material disposed off site. Later, this 'pile cap' was re-mixed with a paddle mixer mounted on an excavator.

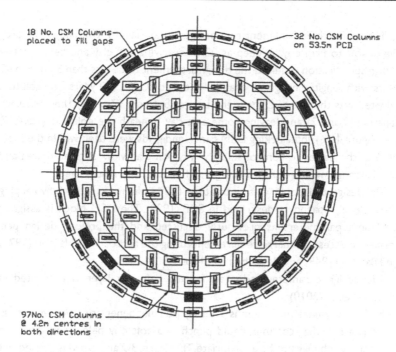

18 No. CSM Columns placed to fill gaps

32 No. CSM Columns on 53.5m PCD

97No. CSM Columns @ 4.2m centres in both directions

Figure 13.4 CSM column layout (after Ameratunga et al. 2009).

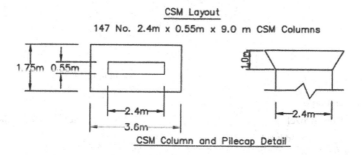

CSM Layout

147 No. 2.4m x 0.55m x 9.0 m CSM Columns

1.75m 0.55m

2.4m

3.6m

1.0m

2.4m

CSM Column and Pilecap Detail

Figure 13.5 Details of CSM and CSM cap.

CASE HISTORY 13.2 (MUTTUVEL ET AL. 2010)

This case history summarizes a DSM project at a bridge approach where the design was based on the equivalent block approach rather than individual columns. The column spacing and length have been selected such that treated ground behave as a block.

The approach embankment to a bridge is underlain by a variably thick profile of soft to firm clay, as deep as 22 m, overlying stiff clay and bedrock. The design specification required the design settlement to be less than 50 mm in 40 years with longitudinal grade change limited to 0.5% in 10 years. The solution adopted was the use of DSM at the bridge abutment (Zone 1), then gradually tapering (Zone 2) and leaving some soft clay beneath the columns (Zone 3) (see Figure 13.6). Interlocking DSM panels and grids were used within the footprint of the batter to control stability. A typical plan view of treated area and panel arrangement are shown in Figures 13.7 and 13.8, respectively.

The design philosophy was to treat the DSM columns as a block by keeping them closely spaced as described in SGF (1997). The DSM block is assigned equivalent parameters as previously discussed and the preload design performed to assess surcharge requirements using the approach by Hansbo (1979) and Barron (1948) and described in Chapter 9.

The design parameters adopted for the soft clay layer is presented in Muttuvel et al. (2010).

In Zone 2 where the columns are only partially penetrating the weak soils, there is a risk that columns could punch into soft clay below and the DSM block approach may not be appropriate. Therefore, 3D analysis was carried out using PLAXIS 3D (2007).

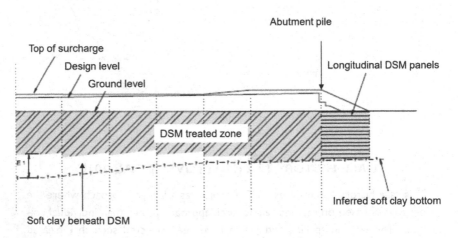

Figure 13.6 DSM zones (adapted from Muttuvel et al. 2010).

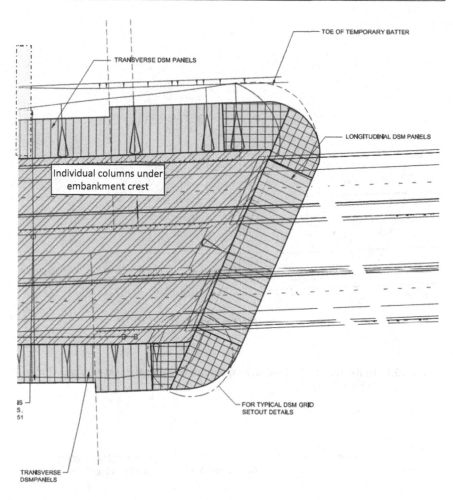

TOE OF TEMPORARY BATTER

TRANSVERSE DSM PANELS

LONGITUDINAL DSM PANELS

Individual columns under embankment crest

B-B

IS
S.
51

FOR TYPICAL DSM GRID
SETOUT DETAILS

TRANSVERSE
DSMPANELS

Figure 13.7 Plan view of treated area (typical).

In Zone 1 where the embankment height was 3.6 m to 4.5 m, the predicted settlement was 115 mm at the end of primary under a fill load of 4.5 m, which was close in agreement with settlement measurements during construction.

In Zone 2, the results of Plaxis 3D indicated that settlement is linear with applied load up to 4 m of fill but slight yielding occurs when the fill height is further increased by 1 m as shown in Figure 13.9. The minor punching does not affect the overall settlement and therefore DSM treatment. This allows the conclusion that the equivalent block approach could also be used for the partially penetrating zone in this instance.

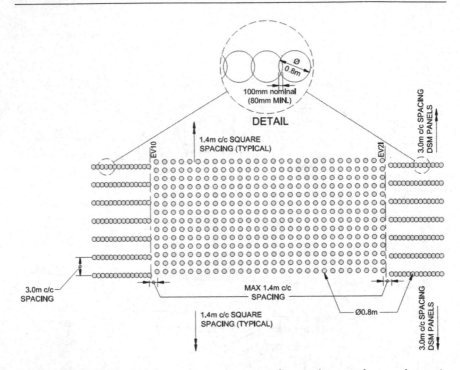

Figure 13.8 Individual and panel arrangement of typical treated area shown in Figure 13.7.

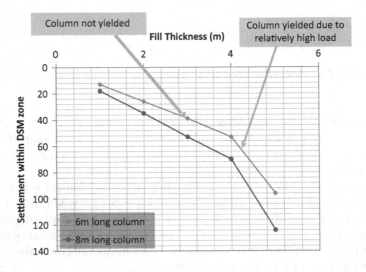

Figure 13.9 Behavior of tapered columns under varying load for selected spacing.

13.3 ENVIRONMENTAL CONSIDERATIONS

Generally, DSMs are more environmentally friendly than some other types of ground improvement. This is because additives are generally kept within the ground and/or contained within site. As existing soil is used as part of the DSM operation, there is a significant reduction in bringing in or disposal of borrow materials. The latter reduces the carbon footprint significantly because of the reduction in transport activities and also reduces treatment requirements if the soil is acidic.

One of the major advantages for a site located in an urban area is the reduction in traffic control and inconvenience to commuters, residents and business owners in the vicinity. Vibration is also minimal, i.e., impact to nearby buildings/structures and utilities is therefore reduced.

It can be considered as environmentally friendly as non-toxic binders are used and there is minimal overflow, which can be controlled.

13.4 CONSTRUCTION ISSUES

Typical construction issues related to the use of heavy machinery and drilling through soils are equally valid for DSMs. Some of these include:

- As the equipment is large and heavy, a properly designed working platform may be needed on land projects.
- If obstructions are encountered such as boulders, DSMs would not be able to penetrate. If the obstructions are shallow, extraction and removal of such obstacles is possible with replacement by sand or similar materials.
- Noise and vibration are generally not major issues at DSM sites.
- There will always be overflow from DSM operations (although much less than for other types of ground improvement) and measures need to planned and implemented to contain such overflow which can be a major issue in acid sulfate soils or contaminated land.
- One of the main construction issues related to DSM is quality control (QC) and quality assurance (QA). There are several documents dealing with QA/QC and the reader is referred to the following:

  - FHWA (2013)
  - Kitazume and Terashi (2013).

  SHRP2 (2014) lists components of QC/QA monitoring programs as reproduced in Tables 13.3–13.5.

Advantages and disadvantages associated with DSM are presented in Table 13.6.

Table 13.3 Typical existing QC/QA procedures and measurement items (adapted from SHRP2 2014)

QC or QA	Material or process	Items
QC	Material related	• Procedures: Sub soil investigations, boreholes and CPT, DMT, pressuremeter, plate load testing, etc. and collection of disturbed and undisturbed samples • Measure strength and modulus
QC	Material related	• Procedures: Laboratory mix trials, unit weight of slurry, field test columns, wet grab sampling, strength testing, hydraulic conductivity testing • Measure strength, permeability
QC	Process control	• Procedures: Installation monitoring, surveying • Measure: Blade rotation rate, penetration/withdrawal rate, binder mix quantity, verticality, injection pressure, geometry of improved soil
QA	Material related	• Procedures: Core sampling, wet grab sampling, penetration/pullout test, extraction/excavation, geophysical testing • Measure degree of mixing/uniformity, strength, permeability

Table 13.4 Performance criteria used in QC/QA monitoring programs (after SHRP2 2014)

Topic	Items
Material parameters	Strength parameters, bearing capacity, stiffness parameters, permeability
System behavior	Settlement, lateral deformation, structural failure of columns

Table 13.5 Emerging QC/QA procedures and measurement items (after SHRP2 2014)

Topic	Items
Material related	Latest applications of geophysics
Process control	None noted

Table 13.6 Advantages and disadvantages of DSM

Factor	Advantages/disadvantages
Method	Disadvantage – Considered simple in the design, especially at feasibility stages, but selection of parameters ultimately achievable in the field is difficult, particularly with dry soil mixing.
Spoil	Advantage – Very little spoil is generated in the installation process except slurry overflow, which could be contained.
Vibration	Advantage – Very little vibration generated and therefore could be easily used in built up areas.
Speed	Advantage – Speed in construction is relatively high in comparison to other methods such as stone columns.
Time	Advantage – DSMs are usually designed in combination with preloading with or without vertical drains particularly with tapered columns. For full depth only nominal surcharge is required. Columns act as conduits for pore pressure dissipation. Need curing time before loading.
Complexity of analysis	Advantage – Analysis is generally considered not complex.
Cost	Less expensive compared to some other ground improvement techniques such as piled embankments but more expensive than preloading.
Preload height	Advantage – Generally, the height of preload could be reduced when DSMs are used if available time for preloading is limited.
Edge effects	Disadvantage – Could create differential settlement issues adjoining a non-DSM area and therefore transition should be taken into account in the design.
Stability	Advantage – Improvement in strength allows faster construction and/or less high strength geotextiles to manage stability.
Remedial measures if DSMs not properly installed or inadequate	Advantage – Additional DSM columns could be installed if necessary. As discussed in Chapter 9, the same outcome cannot be achieved for wick drains.
Depth of improvement	The current state of art limits the depth of installation of DSM/depth of improvement to about 45 m (CDMA 2015).
Working Platform	Advantage – Generally, the equipment used is light and therefore working platform requirements are less onerous compared to piled embankments.
Homogeneity	Disadvantage – Cannot expect uniformity in improvement because of variabilities involved. However, as the behavior of DSM is generally as a block, adverse effects are generally low. Advantage – Can treat layered soil.

(Continued)

Table 13.6 (Continued)

Factor	Advantages/disadvantages
Obstructions	Disadvantage – Cannot procced if large obstructions such as boulders occur.
Soil characteristics	Disadvantage – Full strength may not be gained if acidic conditions encountered.
QA/QC	Disadvantage – Widely recognized quality assurance program not available (SHRP2 2014).

13.5 DESIGN EXAMPLE

13.5.1 Design example 13.1

In general, the following steps could be adopted to carry out DSM design, mainly for the embankment type situation. For other buildings or structures, similar reasoning could be adopted.

- Step 1: Establish project requirements;
- Step 2: Establish geotechnical model for the existing site conditions;
- Step 3: Establish trial DSM property values;
- Step 4: Establish DSM geometry including general layout, and center and shear wall panel replacement ratios considering settlement and stability requirements

Step 1: Establish project requirements

Project requirements are typically defined by the asset owner as part of project specification so that the design needs to meet such requirements. These requirements are governed by many factors including post construction settlement and minimum safety factor in the short and long term.

A proposed highway is underlain by a 12.5 m thick deposit of soft clay, overlying a layer of dense sand. The ground water table is 0.5 m below the existing ground level. As the preliminary analysis indicated that the factor of safety against instability and because the predicted settlement significantly exceeds the project criteria, DSM was adopted as the ground improvement technique for this section of the highway alignment (Figure 13.10).

Step 2: Establish geotechnical model for the existing site conditions

Adopted geotechnical models are shown in Table 13.7.

Step 3: Establish trial DSM property values

Design parameters including strength to meet design requirements shall be developed in consultation with DSM contractors considering the material type and their experience with similar materials and any laboratory and

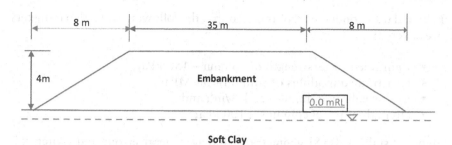

Figure 13.10 Embankment geometry.

Table 13.7 Geotechnical design parameters

Sublayer	1	2	3	4	5	6	7
RL at the bottom of layer (m)	−0.5	−1	−3	−4.5	−7	−12.5	−16.5
Average depth (m)	0.25	0.75	2	3.75	5.75	9.75	14.5
Sublayer thickness (m)	0.5	0.5	2	1.5	2.5	5.5	4
Geotechnical Unit	2b	2b	2a	2a	2a	2a	5b
Bulk unit weight (kN/m³)	14.5	14.5	14.5	14.5	14.5	14.5	16.5
OCR	100.0	13.6	4.0	3.0	2.2	1.6	−
Total stress (kPa)	3.6	10.9	29.0	54.4	83.4	141.4	214.3
Effective stress, σ' (kPa)	3.63	8.42	14.29	22.49	31.87	50.63	76.91
Preconsolidation stress (kPa)	362.5	114.5	57.1	67.5	70.1	81.0	−
$C_c/(1+e_0)$	0.35	0.35	0.35	0.35	0.35	0.35	−
$C_r/(1+e_0)$	0.05	0.05	0.05	0.05	0.05	0.05	−
$C_{a\varepsilon}$	0.015	0.015	0.015	0.015	0.015	0.015	−
c_u (kPa)	30	15	10	12	13	16	60
c_h (m²/yr)	2.5	2.5	2.5	2.5	2.5	2.5	−

(C_c = compression index; C_r = recompression index; $C_{a\varepsilon}$= creep strain rate; c_u = undrained shear strength; c_h = coefficient of consolidation in horizontal direction)
(Note: A unit weight of 20 kN/m³ can be adopted for embankment fill.)

field trial tests conducted. For this example, the following strength parameters are selected:

- Undrained shear strength of column – 150 kPa;
- Constrained modulus of column – 30 MPa;
- Unit weight of column – 20 kN/m³; and
- Strength of panel columns – 100 kPa;

Step 4: Establish DSM geometry including general layout and center columns, and panel area replacement ratios

Stress transferred to soil and columns can be assessed using equations provided in this chapter. The area replacement ratio of DSM columns located under the embankment center can be selected such that columns do not yield under an embankment height of 4 m, adopting Eq. (13.2). A creep factor of 0.75 can be adopted for the clay. In the assessment of ultimate stress of columns, the following can be adopted in the assessment of horizontal stress (SGF 1997):

- Initial horizontal stress: Equal to vertical stress; and
- Increase in horizontal stress due to load transferred to soil as 50% of imposed load in soil.

Based on the above, the minimum area replacement ratio to reduce the risk of column yielding is approximately 0.296. In order to achieve this area replacement ratio, the following arrangement can be adopted:

- Column diameter: 800 mm;
- Columns are arranged at 1.4 m spacing in a triangular pattern.

Adopting the assessed area replacement ratio, the equivalent undrained shear strength, compression parameters and OCR can be assessed for each layer as discussed previously. The assessed settlement during construction is about 100 mm and post-construction settlement is less than 5 mm.

With this arrangement within the center of embankment, the DSM panel within the batter (refer to Figure 13.3) has been considered to achieve a short-term and long-term factor of safety of 1.2 and 1.5, respectively. The slope stability assessment can be carried out using computer program Slope/W. To achieve a short factor of safety of 1.2, the following DSM panel arrangement is required:

- Panel spacing is 2.5 m;
- Column overlapping within panel is 75 mm;
- Column diameter is 800 mm.

The long-term factor of safety for the adopted DSM arrangement under the center of embankment and DSM panels within the batter as discussed above is 1.5.

13.6 SUMMARY

This chapter introduces deep soil mixing and provides a comprehensive coverage of DSM design. Two case histories have been included to emphasize the popularity of DSM as a ground improvement method for soft soil. Advantages and disadvantages in comparison to other ground improvement techniques have been listed. Special attention has been given to quality control and quality assurance aspects, which are very important in delivering DSM projects. A detailed step-by-step procedure to design is provided including an example.

REFERENCES

Ameratunga, J., Brown, D., Ramachandran, M. and Denny, R. (2009). 'Ground improvement for a large above ground storage tank using cutter soil mix columns.' *Proc 17th Intl Conf. on Soil Mechanics and Geotechnical Engineering*, Alexandria, Ed. M. Hamza, M. Shahien and Y. El-Mossallamy, IOS Press, 2280–2283.

Barron, R.A. (1948). 'Consolidation of fine-grained soils by drain wells.' *Transaction of the ASCE*, 113(1), 718–742.

BS EN 14679 (2005). *Execution of Special Geotechnical Works – Deep Mixing*. 56p.

CDMA (Cement Deep Mixing Association) (2015). *Cement Deep Mixing*. www.cdm-gr.com/books/pdf/cdm_en.pdf. Last visited August 2019.

EuroSoilStab (2002). 'Development of design and construction methods to stabilise soft organic soils.' Design Guide Soft Soil Stabilisation, CT97-0351, Project No. BE 96-3177, 95p. Ministry of Transport Public Works and Management.

FHWA (2013). *Federal Highway Administration Manual: Deep Mixing for Embankment and Foundation Support*, Publication no. FHWA-HRT-13-046.

Fiorotto, R., Schöpf, M. and Stötzer, E. (2005). '*Cutter Soil Mixing (C.S.M.) Figure 10.17 Stone column treated soft soil supporting fill embankment. An innovation in soil mixing for creating cut-off and retaining walls.*' Proceedings of the 16th International Conference on Soil Mechanics and Geotechnical Engineering, Millpress Science Publishers/IOS Press, Osaka, Japan.

Hansbo, S. (1979). 'Consolidation of clay by band-shaped prefabricated drains.' *Ground Engineering*, 5, 16–25.

Kitazume, M. (2017). 'Deep mixing method, the Japanese experience and recent advancement.' *Advance in Concrete Technology* by Hong Kong Concrete Institute.

Kitazume, M. and Terashi, M. (2013). *The Deep Mixing Method*. CRC Press, New York.

Ladd, C.C. and DeGroot, D.J. (2004). '*Recommended practice for soft ground characterization: Arthur Casagrande lecture.*' *Proceedings 12th Panamerican Conference on Soil Mechanics and Geotechnical Engineering,* June 2003, paper revised 9 May 2004.

Muttuvel, T., Kelly, R. and Chan, K. (2010). '*Transition design of bridge approach embankment over soft ground using dry soil mixing technique.*' *Proceedings of 7th Ground Improvement Techniques GI10 Seoul,* South Korea, 237–244.

PLAXIS 2D (2019). *Reference Manual.* Bentley Systems, Incorporated, US.

PLAXIS 3D Foundation (2007). *User Manual,* Version 2.2, Delft, Netherlands.

SGF (1997). SGF Report 4:95E. *Lime and Lime Cement Columns, Guide for Project Planning, Construction and Inspection.* Swedish Geotechnical Society.

SHRP2 (2014). *Geotech Tools: Geo-construction Information & Technology Selection Guidance for Geotechnical, Structural, & Pavement Engineers.* http://geotechtools.org/

Chapter 14

Basal high strength geotextiles

C. Lawson

14.1 INTRODUCTION

Soft foundation soils suffer from two fundamental problems when fill is placed on top; low shear resistance and high compressibility. For the construction of embankments and placement of fills, several techniques are used to enhance stability and control settlements arising from the presence of soft foundation soils.

Figure 14.1 shows embankment construction techniques commonly employed to enhance stability over soft foundation soils. These are stage construction, berm construction and basal reinforced construction.

Stage construction (Figure 14.1a) is an historical, conventional technique where thin layers of embankment fill are placed and then left in-place until the soft foundation has consolidated and can support the subsequent layers. This process continues until the required embankment height is finally reached. The advantage of this technique is that it is relatively low cost, but its major disadvantage is that it can take a long period of time to complete – one to two years in many cases. With today's emphasis on speed of construction, this technique is now rarely used.

Berm construction (Figure 14.1b) is also an historical technique where earth berms are constructed on the sides of the embankment to spread the vertical loads by increasing its base area footprint. The advantage of this technique is that stability is enhanced and thus embankments can be constructed quicker; however, the disadvantage is that the technique takes up extra land area, which could be expensive to acquire.

Basal reinforced construction (Figure 14.1c) utilizes basal high strength geotextile reinforcement to provide additional stability to the soft foundation. The strength of the reinforcement at relatively low strain improves the effective bearing capacity of the soft foundation soil compared to where no basal reinforcement is used. The advantage of this technique is that it enables the embankment to be constructed higher and/or with steeper side slopes than would be the case with no basal reinforcement. (It should be noted that a geotextile separator is used for the cases in Figures 14.1a and 14.1b whereas geotextile reinforcement is used for the case in Figure 14.1c.)

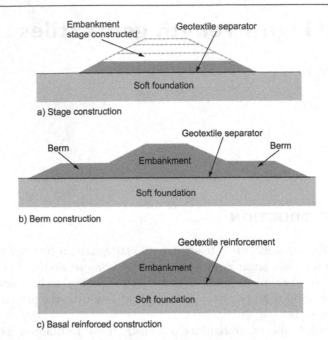

a) Stage construction

b) Berm construction

c) Basal reinforced construction

Figure 14.1 Embankment construction to enhance stability.

Figure 14.2 shows embankment construction techniques commonly employed to control settlements over soft foundations soils. These are to construct the embankment using preloading and wait for settlements to occur, use wick drains to accelerate settlements, and use columns or piles to prevent settlements from occurring. These techniques can all utilize basal high strength geotextile reinforcement for stability purposes. An additional technique, not discussed here, involves the excavation of the soft foundation and its replacement with good quality fill. This technique is discussed in Chapter 7.

The application of a preload to accelerate the rate of consolidation of soft foundations (Figure 14.2a) is a common technique to reduce post-construction settlements. It involves the placement of an additional fill height to precompress the soft foundation, and this fill is then removed at a later point in time before the surface features are constructed on top of the embankment. Preloading is discussed in detail in Chapter 8. In this application basal geotextile reinforcement is used to provide short-term stability to the preloaded embankment.

Ground improvement involving the application of wick drains can be used to accelerate the rate of consolidation of soft foundations (Figure 14.2b). Here, the spacing between the wick drains and the thickness and hydraulic conductivity of the soft foundation all affect the resulting rate of consolidation. Normally, it is planned for most of the settlement to occur

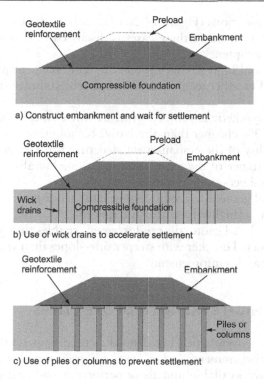

a) Construct embankment and wait for settlement

b) Use of wick drains to accelerate settlement

c) Use of piles or columns to prevent settlement

Figure 14.2 Embankment construction to control settlements.

during the construction phase of the project so that later (post-project) settlements are minimized. The use of wick drains is discussed in detail in Chapter 9. It is common to include basal geotextile reinforcement for this application where the reinforcement provides short-term stability for the embankment during the consolidation period. To further accelerate consolidation, preloading can also be included.

Ground improvement using piles or columns can be used to prevent or minimize settlements from occurring beneath embankment fills (Figure 14.2c). This technique is also discussed in Chapters 10 and 11 from the perspective of the different piles and columns used. Basal geotextile reinforcement is commonly placed across the tops of the columns or pile caps in order to enable the embankment fill to effectively arch between the adjacent piles and columns. This reinforced embankment structure is normally referred to as a 'basal reinforced piled embankment' and the piles in combination with the basal geotextile reinforcement and an initial granular fill layer as a 'load transfer platform'.

The cases that utilize basal high strength geotextile reinforcement are where it is used alone to enhance embankment stability (Figures 14.1c and 14.2a), where it is used in conjunction with wick drains to enhance embankment stability (Figure 14.2b) and where it is used in conjunction with pile and column foundations (Figure 14.2c). The role of the basal high strength geotextile reinforcement in these cases is discussed in detail in subsequent sections of this chapter.

The use of basal high strength geotextile reinforcement provides advantages compared to other techniques. These maybe summarized as follows.

1. Basal geosynthetic reinforced structures are cost effective. They can be 30% to 50% cheaper than alternative techniques.
2. The stability of the embankment structure is no longer based on the shear resistance of the foundation soil alone but also on the presence of the basal geotextile reinforcement.
3. With basal geotextile reinforcement land acquisition can be kept to a minimum (a smaller base area is required).
4. The use of basal geotextile reinforcement enables the embankments to be constructed quicker with steeper side-slopes than would be the case without basal reinforcement.

14.2 BASAL HIGH STRENGTH GEOTEXTILES

Geotextiles (or geosynthetics) are defined as textile (or synthetic) materials used in a soil environment. Today, many different types of geotextiles and geosynthetics are available and these perform a wide range of functions when included in soils. Koerner (2012) and Shukla (2011) are good textbooks on geotextiles and geosynthetics in general. While different geosynthetics behave differently in soil environments, it is the ability of the high strength geotextiles (woven geotextiles and geogrids) to carry tensile loads at defined strains over long periods of time that make them ideal as soil reinforcement materials.

While geotextile reinforcements must carry tensile loads at defined strains over different periods of time (depending on required design life), they must also form an efficient bond with the surrounding soil and have good durability. For basal reinforcement beneath embankment fills the tensile loads generated in the geotextile reinforcement can be relatively high requiring the use of high strength geotextiles. Almost exclusively, these high strength geotextiles comprise high modulus polyester (PET) yarn as their load-carrying component because of its high strength, high-tensile stiffness and good durability. Figure 14.3 shows the two types of basal high strength geotextile reinforcement that are used beneath embankment fills, woven polyester geotextiles (Figure 14.3a) and polyester geogrids (Figure 14.3b).

It is common for high strength geotextiles to have higher strengths in their longitudinal direction (sometimes termed 'machine direction') than in their cross direction (sometimes termed 'cross-machine direction'). Where this occurs, these types of reinforcements are normally termed 'unidirectional reinforcements' as the reinforcement function essentially acts in one

a) Woven polyester geotextile   b) Polyester geogrid

Figure 14.3 Basal high strength geotextile and geogrid reinforcements.

Table 14.1 Summary of initial ultimate properties of basal high strength geotextile and geogrid reinforcements.

	Woven polyester geotextiles	Polyester geogrids
Initial ultimate tensile strength - longitudinal direction	100–2000 kN/m	100–800 kN/m
Strain at maximum load - longitudinal direction	10–12%	10–15%
Bond coefficient	0.8–0.80	0.8–0.95

direction only. In practice, this provides the most efficient form of reinforcement.

Table 14.1 provides a summary of the mechanical properties of basal high strength geotextile and geogrid reinforcements for comparative purposes. The initial ultimate tensile strengths of woven polyester geotextiles cover a wide range from 100 kN/m to 2,000 kN/m, while those for polyester geogrids can range from 100 kN/m to 800 kN/m. Strain at maximum tensile load ranges from 10% to 15% for both types. The bond coefficient (how efficient the reinforcement bond is with the surrounding soil) is high for these materials in the range 0.8 to 0.95 for both types of reinforcement depending on the type of surrounding soil.

While Table 14.1 demonstrates that high strength geotextile and geogrid reinforcements provide a wide range of available initial ultimate tensile strengths, from a reinforced soil design perspective, what is important is the allowable long-term design strength of the geotextile reinforcement and this is not the same as the initial ultimate tensile strengths listed in Table 14.1. An internationally recognized procedure exists to convert the characteristic initial ultimate tensile strength of the geotextile reinforcement to its allowable long-term design strength, and vice versa, by accounting for the various processes that cause loss of tensile strength over time. This procedure is shown graphically in Figure 14.4. The procedure accounts for different reinforcement design lives, reinforcement weathering prior to covering with fill, reinforcement installation in different fill types, changes in reinforcement

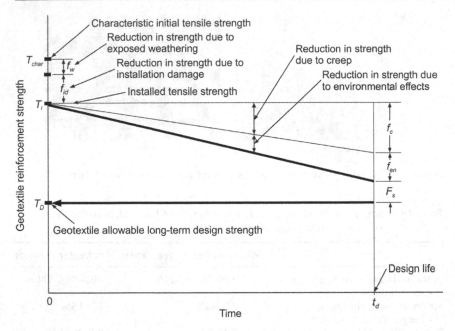

Figure 14.4 Standard procedure for determining the allowable long-term design strengths of geotextile and geogrid reinforcements.

strength associated with creep and environmental effects over time as well as other safety factors that also may be applied to the geotextile reinforcement. In general, for basal reinforced embankments, the allowable long-term design strength of the reinforcement will be typically between 30% to 60% of the characteristic initial ultimate strength of the geotextile reinforcement, depending on the required design life.

Most recognized international design codes, e.g., BS8006-1:2010, represent the procedure shown in Figure 14.4 as a relationship as follows:

$$T_D = \frac{T_{char}}{\left(f_w f_{id} f_c f_{en}\right) F_s} \qquad (14.1)$$

where the variables are shown in Figure 14.4.

Figure 14.4 and Eq. (14.1) show the change in reinforcement strength over time related to five factors; four of these are related to the geotextile reinforcement directly (f_w, f_{id}, f_c, and f_{en}) and the fifth (F_s) is an additional safety factor that may be applied depending on the type of the design procedure used and the quality of the documentation related to the four reinforcement factors.

The reduction factor due to exposed weathering, f_w, accounts for any loss of strength the reinforcement undergoes prior to it being covered over by fill. This normally relates to how the reinforcement is stored on site and the time it takes to cover the reinforcement with fill after it is rolled out. Geotextiles can be subject to the detrimental effects of exposure to ultraviolet (UV) light with the results varying according to the type of geotextile, the level of UV intensity and the exposure time. Woven polyester geotextiles are prone to exposure to UV light and thus should be covered over as soon as possible after rolling out on site. In high-intensity UV climates typically f_w = 1.05 to 1.1 for exposure periods up to 1 month for woven polyester geotextiles. For polyester geogrids that have a PVC protective coating, these are immune to the effects of short-term UV exposure.

The reduction factor due to installation damage, f_{id}, accounts for any loss of strength the reinforcement undergoes when fill is placed and compacted against it. The amount of damage caused is related to the type of geotextile reinforcement being used, the grading of the fill material, the level of compaction and the conditions on either side of the reinforcement. The effect of installation damage is not time-related (like creep and environment) but occurs immediately on installing the geotextile reinforcement. For woven polyester geotextiles f_{id} can range from 1.1 to 1.6 depending on the specific conditions, while for polyester geogrids f_{id} can range from 1.1 to 1.3. Because of this large variation in the possible value of f_{id}, it is important that the assessment of reinforcement behavior is based upon the results of actual installation damage testing of the specific reinforcement under consideration.

The reduction factor due to creep, f_c, accounts for any loss in strength or increase in strain the reinforcement undergoes over its design life. The effect of creep varies with time as shown in Figure 14.4 (the longer the time, the greater the creep). As well as being related to time, creep is also related to temperature. For ground conditions in tropical and equatorial regions ambient ground temperatures approximate 20°C and thus creep testing is normally carried out at this 'standard' temperature. This test temperature also provides a conservative temperature environment for creep testing for temperate environments. Typically, high modulus polyester reinforcements exhibit f_c = 1.35 to 1.5 at 10 years design life and f_c = 1.45 to 1.9 at 100 years design life. Because of the large variation in values and the critical role creep plays in the assessment of reinforcement design strength, quality independent testing of geotextile reinforcements are required.

The reduction factor due to environmental effects, f_{en}, accounts for any loss in strength the reinforcement undergoes over its design life while it is installed in its soil environment. The effect of the soil environment also varies with time as shown in Figure 14.4. The vast majority of soil environments are relatively benign while geotextile reinforcements are environmentally stable over time. For high modulus polyester geotextile reinforcements f_{en} = 1.0 for 10 years or less design life, with f_{en} = 1.1 for 50 years design life. Good quality independent testing over extended time periods may show better results than that quoted here.

The factor of safety, F_s, shown in Figure 14.4 and Eq. (14.1), is applied according to the specific type of design procedure being used. For design procedures based on limit state principles, such as BS8006-1:2010, $F_s = 1.0$ as it is accounted for in the application of various load and material factors in the design procedure. For working stress and the global factor of safety design procedures, which are common in the USA, $F_s = 1.3$ to 1.5 depending on the failure mode being evaluated.

It is crucial, that to make a quality assessment of the long-term design strength of the geotextile reinforcement quality documentation must exist regarding the effects of weathering, installation damage, creep and soil environment. This documentation must also support the time-related nature of the reinforcement behavior.

14.3 BASAL HIGH STRENGTH GEOTEXTILES USED TO ENHANCE EMBANKMENT STABILITY

14.3.1 Design

Figure 14.1c shows the technique where basal high strength geotextile reinforcement is used to enhance the stability of embankments constructed over soft foundations. The basic mechanics of this technique are one of improved bearing capacity when the basal reinforcement is used. Figure 14.5 shows the comparison between an unreinforced embankment and a basal reinforced embankment in terms of a bearing capacity approach. For the unreinforced embankment case (shown on the left-hand side of Figure 14.5), the soft foundation is not only subject to the vertical loading of the embankment fill but also outward horizontal shear stresses beneath the side slope of the embankment. In terms of the magnitude of the bearing capacity factor to be used for the foundation layer, N_c, this specific combination of stresses results in a reduced bearing capacity factor depending on the magnitude of the horizontal outward shear stresses. For the basal reinforced embankment case (shown on the right-hand side of Figure 14.5), the soft foundation is subject to the vertical loading of the embankment fill, but also there is an inward horizontal shear stress on the surface of the soft foundation due to the presence of the basal reinforcement. Now, in terms of the magnitude of the bearing capacity factor to be used for the foundation layer, N_c, this specific combination of stresses results in an increased bearing capacity factor depending on the magnitude of the horizontal inward shear stresses. Comparison between these two cases demonstrates the improvement in effective bearing capacity by using basal reinforcement.

In practice, this improvement in effective bearing capacity by using basal reinforcement normally results in the embankment being constructed higher and/or with steeper side slopes than would be the case where no basal reinforcement is used. While the use of basal reinforcement provides these two

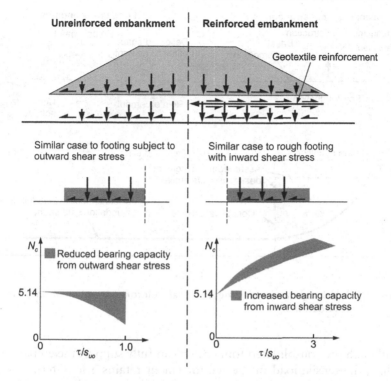

Figure 14.5 Basic mechanics of basal reinforced embankments on soft foundations (after Jewell 1996 – s_{uo} denotes undrained shear strength near
soft foundation surface).

performance benefits, close attention is still required to the rate of application of the embankment fill to maintain stability during construction.
Chapters 8 and 9 provide guidance on the appropriate rate of fill application
on soft clay foundations.

The tensile load generated in the basal reinforcement is not constant, but
changes over time, as shown in Figure 14.6. During embankment construction the load in the basal reinforcement increases until it reaches a maximum T_r when the embankment has reached its maximum height. Following
this, the soft foundation consolidates under the loading of the embankment
fill and gains in shear strength. As it gains in shear strength, the soft foundation can support more and more of the embankment loading and thus the
tensile load in the reinforcement reduces. This process continues until the
soft foundation has fully consolidated, over time t_d, and can fully support
the embankment loading without need for the basal reinforcement. Thus,
the active reinforcement is required only for the time period t_d and this is
commonly referred to as the 'short-term condition,' but in practice it may
take up to ten years for the soft foundation to consolidate.

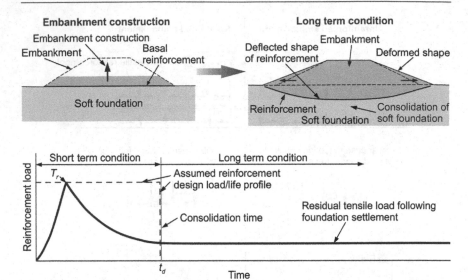

Figure 14.6 Load regime over time in basal reinforced embankments on soft foundations.

Although the consolidated foundation can fully support the embankment loading, the tensile load in the reinforcement retains a long-term residual value due to the deflection the reinforcement undergoes during foundation consolidation. This residual load remains relatively constant with time during the remainder of the design life of the embankment (commonly termed the 'long-term condition'). This residual tensile load is typically 10% to 20% of the maximum tensile load T_r.

From a basal reinforcement design perspective, the two parameters of importance are the maximum tensile load generated in the reinforcement T_r and the time period it takes for the soft foundation to consolidate and fully support the embankment loading t_d. While it may be difficult to accurately determine the actual variation in reinforcement tensile load over this short-term condition, to simplify the design approach it is common to conservatively assume that the load remains constant at T_r over the construction and consolidation period t_d (the dashed line shown in Figure 14.6).

Recognized international design codes such as BS8006-1:2010 relate the reinforcement long-term design strength T_D to the maximum generated tensile load T_r as follows:

$$T_D = f_n T_r \qquad (14.2)$$

where f_n is a factor normally referred to as a 'consequence of failure' factor and is related to the critical nature of the reinforced structure. Most current design codes use $f_n = 1.1$ for basal-reinforced embankment applications.

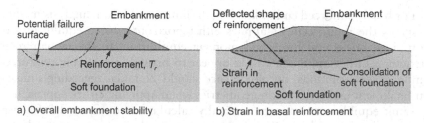

a) Overall embankment stability
b) Strain in basal reinforcement

Figure 14.7 Design limit states governing overall embankment stability and strain in basal reinforcement.

To design the basal reinforcement in the embankment there are two limit states that must be satisfied. These are shown in Figure 14.7 where the maximum tensile load in the reinforcement ensures adequate overall embankment stability over the reinforcement design life (Figure 14.7a) as discussed above, and where the maximum allowable strain in the reinforcement is limited to control embankment spreading (Figure 14.7b). Thus, there is a maximum tensile load requirement and a maximum tensile strain requirement for the basal reinforcement.

14.3.2 Calculation methods for determining maximum reinforcement load T_r

A number of methods exist to determine the maximum tensile load in the basal reinforcement. These methods fall into one of three categories – plasticity chart methods, limit equilibrium methods and continuum methods.

14.3.2.1 Plasticity chart methods

The plasticity chart method is based on bearing capacity (plasticity) solutions where the basal reinforced case is assumed to provide a horizontal stabilizing force compared to the unreinforced case, which is assumed to provide a horizontal destabilizing force (e.g., see Figure 14.5). Plasticity solutions exist in chart form for common soft foundation geometries (e.g., Jewell 1996).

This method provides a simple approach for the determination of the maximum tensile load in the basal reinforcement. It is normally used if a quick, initial estimate is required. However, for more detailed design the following two methods are normally used.

14.3.2.2 Limit equilibrium methods

Limit equilibrium ('slip-circle') methods are commonly used to assess the stability of geotechnical structures. Today, this is carried out with user-friendly computer software.

For basal reinforced embankments, the limit equilibrium analysis method assumes the reinforcement applies either horizontal force equilibrium or combined horizontal force and moment equilibrium to the embankment structure. Either approach would appear to provide acceptable reinforced embankment solutions, but some care should be taken regarding unusual embankment and foundation geometries when applying either approach.

Limit equilibrium software normally calculates the required safe basal reinforcement load by having the designer initially input an assumed reinforcement load and the program conducts a stability search and outputs a resulting factor of safety. The designer subsequently updates another reinforcement load and/or makes changes to the embankment geometry and the process is repeated until the resulting calculated factor of safety meets the stability requirements of the designer.

Limit equilibrium analyses assumes that a perfect bond develops between the reinforcement and the adjacent soil, and this may indicate higher reinforcement loads and levels of stability than may be achievable in practice. Engineering judgement is required where limit equilibrium analyses are demonstrating large increases in stability and calculating excessively high reinforcement loads.

14.3.2.3 Continuum methods

Continuum methods (finite element and finite difference methods) have become fundamental to the design and analysis of geotechnical engineering structures. Their ability to not be limited by soil geometry, use of complex soil models, use of different boundary conditions and adaptation of different structural elements makes these methods very powerful. However, with this power comes the need to use quite sophisticated input parameters if the method is to produce realistic results. Today, with the graphics outputs of these continuum methods computer programs it is possible to observe many different aspects of behavior. For example, it is easy to observe the reduction in lateral deformations due to the presence of the high strength geotextile reinforcement at the base of embankments.

When using continuum methods to model embankment stability with basal high strength geotextile reinforcement it is necessary only to use a simple soil material model, such as a Mohr-Coulomb model. The use of soil models of increased sophistication is normally not warranted for simple stability analyses but may be of use where a more detailed analysis of the soft foundation is to be performed.

Different continuum methods programs represent the basal geotextile reinforcement in different ways, and this must be understood to model the basal reinforcement in the correct manner. Some programs have actual geotextile/geogrid elements where the tensile stiffness (in kN/m) is input directly. Other programs may use other structural elements for the basal geotextile reinforcement where the elastic modulus (in MPa) is input for the reinforcement. In this latter case an assumption of geotextile thickness must be made for ready conversion between reinforcement tensile modulus and elastic modulus. The author has found that an assumed geotextile reinforcement thickness of 3 mm or 4 mm is appropriate for modelling using these different structural elements. The nature of the bond between the reinforcement structural elements and the surrounding soil must also be understood because the structural elements may require these bond values to be input in specific ways to model actual field behavior.

14.3.3 Maximum allowable strain in basal reinforcement

To prevent the embankment from undergoing excessive deformation it is common to limit the maximum level of strain in the basal reinforcement (see Figure 14.7b). Commonly, most design codes prescribe a maximum allowable reinforcement strain of 5% as this has been found to provide acceptable performance in practice. The application of this maximum allowable strain is shown in Figure 14.8 where it is equated to the allowable reinforcement design strength T_D. From this, a suitable basal reinforcement can be determined to achieve the allowable design strength T_D at a maximum strain of 5%.

14.3.4 Construction details

Prior to placement of the geotextile reinforcement the foundation surface should be prepared by removing any tall vegetation, e.g., trees and shrubs. Low-lying grass can be left in place and undisturbed because its root matter

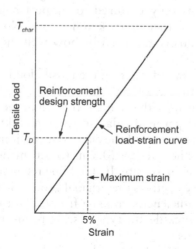

Figure 14.8 Relationship between reinforcement maximum allowable design strength T_D and maximum allowable tensile strain.

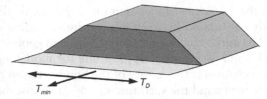

Direction of tensile loads at base of embankment

Direction of installing geotextile reinforcement

Figure 14.9 Direction of loads carried by basal geotextile reinforcement and optimal direction of installation.

can help to bind the surface of the soft foundation. Tall grass should be removed, and any holes should be filled to make a more even surface.

The direction of the tensile loads at the base of the embankment governs the way the basal geotextile reinforcement is to be installed (see Figure 14.9). The calculated design strength of the geotextile reinforcement T_D is required to act in the direction across the width of the embankment. This is the primary strength direction for the basal reinforcement and in this direction, there should be no geotextile joints (joints act as a point of tensile weakness). In the direction along the length of the embankment the basal reinforcement is only required to support a minimal tensile load (T_{min}), typically of the order of 20 kN/m to 50 kN/m. The magnitude of these longitudinal loads is governed by how well the embankment filling process is controlled.

The most efficient way of meeting the tensile load regime at the base of the embankment is to use a single layer of unidirectional high strength geotextile reinforcement (with the high strength in the longitudinal geotextile direction) placed perpendicular to the embankment alignment (see Figure 14.9). Along the length of the embankment, joints can be effected by using either a geotextile overlap (0.5 m to 1.0 m depending on ground conditions) or by seaming on site. It should be recognized that sewn seams never have the same strength as the original geotextile material with typical sewn seam strength efficiencies ranging from 25% to 75% depending on the strength of the geotextile, the type of seam being employed and the type of sewing yarn being used.

The initial layer of embankment fill should be placed on top of the geotextile reinforcement in a careful manner. Mechanical equipment should be limited in weight and not run directly on the geotextile reinforcement. Fill should be placed on the existing fill platform before it is spread over the exposed geotextile by use of a light weight dozer. Where the foundation is particularly soft a swamp dozer may be used for spreading the initial embankment fill ensuring only low vertical stresses are applied to the soft foundation.

The type of fill used immediately above the geotextile reinforcement should be granular (i.e., frictional) in nature and preferably free-draining. Further, there should be no large rocks in this initial fill layer, and the fill grading should be consistent with the factor for installation damage effects f_{id} used to determine the characteristic initial tensile strength T_{char} of the geotextile reinforcement in Eq. (14.1).

Compaction of the initial embankment fill layer may be limited because of the softness of the foundation. However, subsequent fill lifts should be compacted to 90% to 100% Standard Proctor density levels.

14.4 BASAL HIGH STRENGTH GEOTEXTILES USED IN CONJUNCTION WITH WICK DRAINS

14.4.1 Design

Wick drains, also known as Prefabricated Vertical Drains (PVDs), are commonly used to accelerate the consolidation of soft foundation soils. Figure 14.2b shows the technique in general for embankment construction with Chapter 9 describing it in detail. The benefits of the use of wick drains are twofold. First, settlements can be designed to occur relatively quickly within the construction timeframe thereby minimizing later costly maintenance programs. Second, the undrained shear strength of the soft clay foundation increases more quickly (compared with not using wick drains) thus providing earlier improvement in foundation stability. Typical consolidation time-frames when using wick drains can range from 6 months to 2 years depending on drain spacing and foundation hydraulic conductivity versus 10 to 20 years if no wick drains are used.

The combination of wick drains and basal geotextile reinforcement interact together to provide an efficient method of embankment construction over soft clay foundation soils. Figure 14.10 shows this interaction and compares it to the case where no wick drains are used. The use of basal reinforcement without wick drains has already been discussed in Section 14.3.1 and is shown in Figure 14.10. Here, the reinforcement tensile load reaches a maximum value T_r on completion of embankment filling and then reduces over time as the soft foundation consolidates until the foundation can fully support the embankment loading by itself.

When wick drains are used, an identical loading process occurs, but the maximum tensile load generated in the basal reinforcement does not reach

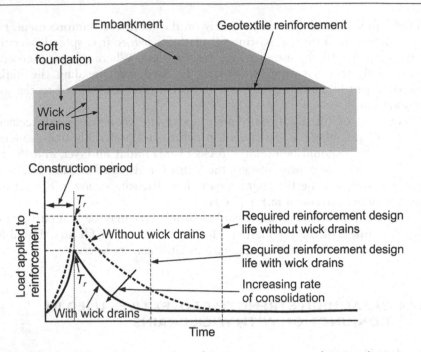

Figure 14.10 Load regime in basal reinforcement over time showing the interaction between wick drains and basal geotextile reinforcement.

the same maximum level without wick drains, Figure 14.10. The reason for this is that the wick drains are already enabling foundation consolidation to occur during the embankment filling process and thus the maximum tensile load generated in the basal reinforcement is reduced. The presence of the wick drains enables a quicker rate of consolidation and thus a shorter time for fully supported foundation conditions to occur. So, the design life over which the basal reinforcement is required is reduced compared to where no wick drains are used. Increasing the rate of consolidation of the soft foundation, by using closer spaced wick drain layouts, can reduce further the required design life of the basal reinforcement. Thus, with the use of wick drains, a lower maximum tensile load in the basal reinforcement can be achieved as well as a shorter required design life. This enables a more cost-effective basal reinforcement solution to be chosen.

While the use of wick drains enables accelerated consolidation of the soft foundation, care is still required regarding the rate of fill placement for the embankment to ensure stability during construction. Chapters 8 and 9 cover this in further detail.

14.4.2 Calculation methods

Two different calculation approaches are used to analyze basal reinforced embankments with wick drains. These involve either decoupling the consolidation component from the strength and stability component (by using a separate consolidation model and a stability model) or using a coupled consolidation/stability model by numerical analysis.

14.4.2.1 De-coupled consolidation and stability analysis

De-coupling the consolidation component from the stability component results in an inherent inaccuracy of the analysis method. However, this approach can be undertaken where the more sophisticated coupled analysis tools (continuum methods) are not readily available or where the available soil input parameters are basic in nature.

Use of two uncoupled analysis methods, one for consolidation and one for stability, results in a lack of compatibility between the two. Thus, considerable engineering judgement is required to relate the two different methods in practice.

Consolidation analysis incorporating wick drains is carried out today using computer software, and these are readily available in varying degrees of sophistication. However, the accuracy of these programs in predicting consolidation depends on the accuracy of the input parameters, specifically the horizontal coefficient of consolidation of the soft foundation and how this changes with depth, location and degree of consolidation. The horizontal coefficient of consolidation is discussed in Chapter 9. This analysis enables the degree of consolidation to be determined at different points in time and provides an estimate of the time to end of primary consolidation.

Stability analysis incorporating basal geotextile reinforcement is also carried out using computer software based on the limit equilibrium approach (see Section 14.3.2.2). However, considerable engineering judgement is required when assessing the increase in undrained shear strength resulting from the presence of the wick drains at different points in time during the consolidation process. This is important because foundation undrained shear strength is a key input parameter when assessing the stability of the basal reinforced embankment and the tensile loads generated in the basal reinforcement.

Generally, an uncoupled analysis can estimate the required design life of the basal reinforcement fairly well, provided accurate values of foundation coefficient of consolidation are used. However, it requires considerable engineering judgement to determine the maximum tensile load in the basal reinforcement (T_r) due to the presence of the wick drains (see Figure 14.10). For simplicity, it has been common practice to assume that the use of wick drains does not reduce the maximum basal reinforcement tensile load; with the maximum tensile load being the same as that for when no wick drains are used. This is a conservative approach when assessing the required design tensile strength of the basal reinforcement.

14.4.2.2 Coupled consolidation and stability analysis

To perform a coupled consolidation and stability analysis a sophisticated numerical analysis by use of continuum methods is required. The major advantage of this approach is that changes in foundation consolidation can be immediately reflected in the stability of the embankment and the tensile loads generated in the basal reinforcement. Today, this can be performed relatively easily using finite element and finite difference computer software. However, this type of numerical analysis requires the choice of an appropriate soil numerical model and the input of relatively sophisticated soil parameters to provide a result that makes sense. Here, engineering experience is vital.

When carried out correctly, coupled consolidation and stability analyses can accurately model the changes in basal reinforcement tensile loads over time as shown in Figure 14.10, but this is dependent on the accuracy of the input parameters used.

When using wick drains the maximum allowable strain in the basal reinforcement should also conform to that discussed in Section 14.3.3.

14.4.3 Construction details

In general, the overall method of construction is the same as where no wick drains are used. However, there are some specific differences with the way the base of the embankment is constructed when using wick drains. The construction at the base of the embankment is commonly termed an 'upside down' construction since the basal reinforcement is not located at the base of the construction but rather located above the drainage blanket, see Figure 14.11. This prevents the geotextile reinforcement from being punctured during the wick drain installation process.

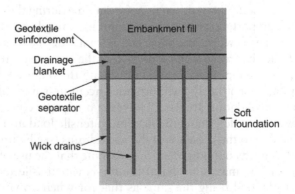

Figure 14.11 'Upside down' construction of basal geotextile reinforced embankment with wick drains.

Large vegetation (trees, tall plants, etc.) should be cleared from the site but low-lying vegetation (grass, etc.) can be left in place. Holes should be filled to provide a more-even surface. Next, a geotextile separator should be placed over the area followed by a granular layer of adequate thickness to construct a stable working platform to support the wick drain installation rigs (see Figure 14.11). Following installation of the wick drains, further granular material is placed to complete the horizontal drainage blanket for the wick drains. Finally, the geotextile reinforcement is placed across the top of the drainage blanket in the location as shown in Figures 14.9 and 14.11 and by the same layout as described in Section 14.3.4. The embankment fill is then placed and spread on top at a safe rate of fill placement.

14.5 BASAL HIGH STRENGTH GEOTEXTILES SUPPORTED ON PILES AND COLUMNS

14.5.1 Design

Piles and columns can be used to support a basal reinforced embankment (known as a basal reinforced piled embankment) to minimize the amount of embankment settlement occurring as shown in Figure 14.2c. This technique, shown in detail in Figure 14.12, is different from the previous two techniques already discussed in this chapter in that piles or columns are used to prevent settlement of the embankment. The basal geotextile reinforcement spans across the pile caps ensuring all the embankment loading is transferred to the pile caps (and then to the piles) and not to the soft foundation in between. This technique enables embankments to be constructed at any rate and have side slopes independent of the shear strength of the soft foundation.

Basal reinforced piled embankments are typically used where differential settlements and stability cause problems, namely:

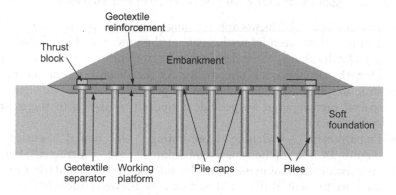

Figure 14.12 Basal reinforced piled embankment technique.

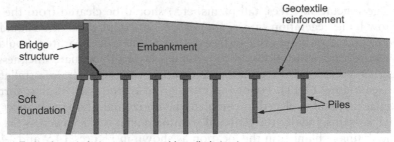

a) Embankment abutments approaching piled structures

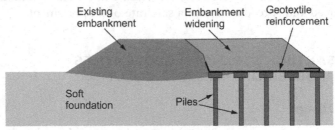

b) Embankment widening preventing differential settlements

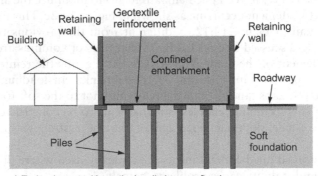

c) Embankment with vertical walls in a confined area

Figure 14.13 Applications for basal reinforced piled embankments.

- Embankment abutments approaching piled structures (Figure 14.13a). To prevent differential settlements at the junction of piled structures (e.g., bridges), it is common to use a basal reinforced piled foundation beneath the approach abutments. As the distance from the bridge structure increases, the embankment piles are shortened in length and spaced further apart to provide a smooth transition to the existing embankment.
- Embankment widening preventing differential settlements (Figure 14.13b). Where new embankments are constructed abutting an existing embankment a basal reinforced piled foundation can be used to prevent differential settlements between the old and new embankments.

- Embankment with vertical walls (Figure 14.13c). When embankments are constructed in confined spaces it is common for them to have vertical retaining walls. To prevent the retaining walls from rotating inwards, and over-stressing the walls, a basal reinforced piled foundation is used.
- Preventing large embankment settlements in general. In some cases, it may be necessary to prevent large embankment settlements from occurring, which can be the case when constructing embankments over peat.
- Where speed of construction is paramount. Basal reinforced piled embankments can be constructed at any rate; thus, the technique is ideal where speed of embankment construction is paramount.

A variety of different piles and columns have been used for basal reinforced piled embankments. Examples are listed in Table 14.2. Piles, which are relatively small in diameter, normally utilize pile caps (as shown in Figure 14.12) to attract the embankment loadings. Columns, being larger in diameter, generally don't require caps. It is common to design the piles and columns as end-bearing units, which means they are extended down to a firm foundation stratum. In some cases, this has required piles to be installed at over 30 m in depth.

To properly design a basal reinforced piled embankment, an assessment is made of the pile group capacity, pile group extent, overall stability, as well as the maximum tensile loads acting in the basal geotextile reinforcement. Tensile loads are generated in the basal reinforcement due to the vertical arching at the base of the embankment fill and the horizontal outward thrust of the embankment fill (Figure 14.14). Both these tensile loads are generated simultaneously.

For this application, the basal reinforcement is required to carry the generated tensile loads for the full design life of the structure. For example, if the basal reinforced piled embankment is to be designed for a 100-year design life, then the basal reinforcement will also have to carry the generated

Table 14.2 Different piles and columns used for basal reinforced piled embankments

Type	Comments
Piles:	High capacity, therefore maximum spacings used.
Pre-cast concrete piles	Low capacity, used where consistent high
Timber piles	groundwater levels occur.
Columns:	Wet or dry installation.
Concrete/cement columns	See Chapter 10 for further details.
Stone columns	See Chapter 10 for further details.
Confined stone columns	See Chapter 11 for further details.
Semi-rigid columns	

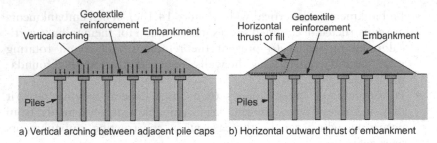

Figure 14.14 The two embankment conditions causing tensile loads in basal geotextile reinforcement.

tensile loads for 100 years. This makes this application different to the applications described earlier in this chapter where the reinforcement load was only required to be carried until the soft foundation had consolidated. Basal reinforced piled embankments do not rely on the soft foundation undergoing significant consolidation and gaining strength.

14.5.2 Calculation methods

Various methods have been developed to calculate the tensile loads and strains generated in the geotextile reinforcement of basal reinforced piled embankments. The most common methods utilize an analytical model to describe the arching occurring between adjacent pile caps and then a subsequent calculation of basal reinforcement tensile loads and strains. The reinforcement loads due to the horizontal outward thrust of the embankment fill is normally determined based on a Rankine analysis.

14.5.2.1 Tensile loads due to vertical arching between adjacent pile caps (Figure 14.14a)

The distribution of vertical stress at the base of a piled embankment is a complex foundation interaction problem. Figure 14.15 shows a general model of this condition. The rigid piles with pile caps attract a greater proportion of the embankment vertical stress, p'_c, due to the fill arching between the adjacent pile caps. The unarched component of the fill, of vertical stress p'_f, acts on the soft foundation between the adjacent pile caps. The role of the basal reinforcement when it undergoes deflection due to the action of the unarched vertical stress p'_f is to transfer this vertical stress to the adjacent pile caps and not directly to the soft foundation. This deflection of the basal reinforcement generates tensile loads and strains in the geotextile.

The general model shown in Figure 14.15 represents a three-dimensional (3D) problem and consequently the appropriate solution should be based on a 3D analytical methodology. Table 14.3 lists three of the most common 3D methods used to determine the tensile loads and strains in the basal

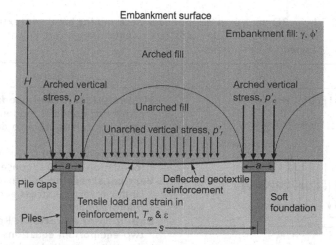

Figure 14.15 General model used to determine geotextile tensile loads due to arching at base of embankment fill.

reinforcement resulting from embankment arching between adjacent pile caps or columns. The BS8006 method, which was originally developed in the 1980s, is the oldest and simplest approach and is in common use in many English-speaking countries. The EBGEO (2011) and CUR (2016) methods are more recent and more rigorous in nature.

Typical strains in the basal reinforcement are quite low, commonly ranging from 3% to 4%.

While analytical models have been the most common design tool for basal reinforced piled embankments, numerical (i.e., continuum) modelling has also been performed in specific cases. Numerical modelling has the advantage over analytical models in that decoupling of the basal reinforcement (to determine its tensile loads) from the stresses and strains at the base of the embankment is not required and both load cases shown in Figure 14.14 can be handled in a single process. However, there are specific aspects of numerical modelling where close attention to detail must be paid. First, where individual pile caps exist, a proper 3D analysis should be performed to arrive at the appropriate loads and strains. Applying 2D analyses or 2D axisymmetric analyses tends to overestimate the amount of arching and underestimate the tensile loads and strains in the basal geotextile reinforcement. Second, numerical methods tend to be very specific and while they can model specific problems they are not adept as a general design tool.

14.5.2.2 Tensile loads due to horizontal outward thrust of embankment (Figure 14.14b)

In the direction across the width of the embankment the basal reinforcement must counteract the horizontal outward thrust of the embankment fill.

Table 14.3 Three 3D analytical methods for determining basal geotextile reinforcement tensile loads due to embankment arching at the base of piled embankments

Method	Features
BS8006-1:2010 method 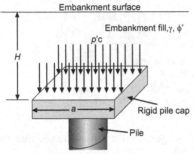 Arching on top of rigid pile cap: $$\frac{p'_c}{\sigma'_v} = \left(\frac{C_c a}{H}\right)^2 \qquad \sigma'_v = \gamma H$$	**3D analytical model used:** Positive projecting conduits model. Empirical relationship to describe arching. **Foundation support:** Assumes no foundation support (conservative). **Vertical stress on reinforcement:** Uniform vertical stress acting on reinforcement. **Equilibrium equations:** Reduces to two equilibrium equations depending on piled embankment geometry. **Deflected reinforcement shape:** Approximates a parabola.
EBGEO (2011) method 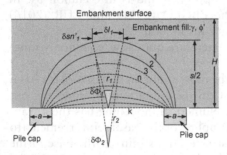	**3D analytical model used:** Multi-shell arching model. **Foundation support:** Accommodates foundation support by use of Winkler spring constants. **Vertical stress on reinforcement:** Triangular vertical stress acting on reinforcement. **Equilibrium equations:** Single complex equilibrium equation (graphical solution available). **Deflected reinforcement shape:** Partial differential equation (with graphical solution).
CUR (2016) method 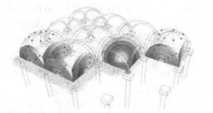	**3D analytical model used:** Concentric arches model. **Foundation support:** Accommodates foundation support. **Vertical stress on reinforcement:** Inverted triangular or uniform vertical stress acting on reinforcement depending on conditions. **Equilibrium equations:** Necessary to solve several complex equilibrium equations. **Deflected reinforcement shape:** Partial differential equation.

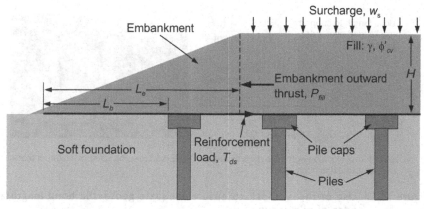

$$T_{ds} = P_{fill} = 0.5 K_a \left(\gamma H + 2w_s \right) H$$

Figure 14.16 Geotextile tensile load due to horizontal outward thrust of embankment fill

In Figure 14.16 the horizontal outward thrust P_{fill} is resisted by a tensile load in the basal reinforcement T_{ds}. A simple Rankine earth pressure analysis can be used to determine the magnitude of the basal reinforcement load T_{ds} as is shown in Figure 14.16.

14.5.2.3 Overall tensile loads in the basal geotextile reinforcement

In the longitudinal direction along the length of the embankment the tensile loads are due to those generated by the unarched fill, i e , $(T_r)_{long} - T_{rp}$. In the transverse direction across the width of the embankment the tensile loads are due to that generated by the unarched fill plus the horizontal outward thrust of the embankment fill, i.e., $(T_r)_{trans} = T_{rp} + T_{ds}$. Thus, the tensile loads in the transverse direction across the embankment will always be greater than those along its length.

From these tensile loads, the required basal reinforcement design strengths T_D along the length and across the width of the embankment can be determined using Eq. (14.2).

14.5.3 Construction details

Construction of the basal reinforced piled embankment structure shown in Figure 14.12 involves the following. The site is first cleared of any large vegetation (e.g., trees, large plants, etc.) and other objects. Low-lying grass vegetation can be left in place.

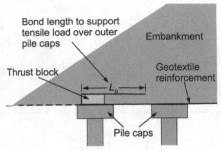

a) Use of thrust block on outer pile caps to enable   b) Example using gabions as thrust block
required geotextile bond length

Figure 14.17 Use of thrust block to ensure adequate geotextile bond length at
edge of pile group

A working platform is then constructed across the base of the embankment to support the piling installation equipment (see Figure 14.12). This working platform normally consists of a geotextile separator with a granular fill layer placed on top to the required thickness where it can support the piling equipment loadings.

The piles or columns are then installed at the design spacings and to the depth of a competent stratum. When piles are used, it is normal practice to include concrete pile caps to increase the surface area of influence of the piles. Pile caps should have a smooth upper surface with no sharp edges that might unduly damage the basal geotextile reinforcement.

The tensile loads generated in the basal reinforcement must be continuous even across the outer pile caps beneath the embankment. To achieve this a suitable reinforcement bond length L_b must be provided beyond the outer piles (e.g., see Figure 14.16). However, extending the geotextile reinforcement to the toe of the embankment may not achieve this bond resistance. Consequently, an alternative approach is commonly used where the reinforcement is placed around a thrust block located on the outer row of piles and continued inward within the embankment fill to arrive at a suitable bond length, Figure 14.a. This thrust block must provide the required passive resistance and normally consists of a row of gabions, as shown in Figure 14.b, or concrete posts, etc. laid across the outer pile caps.

While any granular fill type can be used immediately above the geotextile reinforcement, there have been cases where a highly granular fill layer has been used to enhance the arching at the base of the embankment. The thickness of this layer has ranged from 0.5 m to 1.0 m depending on the spacing between adjacent pile caps. Once this initial layer has been placed, normal embankment fill can be used to complete the embankment.

14.6 SUMMARY

Basal high strength geotextile reinforcement can be used to enhance the stability of embankments and fills constructed over soft foundation soils and can be used to control their settlements when used in combination with another ground improvement technique, e.g., wick drains or piles and columns.

High strength geotextile reinforcement composed of high modulus polyester yarns provides an effective means of reinforcing embankments constructed over soft clay foundations. Recognized procedures exist to convert the design strength of the reinforcement to its required initial tensile strength. Such procedures account for installation damage, creep and environmental effects and must be supported by well-documented reinforcement evaluation programs.

The use of basal high strength geotextile reinforcement to construct stable embankments over soft clay foundation soils is a common geotechnical engineering technique for which standard design procedures exist. Basal reinforced embankments can be constructed higher and have steeper side slopes than embankments without basal reinforcement. The maximum basal reinforcement loads occur in the direction across the width of the embankment with relatively small loads occurring along the length of the embankment. The most efficient use of the geotextile reinforcement is to use a single layer of unidirectional reinforcement laid across the width of the embankment in one continuous piece (with no joints).

The use of a combination of basal geotextile reinforcement and wick drains provides an effective consolidation and stability solution for embankments constructed over soft foundations. The wick drains accelerate the rate of foundation consolidation and hence shear strength gain. The basal reinforcement provides the early stability for the embankment but is only required over a relatively short period of time due to the use of the wick drains. This technique provides additional economies in the amount of basal geotextile reinforcement required.

Basal geotextile reinforced piled embankments are used where differential settlements are to be minimized. With this technique, embankments can be constructed at any rate and at any side slope without consideration of foundation undrained shear strength. For this application, basal reinforcement loads occur both across and along the length of the embankment. Consequently, the most efficient basal reinforcement layout is to use two layers of unidirectional reinforcement installed at right angles to each other.

REFERENCES

BS8006-1 (2010). *Code of practice for strengthened/reinforced soils and other fills*, British Standards Institution, London, UK.

CUR (2016). *Design Guideline Basal Reinforced Piled Embankments*. Edited by van Eekelen, S.J.M. and Brugman, M.H.A., CRC Press, The Netherlands.

EBGEO (2011). *Recommendations for Design and Analysis of Earth Structures Using Geosynthetic Reinforcements*. German Geotechnical Society, Germany.

Jewell, R.W. (1996). *Soil Reinforcement with Geotextiles*. CIRIA Special Publication 123, Construction Industry Research and Information Association, UK.

Koerner, R.M. (2012). *Designing with Geosynthetics*, 6th edition. Xlibris Corporation, New York.

Shukla, S.K. (2011). *Handbook of Geosynthetic Engineering*, 2nd edition. ICE Publishing, London.

Chapter 15

Mass stabilization

A. O'Sullivan

15.1 INTRODUCTION

Mass stabilization refers to the stabilization of weak soils through the introduction of a binder, such as cement, lime, fly ash or gypsum to make them stronger and less compressible. Mass stabilizing is a shallow treatment limited to depths up to 5 meters and involves 100% treatment of a targeted zone. It is generally used as a ground improvement technique in the construction of road and rail embankments, ports and harbors, and in land development including reclamation sites and landfills. Stabilization can be undertaken either in-situ or ex-situ, giving rise to the terms in-situ (mix in place) and ex-situ (removal, mix and replace) mass stabilization (or IMS and EMS). IMS is generally used in soils with high moisture content viz., clays, organic soils, peat and contaminated soils where the in-situ soil strength is generally less than 20–40 kPa. Here, the stabilizing agent is applied in dry form and utilizes the available moisture in the soil for hydration. Where the extent of excess moisture in the soil is less and the in-situ shear strength is greater, the binder can be added in grout or slurry form. Stabilized material offers higher strength and lower compressibility and hydraulic conductivity. A schematic section of in-situ mass stabilization is shown in Figure 15.1.

The objectives of mass stabilization for soft clays could be summarized as follows:

1. Increase strength;
2. Improve bearing capacity;
3. Reduce compressibility;
4. Reduce plasticity;
5. Reduce expansivity (more relevant to stiffer soils);
6. Reduce permeability;
7. Reduce sensitivity; and
8. Remediation (encapsulation) of contaminated sites using stabilization and solidification.

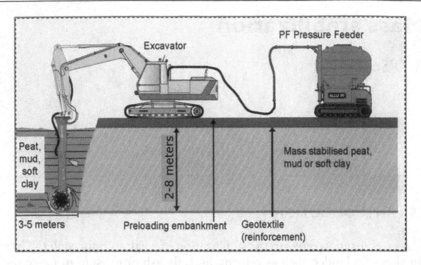

Figure 15.1 Schematic section of in-situ mass stabilization machinery in operation (courtesy Andy O'Sullivan).

According to Kiukkonen (2014), mass stabilization of clays was initially developed in Finland in the 1990s, although the precursor, column stabilization, commenced in the 1960s in Sweden and Japan. It is now considered a practical and economical method of ground improvement for shallow weak soils. Although it is most popular in Europe, other countries such as New Zealand have adopted the technique for the local conditions.

The construction process of in-situ mixing can be challenging, particularly around dealing with achieving an efficient mixing process, spatial control and dealing with underground obstructions. Furthermore, this process relies on achieving strength through the stabilization process and without compactive effort. Therefore, an understanding of water contents, binder selection, strengths and quality control are important considerations.

Mass stabilization ex-situ is mostly popular when the materials are to be used as good-quality engineered fill. It involves removing soil and transporting to another area or site to stockpile where soil mixing with a binder could be carried out. Various pugmill mixing systems are available and rely on efficient mixing and controlled binder distribution to the treated soil. Some mixing trials have demonstrated that mixing efficiency is as important as the binder quantity, as demonstrated by Hiway Stabilizers at the Tui Mine remediation site in Waikato, New Zealand (Quickfall et al. 2013). An example of this system using the Reterra system developed by the Japanese firm Komatsu for EMS is shown in Figure 15.2.

Ex-situ stabilization mostly occurs where the material is well below optimum moisture content or too stiff to allow for effective in-situ treatment or when dredged materials need to be improved. This is the case close to ports where dredging is carried out for maintenance of shipping lanes (Burgos

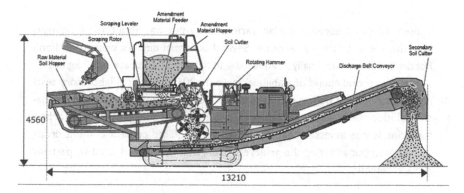

Figure 15.2. Schematic showing the Reterra system developed by Komatsu (courtesy Hiway Geotechnical Ltd).

et al. 2007). Even if contaminants are present in the dredged materials, so long as the materials are placed in a contained area, effective stabilization treatment can still be undertaken.

The ex-situ stabilization process has the advantage of being a more controlled process and higher levels of quality control are generally more easily achieved. Furthermore, compaction of the treated material can lead to higher strengths. Higher production outputs and efficiencies would generally make ex-situ a more cost-effective option than in-situ mixing depending on the design and engineering criteria.

There are not many technical case studies reported by the industry apart from information in company/contractor websites. We report below a case study briefly and the reader may refer to the original paper.

CASE STUDY 15.1   JÄTKÄSAARI IN HELSINKI (AFTER LAHTINEN ET AL. 2014)

The site was formerly occupied by an old container harbor in Helsinki, Finland. Since the harbor was moved in 2008 to a different location, the site was available for other developments. Four temporary basins were constructed using mass stabilization to improve the contaminated dredged sediments.

Mass stabilization used at the site serves a different purpose than in most other sites as the stabilized mass, after the curing period, is temporarily allowed to be in store but is used as construction materials for various other projects in and around the city of Helsinki. By 2014, two rounds of stabilization had occurred where the basins have been emptied every year, i.e., after sufficient curing time in each case, and refilled with new dredged sediments.

Prior to the initial treatment in 2011, a detailed assessment of the required binder content had been carried out using laboratory testing. The moisture content of the dredged materials varies widely but was about 100% on average with the pH sitting at about 8. Several potential binders were tested and Portland cement was finally used for the initial works at a rate of 60 kg/m³.

For the second round of stabilization, i.e., in 2012, additional laboratory testing was carried out before the stabilization works commenced to seek alternative solutions because of the high price of cement and considerably high carbon footprint. It was assessed that the amount of required cement content drops by 50% without affecting the achieved strength if fly ash was used to partially replace cement.

Other types of ground improvement closely related to mass stabilization are soil mixing and deep soil mixing (DSM). The former generally refers to shallow surface soil improvement such as for subgrade improvement, and DSM refers to improving the soils in a deep stratum. DSMs are constructed as columns and panels as discussed in Chapter 13. The main differences between the mass stabilization and DSM are:

- Machinery – Generally an excavator based equipment and therefore readily available, and mobilization costs plus moving from site to site, for both in-situ and ex-situ, is generally easy and less expensive.
- Depth of stabilization – Limited to 5 to 7 m in mass stabilization under favorable conditions but optimal results are obtained when the depth of improvement is limited to 3 to 5 m (Forsman et al. 2015a). As previously mentioned, the depth is limited because stabilization is usually carried out using an attachment to an excavator, although there is no limitation on depth from the point of view of design. Deeper stabilization would need very heavy machinery and the costs increase substantially. In such situations, DSM columns are likely to be more appropriate and cost effective.
- Design strength – Mass strength in mass stabilized material is less than the strength developed in DSM columns/panels. Figure 15.3 illustrates the effect of stabilization on the unconfined compressive strength of a clay. Compared to soft clay, the shear strength is increased and the strain to achieve peak strength is lesser. Compared to DSM, the strength achieved would be less and the strain at peak shear strength is greater.
- Coverage – Nearly 100% for mass stabilization but significantly less for DSM.

Mass stabilization is sometimes used in combination with other ground improvement methods. A simple example is the use of mass stabilization to create a working platform for heavy machinery that is used for deeper

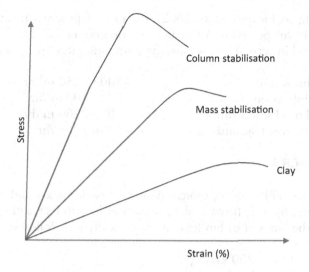

Figure 15.3 Effect of on shear strength due to stabilization.

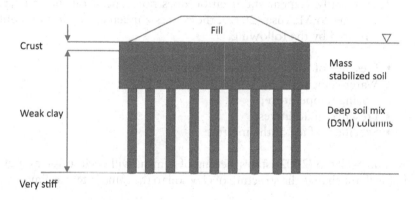

Figure 15.4 Example where mass stabilization is used in combination with DSM (courtesy Hiway Geotechnical Ltd).

ground improvement such as semi rigid inclusions. Another example is the mass stabilization to maximum depth possible using the site machinery followed by DSM as a schematic shown in Figure 15.4. This was undertaken at a site at Dargaville in the Kaipara District of New Zealand (O'Sullivan et al. 2013), which was underlain by 20 m of very soft alluvial soils where IMS was used to provide a stable working platform for the DSM plant.

According to EuroSoilStab (2002), the rate of progress in-situ is of the order of 100 m³ per hour. According to Forsman et al. (2015b), based on data gathered in typical conditions, the following rates are applicable:

Peat stabilization	~100 to 150 m³/hr
Clay stabilization	~80 to 100 m³/hr
Dredged mud stabilization	~100 to 200 m³/hr
Hard clay or silt stabilization	~50 to 80 m³/hr

15.2 BINDERS

According to Wilk (2013), typical binders used for mass stabilization are cement, lime, fly ash, furnace slag, industrial by-product and special mixtures, and the amount of binder content typically is as follows:

- Mud – 120 to 200 kg/m³;
- Peat – 150 to 250 kg/m³; and
- Sediment (clay) – 70 to 200 kg/m³.

Cement: Cement reacts with water in the soil matrix to harden the soil. Theoretically, cement stabilization does not depend on the soil type. According to Makusa (2012), the process of cement hydration could be affected by the following:

- Presence of foreign matters or impurities;
- Water-cement ratio;
- Curing temperature;
- Presence of additives;
- Specific surface of the mixture.

As EuroSoilStab (2002) states 'Setting of cement will enclose soil as a glue but will not change the structure of clay soil to the same extent as lime does.'

Lime: Depending on local market conditions, lime can be significantly less expensive than cement and therefore it is appealing to a contractor when compared to cement. It reacts with pozzolanic materials in the presence of water to produce soil compound. Two types of lime are available, viz. quicklime (CaO) and hydrated lime ($Ca(OH)_2$), and both could be used for lime stabilization, although quicklime when reacting with water in the soil becomes hydrated lime.

Fly ash: Fly ash is a product of coal-powered generation plants and quite cheap and readily available because it is essentially a waste material from the power generation activity. According to Makusa (2012), it has little cementitious properties compared to lime and cement, and therefore a small amount of activator may be needed. EuroSoilStab also states that ash is generally not very reactive on its own but may reduce the cost of a blended product.

Blast furnace slag: It is a non-metallic product from the process of production of iron. According to Makusa (2012) the chemical composition is similar to that of cement but is non-cementitious on its own. As EuroSoilStab (2002) states it has to be activated with lime and cement to achieve a faster reaction and may be regarded as a low cost substitute for cement. Other types of binders are briefly discussed by EuroSoilStab (2002).

Where organic contaminants occur, some of them could retard/obstruct strength gain in the stabilized soil (Forsman et al. 2015b).

15.3 DESIGN CONSIDERATIONS

Generally, the design of mass stabilization is centered around the strength and stiffness to be achieved. It is generally assumed that the mass that is stabilized is homogenous and acts as an elasto-plastic material. Once the requirements of strength needed for the project are established, the types of binders and the binder ratios need to be assessed to provide an economical design. Undrained shear strength obtained from mass stabilization is generally between 30 kPa and 70 kPa and generally does not exceed 100 kPa (Forsman et al. 2015b). Krzeminski and Lai (2019) state that typical undrained shear strength values range from 50 kPa to 250 kPa. EuroSoilStab (2002) recommends a maximum characteristic undrained shear strength of 150 kPa irrespective of the results of laboratory and field tests because of uncertainties in mixing and unequal spread of binder. However, higher values have been designed for and achieved in construction projects. Krzeminski and Lai (2019) discuss required shear strength of 250 kPa where mass stabilization was adopted and achieved in the field to improve to a depth of 6.5 m.

To obtain the design shear strength, it is advisable to initially carry out laboratory tests to assess how effective the stabilization is on the materials to be improved. The results thus obtained need to be factored down to obtain a design value as the field strength developed is likely to be lower than the laboratory strength, the latter obtained under controlled conditions on small samples. Further, site soils are likely to be variable across the site in addition to mixing of soil and binder could also be variable across the site. More realistic values could be obtained by carrying out trials on site.

Once the shear strength to be used in the design is established, based on test results and allowing a factor of safety for variability that could be expected, standard techniques discussed in Chapter 5 could be used to assess settlement and stability. Another parameter required in this assessment is the stiffness of the mass stabilized soil. Correlations between shear strength and modulus as discussed in Chapter 5 could then be used. EuroSoilStab (2002) suggests that the characteristic compression modulus (in confined compression) is 50–100 times the characteristic undrained shear strength, with 50 being for organic soils and 100 for silty clays. It also states that stabilization with binders could be stiffer, i.e., 50–150 times.

Settlement assessment in a mass stabilized soil could be carried out using basic principles discussed in Chapter 3 with the use of the design modulus discussed above. EuroSoilStab (2002) warns that considerable settlements can be derived during the curing period (when the only load is from the working platform) and these settlements should be calculated separately. In fact, EuroSoilStab (2002) recommends that a preloading working platform should be applied immediately after the stabilization work as this compresses the stabilized mass, and increases its strength and stiffness.

The settlement of the soil layers underlying the stabilized mass should be carried out using first principles discussed in Chapter 3. However, in this instance, as the stabilized soil acts as one whole stiff mass, no load spreading should be assumed within the mass.

According to Forsman et al. (2015a), target design strength can be achieved within one to three months.

There are several modes of failure that needs to be investigated for stability. They include the following (FHWA 2017):

- Sliding of the embankment over the mass stabilized block;
- Sliding below the mass stabilized block;
- Failure through the block;
- Overturning of the block;
- Bearing capacity at the toe of the block if lateral loads play a role; and
- Crushing of the toe of the block if lateral loads apply.

CASE STUDY 15.2   COAL EXPORT TERMINAL, AUSTRALIA (TSE ET AL. 2012)

The project is known as Sub-Stages 2AA and 2F of the development of Coal Export Terminal 3 (CET3) in Kooragang Island, NSW, Australia. The extension of the coal stockyard is constructed over the Stage 1 dredge settling ponds where the materials consist of very soft silt and very loose sands. The sub-surface profile consists of 2 to 4 m of dredged sediments overlying 2 to 4 m thick, soft alluvial silty clay. The groundwater level is at or above the dredged sediments.

On-site trials of mass stabilization using dry binder indicated that mass stabilization is possible. Following the trials using cement as the binder, a combination of mass soil stabilization and dynamic replacement techniques were used to support 21 m high coal stockpiles and stacker-reclaimer machinery and berm loads. Soft soil previously stabilized and allowed to cure was used as the platform to progress in a systematic way. It is understood that generally a three-day curing period was sufficient for the mixing rig to walk on the stabilized soil, which was covered by a 0.3 to 0.5 m layer of sand/gravel.

The performance was assessed to confirm whether the design strengths have been achieved by carrying CPTs after stabilization. Where the mass stabilized layer was 5 m thick, a trial embankment 40 m × 40 m was constructed in an area where extremely soft dredged mud was found. The dredged mud consisted of 3 m of c_u = 5 kPa mud overlying 2 m of c_u = 20 kPa alluvial silty clay, which in turn overlies a medium dense sand layer. The 9 m high embankment imparts a load of 180 kPa and was constructed in two stages, first stage to 5 m and left for one month for strength gain.

Observed settlements were significantly less than the predicted settlement of 230 mm under the expected stockpile load of 210 kPa and indicated that only immediate settlement occurred, i.e., behaved as a granular material.

New Zealand reports of mudcrete, which is mud stabilized using cement. The following case study discusses a project recently undertaken that explains the beneficial use of mudcrete.

CASE STUDY 15.3   VIADUCT BASIN, AUCKLAND (PRIESTLY 2019)

The project refers to the redevelopment of the eastern section of the Viaduct Basin in Auckland, New Zealand. A main part of the project was the use of dredged mud for reclamation. To increase the strength and reduce contaminant leachate, the dredged mud was subjected to mass stabilization using Portland cement, making it mudcrete.

Early results of the investigations indicated the high presence of DDT, lead and zinc which means marine disposal will not be possible unless contained. The transformation of mud to mudcrete immobilized contaminants such as heavy metals and some organics. This means disposal is not constrained by treating contaminants. After carrying out laboratory trials with cement, lime, fly ash and a propriety product, Portland cement was found to be the most effective additive. Based on laboratory test results, a target ratio of 20% for cement to marine mud was adopted. The mudcrete assisted the contaminated discharge, first by reducing the leachate strength, and second by reducing the permeability of the mud when it becomes mudcrete.

Dredging and stabilizing were carried out from flat top barges. Dredged materials from hydraulic excavators were discharged to shallow mixing bins on the barges and cement was sprinkled and mixing done by a modified stabilizer fitted onto a hydraulic excavator. Once mixing was complete, the mudcrete was placed using hydraulic excavators, although a clam shell bucket was used when material placement was below sea level.

After one day of placement, mudcrete could be walked on and the construction plant could travel after three to four days. After seven days of curing, shear strengths obtained were greater than 200 kPa compared to the design intent of 100 kPa. Sampling and testing post placement indicated that the mass stabilized is not homogenous, ranging from a well-mixed mass of having a shear strength greater than 400 kPa to no mixing at all, i.e., only soft clay. It was postulated that well-mixed material was confined to the perimeter of each load, the material had a lower strength or was soft clay. The paper concludes that the mass could be described as competent lumps of material in a weakly cemented matrix, or analogous to large boulders placed in a stockpile surrounded by weakly bonded material.

15.4 CONSTRUCTION ISSUES

Although the application of IMS material falls with the category of earthworks, several construction-related issues could crop up and the designer's expectations may not be met. Quality Control and Quality Assurance are most important for mass stabilization projects.

In general, the machinery used for mass soil mixing is relatively light relative to that utilized for DSM installation.

If the mixing volume is large, the potential for heaving, which could affect the adjacent infrastructure, should be considered. Typical measures could be adopted to reduce such effects such as reducing the rate of mixing, progressing of mixing directed away from the boundary and/or trenching. Alterations to the potential application rate are not recommended as these will affect the long-term performance of the stabilized soil mass.

Although in typical ground improvement methods instrumentation is used mostly to measure settlement and lateral displacement and provide alerts on stability using piezometers and inclinometers, in mass stabilization, monitoring of the application process is more important as the design is based on the characteristics of the stabilized mass. Such monitoring includes the volumes of binder and water added as well as typical process monitoring such as speed of rotation and rate of application. Such data assists quality control of the product.

Post construction testing can prove difficult to verify that the design requirements have been achieved due to the variability of the post-treated soil mass. Typically, prescribed testing methods include:

- CPT and SPT testing to assess the shear strength gain and the homogeneity of the soil mass;
- Undisturbed sampling and laboratory testing to assess shear strength and stiffness, and, if necessary, permeability;
- Plate load testing;
- Surface MASW surveys to confirm that the overall design stiffness of the improved soil mass has been achieved.

15.5 ENVIRONMENTAL CONSIDERATIONS

Several environmental issues could be associated with mass stabilization:

- Vibration and noise – As previously mentioned, vibration and noise are not serious concerns for mass stabilization sites as only light machinery is used.
- Environment – In-situ mixing avoids transport and disposal and hence less stress on the environment with significantly less earthmoving movements.
- Quick lime – One of the main environmental issues is related to quick lime if used. Quick lime, if used, may affect the human body, viz. damage eyes and skin. The effects could not only affect the site staff but innocent bystanders such as pedestrians walking near the site.
- Dust generation and management is a necessary consideration.

15.6 ADVANTAGES AND DISADVANTAGES

As with other improvement techniques, there are advantages and disadvantages associated with mass stabilization works and summarized in Table 15.1.

Generally, the compaction of the cement type stabilized soils needs to be carried out with no delay. Otherwise, additional compaction may be needed to overcome the adverse behavior. On the other hand, delay is a necessity for soils stabilized with lime.

15.7 SUMMARY

This chapter covers mass stabilization as applicable to ground improvement projects. A detailed discussion is presented with special attention given to construction. Mass stabilization is a genuine ground improvement method

Table 15.1 Advantages and disadvantages – mass stabilization

Item	Advantages/disadvantages
Complexity	Advantage – Very simple methodology. Basic theory sufficient for design.
Price	Advantage – Relatively cheap compared to other methods where a binder is used such as cutter soil mix, deep soil mix.
Time	Advantage – Improvement within days compared to months when preloading is adopted.
Complexity	Advantage – Very simple methodology and easily understood.
Subgrade	Advantage – Provides a good base for any earthworks.
Settlement	Advantage – Reduces settlement due to development loads significantly.
Acid sulfate soils	Advantage – Avoids the need to treat acid sulfate soils if present on the site compared to excavation and replacement.
Contaminated soils	Advantage – As previously discussed.
Disposal	Advantage – Avoids disposal off site of poor quality materials including contaminated soils and acid sulfate soils, i.e., reduces or eliminates transport, stockpiling, landfilling, etc. leading to very high cost and time of management procedures that need to be implemented, testing to be carried out on soil to be removed, slow progress, and transport to specific landfills when contaminated soils are encountered.
Use as fill	Advantage – As previously discussed, modified soil could be used for engineering applications elsewhere.
Long-term performance	Disadvantage – As the mass stabilization method is relatively new, the long-term performance is yet unknown under field conditions.
Obstacles	Disadvantage – One of the main disadvantages is that the method may not work in certain soil types. This means, laboratory experiments need to be carried out well in advance. Relying on the method could prove to be a disaster if the test results indicate adverse behavior and may affect the overall project planning, management and costs.
In-situ conditions	Advantage – Mass stabilization accepts in-situ conditions and therefore necessity for typical earthworks construction such as dewatering and lateral supports are not deemed necessary, again reducing the overall costs and time.
Noise and vibration	Advantage – As discussed above, very little noise or vibration occurs and therefore the method is very environmentally friendly in these aspects.
Obstacles	Disadvantage – Equipment may not penetrate hard clayey layers and dense sandy layers.
Monitoring	Disadvantage – Difficult to monitor improvement during construction.

that has a lot of advantages compared to other types of ground improvement. In-situ stabilization, which eliminates transporting and disposing elsewhere, is a great advantage and is one of the main reasons for its popularity. Advantages and disadvantages of the methods in relation to other ground improvement methods are also provided. The main disadvantage is the depth limitation because of the light machinery used for construction. It is also accepted that mass stabilization may not work with some types of materials and therefore early laboratory testing needs to be carried out to assess the feasibility.

REFERENCES

Burgos, M., Samper, F. and Alonso, J. J. (2007). 'Improvements carried out in very soft dredged mud soil in the Port of Valencia (Spain).' *Proceedings 14th European Conference*, Eds. Cuéllar J. et al., Madrid, The Netherlands, 234–235.

EuroSoilStab (2002). 'Development of Design and Construction Methods to Stabilize Soft Organic Soils.' *Design Guide Soft Soil Stabilization*, CT97-0351, Project No. BE 96-3177.

FHWA (2017). *Ground Modification Methods Reference Manual – Volume I. Publication No. FHWA-NHI-16-027*. Federal Highway Administration, USDOT, Washington, DC.

Forsman, J., Jyrävä, H., Lahtinen, P., Niemelin, T. and Hyvönen, I. (2015a). *Mass Stabilisation Handbook*.

Forsman, J. Lindroos, N., Jyrävä, H., Niemeläinen, T., Lahtinen, P., Niutanen, V. and Kreft-Burman, K. (2015b). '*Mass stabilisation method – a new handbook for design, execution and quality control.*' *Proceedings of the XVI ECSMGE: Geotechnical Engineering for Infrastructure and Development*, ICE Publishing, 1427–1432.

Kiukkonen, P. (2014). '*Mass stabilisation machinery.*' *Proceedings of the South Baltic Conference on Dredged Materials in Dike Construction*, Rocstock. 195–198.

Krzeminski, M. and Lai, M. (2019). '*Ground modifications using mass soil mixing technique in geotechnical engineering.*' *Proceedings 13th Australia New Zealand Conference*, Ed. H.E. Acosta-Martinez and B.M. Lehane, Australian Geomechanics Society, Perth, 745–752.

Lahtinen, P., Forsman, J., Kiukkonen, P., Kreft-Burman, K. and Niutanen, V. (2014). '*Mass stabilisation as a method of treatment of contaminated sediments.*' *Proceedings of the South Baltic Conference on Dredged Materials in Dike Construction*, Rocstock, 145–155.

Makusa, G.P. (2012). 'State of the art review.' *Soil Stabilization Methods and Materials in Engineering Practice*.

O'Sullivan, A., Soric, D. and Terzaghi. S (2013). 'Stress path approach to settlement analysis of soft ground.' *Proceedings of 19th NZGS Geotechnical Symposium*, Ed. C.Y. Chin, Queenstown, 874–881.

Priestly, S.J. (2019). '*Beneficial use of mudcrete in reclamations.*' *Proceedings 3rd International Conference on Building Information Modelling (BIM 2019) in Design, Construction and Operations*, Seville, Spain, 299–306.

Quickfall, G., Basheer, G., Croucher, B., Jenkins, I.R., Fellows, D.L. and Willson, T. (2013). '*Tui Mine remediation project – geotechnical, environmental and construction challenges.*' *Proceedings of 19th NZGS Geotechnical Symposium*, Ed. C.Y. Chin, Queenstown, 423–430.

Tse, T.K.M., Haines, N., Lai, M. and Hoffmann, G. (2012). '*Ground treatment of dredging settling ponds for NCIG CET3 Coal Stockyard.*' *Proceedings of the International Conference on Ground Improvement and Ground Control*, Eds. B. Indraratna, C. Rujikiatkamjorn and J.S. Vinod, 1687–1695.

Wilk, C. (2013). *Mass Stabilization – Virgin and Contaminated Soil/Sediment.* Presentation at the Remediation Technologies Symposium, ALLU group Inc. www.esaa.org/wp-content/uploads/2015/06/13-Wilk.pdf.

Chapter 16

Observational approach and geotechnical instrumentation

K. Dissanayake and C. Bridges

16.1 INTRODUCTION

Engineering solutions to technical issues have traditionally been based on a single, fully established and robust design, where there is no or very limited intent to modify the designs and the objectives during the process of construction. By contrast, observational approaches can best be implemented to overcome encountered difficulties or unknowns.

Observational approach techniques developed to deal with uncertainties in the design process related to applied soil mechanics, and particularly instrumentation and monitoring, are widely known to geotechnical engineers. However, instrumented observations in the majority of civil engineering projects are often being carried out as a passive tool to assess the performances of critical elements of the design for checking the validity of the original predictions, assumptions and for proving confidence to third parties for the fulfilment of contractual and legal obligations (Patel et al. 2007).

Some potential benefits of this design approach are schematically presented in Figure 16.1 (Nicholson et al. 1999).

Together with soil mechanics, the combined process of *'learn and redesign as you go'* also evolved. This essentially consisted of an active role of predicting, monitoring, reviewing and modifying designs. Professor Peck, in his famous Rankine Lecture, collated these procedures and initiated a framework for the observational design approach (Peck 1969). Its distinct benefits over the conventional design approaches have since led to its framework and principles being extended by a number of engineers and academics. For example, Nicholson et al. (1999) further developed this approach, establishing comprehensive guidelines to implement this in practice while improving its definition. As a result, the observational approach has now been recognized as one of the design methods in design codes, such as Eurocode 7 (BSI 2004) and DIN (2010).

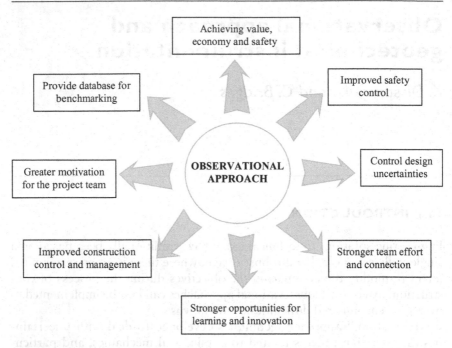

Figure 16.1 Potential benefit of observational approach (adapted from Nicholson et al. 1999).

Depending on the circumstances, the observational approach (OA) could often lead to project savings in terms of time and cost without compromising the intended performance and safety. Moreover, appropriate instrumentation and monitoring could provide the needed assurance concerning safety. The emphasis on prediction, observation, feedback, evaluation and teamwork also creates a strong opportunity for learning.

In addition, the implementation of adequately designed and well-planned instrumentation and monitoring programs may also help engineers and stakeholders in their decision-making process at various stages of their projects. Some of the main benefits of such systems are:

- Characterizing the initial site conditions at an early stage while monitoring common parameters of interest, such as pore water pressure or groundwater, permeability of the soil, suction, contamination detection and slope stability, while establishing baseline readings.
- In-service performance predication by monitoring critical design elements over designated periods during their design lives. These design verification instruments are used to check the performances against predictions.

- Instrumentation can also be used both to enforce the quality of workmanship on a project and to ensure that the work is accomplished conforming project specifications.
- Instrumentation and monitoring data can provide evidence for legal defense of designers and contractors should owners of adjacent properties claim that construction has caused damage.
- Instrumentation is also used for operation and maintenance purpose of civil structures, e.g., asset management.

It is evident that the instrumentation and monitoring is an integral part of the OA. Therefore, greater emphasis is placed here to discuss numerous topics related to it, while briefly outlining the current trends on OA pertaining to soft clay engineering and ground improvement works.

However, there are inherent limitations in this design approach, viz.:

- OA *must not* be used where:
  - there are no or negligible uncertainties in the ground/loading conditions;
  - the design changes are neither possible nor feasible;
  - there is insufficient time to implement and complete the contingency and emergency plans, such as sudden brittle behavior of design elements under ultimate conditions; and
  - observations would be unreliable or difficult to obtain.

- Appropriate safety management systems, governed by relevant legislation and safety criteria, *must be* in place where:
  - OA is used to control the risk of exceeding an ultimate limit state condition; and
  - there is a risk to people and/or sensitive environments associated with the use of OA.

- The overall economy through the implementation of OA could only be achieved if:
  - thorough and high-quality site investigations are carried out;
  - site investigations are substantially completed, and the data are interpreted before the implementation; and
  - all design cases are analyzed covering likely scenarios.

Van Baars and Vrijling (2005) identify general conditions that should be met for successful implementation of OA. However, Spross and Johansson (2017) argue that the decision in choosing between OA and conventional design should be made under a reliability constraint using a probabilistic optimization methodology. The authors' intention in this chapter is to introduce OA as an alternative design approach and, thus, further reading is recommended for details in selecting best approach relevant to the project depending on the circumstances.

16.2 THE OBSERVATIONAL APPROACH (OA)

Most of the geotechnical codes of practice around the world recognize the following approaches for design, verification and construction of geotechnical elements (Nicholson et al. 1999; Spross 2014):

1. Calculating design quantities with analytical, semi empirical or numerical models – i.e., models that are either accurate or err on the safe side;
2. Adopting prescriptive measures – i.e., use of fully empirical design rules when calculation models are unavailable or not required and hence, very conservative;
3. Using experimental methods including in situ load tests – i.e., trial testing on centrifuge, turn table, and laboratory or prototype (e.g., trial embankments) models; and
4. Observational approach (OA) – often adopted where geotechnical behavior is difficult to predict or to overcome unexpected issues during the course of construction.

The first two methods above cope with uncertainties, arising from the variability of material properties and the lack of understanding and knowledge of the behavior of design elements, in two ways; viz., by adopting a usually excessive factor of safety and/or by making assumptions in accordance with general, average experience. As the reliability of geotechnical instrumentation and laboratory testing techniques developed, the third design process using experimental methods became popular in optimizing designs.

Terzaghi and Peck (1967) recognized that the geotechnical designs based on excessive safety factors and unfavorable assumptions are inevitably uneconomical. However, there are no other means for a designer to provide assurance in advance of construction that the soil-supporting structure would not develop unanticipated defects in its lifetime. As a means to overcome this deficiency, a formal framework and principles for OA, i.e., the fourth method of design, was first laid down by Peck in his 1969 Rankine Lecture, primarily based on the application of the observational method pioneered by Terzaghi (1943).

The OA, as its name implies, fundamentally facilitates design changes during construction while establishing a framework for associated risk management. It is based on a flexible design that is customized to suit the actual construction conditions, which is validated by responses from observations and performance monitoring. Nicholson et al. (1999: 12) provided the current and most widely applied definition for this approach:

The observational method (or approach) in geotechnical engineering is a continuous, managed, integrated process of design, construction control, monitoring and review that enables previously defined modifications to be incorporated during or after construction as appropriate. All these aspects have to be demonstrably robust. The objective is to achieve greater overall economy without compromising safety.

Peck (1969) defined two main categories of projects where the OA can be adopted, viz. *ab initio* and *best way out*. The intended use of OA principles from the inception of project is called *ab initio* application, whereas the introduction of OA principles to projects that run into difficulties arising from unexpected circumstances is referred to as the *best way out* approach. The initial framework introduced by Peck (1969) listed below is considered to be generally valid in developing a comprehensive design for both of the above applications:

a. Sufficient exploration to establish at least the general nature, pattern and properties of the deposits, but not necessarily in detail;
b. Assessment of the **most probable** conditions and the **most unfavorable** conceivable deviations from these conditions. In this assessment geology often plays a major role;
c. Establishment of the design based on a working hypothesis of behavior anticipated under the **most probable** conditions;
d. Selection of quantities to be observed as construction proceeds and calculation of their anticipated values on the basis of the working hypothesis;
e. Calculations of values of the same quantities under the **most unfavorable** conditions with the available data concerning the subsurface conditions;
f. Selection in advance of a course of action or modification of design for every foreseeable significant deviation of the observational findings from those predicted on the basis of the working hypothesis;
g. Measurement of quantities to be monitored and evaluation of actual conditions; and
h. Modification of design to suit actual conditions.

However, the establishment of the initial design based on the *most probable* ground conditions has attracted the most concern (Szavits-Nossan 2006). It appears that a considerably high likelihood of unfavorable circumstances occurring during construction attributed to the selection of the *most probable* ground condition in the development of overly optimistic initial design. Case histories have demonstrated that such selection has often led to the implementation of contingency measures causing cost and time overruns, while compromising the safety of construction activities, unless sufficient and reliable data is available from previous works.

To overcome some of these drawbacks, many authors, e.g., Powderham and Nicholson (1996), proposed adaptation of *more probable* (i.e., moderately conservative) ground conditions for the development of initial design for the OA. They also suggested the implementation of '*progressive modification*' to the initial design, where the subsequent design changes will be reviewed and executed sequentially. They also argue that such an approach would lead to initiating necessary changes to the design from a position of known safety via reviewing and establishing safety from the performance observations. It is envisaged that the balance between the risk and the cost can thereby be reviewed and adjusted appropriately.

Most geotechnical engineering projects encounter a wide range of uncertainty in the soil or rock parameters. A typical probabilistic normal distribution of soil parameter results is schematically presented in Figure 16.2. In OA, the *most probable* and *moderately conservative* (i.e., *more probable*) ground conditions are adopted for deformations and load predictions under serviceability limit state (SLS) designs. In contrast, the *most unfavorable* ground conditions are used for ultimate limit state (ULS) designs and robustness checks during risk assessments.

The rationale of selecting geotechnical design parameters and establishing geological design models pertaining to soft soils and ground improvement works are discussed in the previous chapters.

The influence of various uncertainties associated with ground conditions, e.g., loadings, parameters, ignorance, etc., on the predictions of performances, such as embankment settlements and lateral deformations, with respect to the measured values is also presented in Figure 16.2. General experience is that the prediction results of performance parameters, such as deformations, based on initial design are a few times greater than the measured. The differences between predictions and measured values are attributed to uncertainties or ignorance. Such difference can often be reduced to

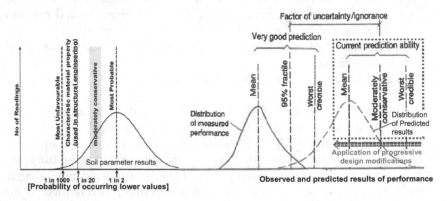

Figure 16.2 Typical probabilistic distributions of soil parameter results, measured performance and predicted results (adapted from Powderham and Nicholson 1996, and Nicholson 2011).

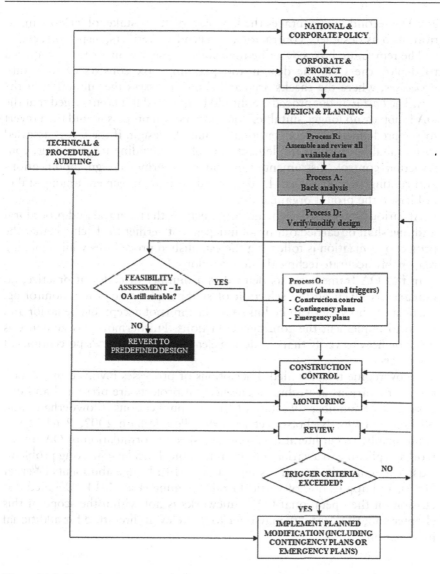

Figure 16.3 Typical operational framework of OA (adapted from Nicholson 2011).

predetermined levels with successful implementation of the OA through progressive design modifications by cautiously revising the soil parameters towards the most probable value.

Figure 16.3 illustrates the typical operational framework proposed by Nicholson (2011) for implementing the Observational Approach. The first box represents the national and corporate policies governing design codes, specifications, quality management systems, health and safety regulations, etc. The second box embraces the key players in the stakeholder's organization, such as designers, contractors, independent verifiers, inspectors, etc.

The remaining boxes describe the robust management structure required to deploy the OA. The design and planning box consists of four sub-processes, where the checks are carried out to assess the suitability of the OA. The *best way out* procedure should be initiated if it is envisaged that the OA is not, or no longer, suitable. The construction process would then revert to the predefined (i.e., conventional or initial) design. If the OA is assessed to be suitable, then the implementation plans including trigger levels, construction control, instrumentation and monitoring, design modifications and contingency plans shall be developed with an agreement among stakeholders in the project organization.

At various stages of design and construction, the technical and procedural auditing shall be carried out by an independent verifier to check whether the project organization is following the established procedures while reaching reasonably accurate technical interpretations.

In this OA framework, as defined in a number of codes of practice, no guidance is given on the selection of suitable parameters for monitoring, particularly how they can be linked to the limits of acceptable behavior and when to implement the contingency actions. Furthermore, no guidance is given on how to verify that the design element fulfils society's perceptions of safety levels.

To overcome this, detailed descriptions of processes involved in OA pertaining to various geotechnical engineering projects are presented and discussed in a number of state-of-the-art publications (Powderham and Nicholson 1996; Nicholson et al. 1999; Powderham 2002; Patel 2012). Additionally, recent literature on project specific formulation of OA frameworks applicable to a variety of different geotechnical engineering problems exist, e.g., embankments (Prastings et al. 2014); bridge abutments (Werner 2013); and application of vertical drains (Chung et al. 2014). Detailed discussion on the operational OA frameworks is not within the scope of this chapter and, thus, readers may refer to the relevant literature for additional information.

16.3 UNCERTAINTIES AND OBSERVATIONAL APPROACH

In 1945 Terzaghi wrote:

> In the engineering for such works as large foundations, tunnels, cuts, or earth dams, a vast amount of effort and labour goes into securing only roughly approximate values for the physical constraints that appear in the equations. Many variables, such as the degree of continuity of important strata or the pressure conditions in the water contained in the

soils, remain unknown. Therefore, the results of computations are not more than working hypotheses, subject to confirmation or modification during construction.

(Peck 2001: 71)

Therefore, the principal objective of the implementation of the OA is to eliminate unnecessary efforts at the initial design phases while preventing unacceptable risk levels caused by uncertainties associated with the construction elements, particularly in the ground, while achieving overall savings within a risk management framework without compromising safety. Therefore, it is indispensable to understand the inherent uncertainties and risks associated with problems dealt by geotechnical engineers.

The uncertainties associated with geotechnical engineering problems are systemic and widespread. However, these can broadly be categorized into two, viz. aleatoric and epistemic uncertainties. Aleatoric uncertainties represent random, unpredictable and spatially variable quantities that govern the design solution and construction, e.g., critical geological features or details that might have missed from the limited field exploration programs, characteristic values of quantities pertinent to the engineering solution, etc. There are three categories of epistemic uncertainties arising due to lack of knowledge on the pertinent quantities, viz. characterization uncertainties, model uncertainties and parameter uncertainties (Spross 2016):

- Characterization uncertainties – Associated with interpreting ground from the results of an exploratory investigation and may depend on the measurement errors and unrepresentative data;
- Model uncertainties – Pertaining to behavior models (e.g., constitutive models), analytical, statistical and numerical methods (e.g., Finite Element (FE) and tools (e.g., computing software); and
- Parameter uncertainties – Relating to the potential variabilities associated with the assignment of properties of interests, which are estimated from test data or empirical factors.

In addition to the above, a number of authors (Christian et al. 1994; Patel et al. 2007; Baynes 2010) have reported a couple of other areas, where a significant number of uncertainties may persist:

- Solution uncertainties – Uncertainties associated with the proposed initial design, such as appropriateness of the treatment method and safety margins, its cost effectiveness, etc.; and
- Construction, instrumentation and monitoring uncertainties – Often related to complexities of the project and its management (e.g., team psychology), human errors and negligence, methods of construction, accuracy, precision and reliability of instrumentation and monitoring works including data processing.

Often, the focus of researchers and project engineers has mainly been on the expert judgement on initial deterministic designs to reduce the uncertainties related to ground characterization, parameter determination and behavioral models by refining and modifying the initial deterministic design based on the interpretation of observations (e.g., Powderham 2002; Finno and Calvello 2005; Baynes 2010). Other researchers, such as Fuentes et al. (2018), have discussed the application of OA on complex issues from the commencement to the end, addressing lessons learnt during the implementation including sets of critical parameters that should be under observation for achieving successful accomplishment.

Some researchers try to incorporate uncertainties associated with soil properties and their spatial distribution using standard statistical or probabilistic approaches, e.g., Christian (2004) and Prastings et al. (2014). They demonstrate that the project outcomes in relation to site measurements and observations could successfully be achieved by carrying out either Bayesian updating or reliability analyses of parameters and observations.

In addition, Whitman in his Terzaghi Lecture in 1981 recognized and opened up a wider application of the OA for managing geotechnical engineering risks using stochastic models via statistical inference (Whitman 1984). Recently, a number of scholars (e.g., van Baars and Vrijling 2005; Spross 2016) successfully demonstrated the ways and means of managing the risks arising from all aspects of project uncertainties collectively, by integrating the design, construction and procurement processes through an OA.

Detailed presentation of stochastic methods in OA pertaining to soft clay engineering ground improvement is out of the scope of this chapter and readers may refer to the latest publications, some of them mentioned here, for additional information. However, the details of risk assessments of the uncertainties associated in these aspects are presented in Chapter 17.

It is obvious from various research publications and presentations that the world trends mostly towards the application of probabilistic methods. Therefore, for success of projects, a good engineer must acquire adequate knowledge of such methods and shall also have sound engineering judgement or intuition without relying too much on mathematical solutions. Renown scientist Gauss once quoted that *'The solution I have already detected and now, it remains to understand how I have achieved it,'* which describes the indispensability of sound engineering judgement (Palmström and Stille 2015: 219).

16.4 GEOTECHNICL INSTRUMENTATION AND MONITORING

'The purpose of geotechnical instrumentation is to assist with answering specific questions about the ground/structure interaction. If there are no questions, there should be no instrumentation' quote Dunnicliff and Powderham (2001: 11), which clearly and simply elaborates the main objective of geotechnical instrumentation.

Due to inherent uncertainties (see Section 16.3), particularly related to soft soil and ground improvement projects, the geotechnical engineering works are perhaps considered to be relatively hazardous and economically risky. In many such projects, e.g., Singapore's Nicoll Highway Project in 2004, failures are often caused by unexpected changes in the strength and deformability of the soil on or in which the structures are built.

Generally, the changes of the soil mass due to redistribution of stresses, resulting from the application of external loads or relaxations, are reflected in geologic units and structures as deformation, strain, stress and pressure. These can now be measured precisely, accurately and, if needed, in real time, as a result of rapid advancement of electronics and communication systems.

Early detection of changes in any of the critical parameters of the ground or the structure is of great importance due to the reasons presented in Table 16.1.

Field observations including measurements assist geotechnical engineers to achieve a number of objectives as outlined above. The Key Performance Indicators (KPIs), types and purpose of some typical geotechnical instruments are presented in Table 16.2.

Instruments adopted for the measurements of in situ moisture, field suction, inclination or tilt, strain and temperature are out of the scope of this chapter and are not discussed. Readers may peruse Dunnicliff (1993) for additional information on instrumentation that measure these parameters.

The geotechnical surveillance of structures is made visually through planned site inspections and using geotechnical instrumentation. The main aim of the next section is to provide general information related to planned site inspections and some typical instruments that can be used to achieve safety and economy on soft soil and ground improvement projects.

16.4.1 Visual monitoring or site inspections

Visual monitoring of geotechnical structures is a fundamental concept for geotechnical engineers and engineering geologists. Peck has stated that 'Only when the eye cannot directly obtain the necessary data is there a need to supplement it by more specialised instruments. Few are the instances in which measurements by themselves furnish a sufficiently complete picture to warrant useful conclusions' Dunnicliff (1982: 18).

Visual observations generally comprise identification of:

- Local bulging or heaving of areas adjacent to embankments or earth fills;
- Initial evidence of sinkholes, local depressions, mud waving;
- Surface cracking on slopes indicating the occurrence of displacement;
- Any transverse cracking of dams, earth fills or embankments;
- Local seepage issues across dams or water retaining structures; and
- Any new seepage areas, wet areas and/or water ponding downstream of dams or water retaining structures etc.

Table 16.1 Common reasons for geotechnical instrumentation (after Dunnicliff 1988, 1993; Durham 2004; Marr 2007)

Reason	Remarks
Define initial site conditions	• Purpose is to characterize initial site conditions; and • Common parameters of interest are pore water pressure, permeability, suction, moisture variations and movements.
Design verification or proof testing	• Purpose is to verify design assumptions, to check that its performance is as predicted and modify the design in later phases; and • Common parameters of interest are vertical and lateral movements, stresses, strains, pore water pressure, seepage, moisture variation, suction changes and deformations.
Construction control	• Purpose is to monitor the effects of construction and assess its rate without initiating any adverse behavior; and • Common parameters of interest are vertical and lateral movements, stresses, strains, pore water pressure, seepage, moisture variation, suction changes and deformations.
Safety	• Purpose is to obtain early warning of impending failures, decide timing for safe evacuation and timing for implementing remedial measures. • This requires quick retrieval, processing and presentation of data so that decisions can be made promptly.
Legal protection and enhance public relation	• Purpose is to provide evidence for legal defense of designers and contractors should owners of adjacent properties claim that the construction has caused damage. Instruments can also be installed to assure the public that their concerns are looked after.
Performance	• Purpose is to monitor the in-service performance of structures over the period of contractual liability or design life; and • Common parameters of interest are pore water pressure and deformations (e.g., vertical and lateral movements).
Investigating anomalous performance	• Purpose is to understand further about the situations of an anomalous or suspected anomalous performance of a geotechnical element.
Research and advancement of knowledge	• Purpose is to gather further knowledge on the behavior of various elements associated with geotechnical works either at laboratories or in the field as trials and thereby, advancing the state of knowledge.

Table 16.2 Some KPIs and typical instruments used to monitor common geotechnical parameters

Performance Indicators	Common purposes of observation	Examples of some typical instruments	Measurement uncertainty Indicative/typical range
Surface deformation: Δx, Δy and/or Δz (or Δl)	– Establish initial site condition – Design verification – Construction control – Safety – Performance – Investigating anomalous performance – Legal protection – Research & development	– Survey monuments markers, prisms or targets for latitude, longitude and elevation measurements – Settlement plates or platforms (or rod and plate settlement gauge) for vertical movement at a point – Liquid levelling or hydraulic profile gauges for vertical movement profile along a line – Settlement cells for vertical movement at a point – Borros anchors for vertical movement at a point – Surface extensometers for distance between surface points	Depends on the method of measurement For example: + Survey leveling – 2 mm/0.5 mm; + Total stations – 1 to 5 mm; + Electronic distance measurements – 0.5 to 3 mm + Global Positioning System GPS 20 mm
Displacement in the ground or in the structure (internal movements) Δx, Δy and/or Δz (or Δl)	– Establish initial site condition – Design verification – Construction control – Safety – Performance – Investigating anomalous performance – Legal protection – Research and development	– Inclinometers for transverse deformation normal to the axis of measurements – Deflectometers for transverse deformation around excavations together with rapid monitoring – Shear strip, shear plane indicator or shear probe for depth and extent of shear plane – Slope extensometers for depth and extent of shear plane – Borehole extensometers for vertical deformation at points	+ Inclinometers – 1 to 2 mm over 10 m depth; + Extensometers – 0.01 to 5 mm + Liquid level gauges – 0.1 to 1 mm

(Continued)

Table 16.2 (Continued)

Performance Indicators	Common purposes of observation	Examples of some typical instruments	Measurement uncertainty Indicative/typical range
		– Liquid level gauges or hydrostatic profile gauges for relative vertical deformation profiles – Fiber optics cable sensors (potential use) for vertical deformation profiles – Fiber optic extensometers for relative vertical movements – Shaped accelerometer arrays for transverse deformation, which uses Micro Electro Mechanical System (MEMS)	
Water and pore water pressure Δl (m) or Pressure (kPa)	– Establish initial site condition – Design verification – Construction control – Safety – Performance – Investigating anomalous performance – Legal protection – Research and development	– Measurement of groundwater level and its seasonal fluctuation – Observation well with dipmeter – Assessing groundwater pressure around the piezometer element: × Open standpipe piezometer with dipmeter × Twin tube hydraulic piezometer, e.g., Bishop piezometer × Pneumatic piezometer × Vibrating wire piezometer × Electrical resistance piezometer × Fibre optic piezometer	+ Dipmeter accuracy goes up to 10 mm + Pneumatic piezometer–0.5% of measuring range + Vibrating wire and electric piezometers – 1% of measuring range + Fiber optic piezometer – 0.25% of measuring range

However, visual inspections have several drawbacks (Stateler et al. 2014), viz.:

- Visual inspections can only provide qualitative assessments and cannot detect subtle changes;
- Often limited to only surface observations and, thus, any changes within the geotechnical structure cannot be observed; and
- Performance parameters of geotechnical structures cannot be evaluated from visual observations alone as they do not produce hard data for comprehensive assessments.

Therefore, a well-designed instrumentation and monitoring program shall overcome the above drawbacks. Peck further elaborated that *there is no substitute for systematic intelligent surveillance. But monitoring and surveillance are not synonymous with instrumentation* (de Rubertis 2018: 1).

16.4.2 Instrumented monitoring

Designs can be validated, verified and updated if field performance could be observed while monitoring the behavior of geotechnical elements such as the site investigation and construction progress. Therefore, monitoring, which is the collection, processing and evaluation of measurements recorded by instruments combined with information from visual observations, often forms an integral part of the design process.

The simplified steps shown in Table 16.3 should be followed for a successful planning and implementation of an instrumented monitoring program.

16.4.3 Selection of instrument type

An uncomplicated instrumentation program is more likely to produce reliable measurements with fewer errors and less cost. Therefore, overall simplicity is considered vital when finalizing a design for instrumented monitoring. However, this does not mean that deploying many instruments is a bad idea. The number of instruments required depends on the number of KPIs requiring measurements to answer questions on the behavior of the geotechnical element. The attributes shown in Table 16.4 are important for a successful instrumented monitoring program irrespective of its scale.

16.4.4 Types of instruments

Geotechnical monitoring is based on the measurement of geotechnical key performance indicators or parameters. The vital geotechnical performance indicators and typical types of currently adopted instruments are also presented in Table 16.2.

Table 16.3 Proposed steps for a systematic approach in planning and
implementing successful geotechnical monitoring programs (after
Dunnicliff et al. 2012; de Rubertis 2018)

Step	Task	Remarks
1	Define the project conditions	Be familiar with planned construction works, geological and hydrological conditions, site constrains and restrictions, etc.
2	Postulate potential failure mechanism that controls the behavior of project elements	Preliminary or initial design should be performed with at least two sets of geotechnical design parameters: • Most probable, and most unfavorable (i.e., worst credible); and • Moderately conservative, and most unfavorable (i.e., worst credible) And the design should cover all the likely scenarios to be encountered.
3	Define the requirements of geotechnical instrumentation	Every instrument should be selected and located to assist in resolving uncertainties, verify design considerations and quality requirements associated in the design.
4	Identify, analyze, allocate and plan for control of risks	All risks associated with construction should be identified and quantified based on probability of failure and its consequences. Instrumentation planning and priorities should be based on this assessment. Responsibility for each risk should be allocated to a party or more and included in the contract documents.
5	Select key performance indicators (KPIs) or parameters:	Typical geotechnical parameters include: • Groundwater level and pore water pressure in soils • Settlement, lateral movement, heave, crack widths; and • Load and stain in geosynthetics and structural members
6	Predict magnitudes of change	• Predictions are required to establish ranges and accuracies of the instruments to be selected; and • If monitoring is performed for construction control or safety purposes, numerical values of KPI are required to establish trigger levels to take action.
7	Devise remedial action (i.e., contingency plans) or design alterations for OA	Action plans should be deduced and be in place if the observation demonstrates that a remedial action is needed. Prior arrangements should be made to determine how all parties will be forewarned of the planned remedial actions.

(Continued)

Table 16.3 (Continued)

Step	Task	Remarks
8	Assigning the task for construction phase	This is one of the tasks of utmost importance and includes: • Selecting and purchasing instruments; • Installing instruments; • Monitoring; • Maintaining instruments and data; • Interpreting data; and • Implementing actions resulting from the observations and monitoring results.
9	Select instruments	Previous steps should be completed before instruments are selected. Reliability and simplicity should be the key features in the selection criteria. The scale of the geotechnical monitoring program should match the identified risks, scale and complexity of the geotechnical requirements. Overall cost of procurement, calibration, installation, monitoring, maintenance and data processing should be compared.
10	Select instrument locations	Suggested practical approach is: • Establish zones of high risks; • Identified critical sections are regarded as *primary instrumented sections* and instruments are located to obtain comprehensive details of ground/structure interaction and construction performance; • Instruments should be installed at a number of *secondary instrumented sections* for comparison purposes; and • If the behaviors of secondary sections are significantly different from that of primary sections, additional instruments may be installed as the construction progresses.
11	Identify what issues may cause changes to the measured data	All records, including diaries, construction details, visual observation notes, etc., pertaining to the monitoring works shall be maintained to support interpretation of measured data. Monitoring data itself is not sufficient to establish underlying causes for the changes to measured data.
12	Prepare budgets	Budgets should incorporate the cost of current and all future tasks involved in the instrumentation program, and may include: • Calibration; • Installation and maintenance; • Monitoring (data collection); • Interpretation of monitoring data; and • Reporting of conclusions, implementation, and processing and presentation of monitoring data.

(Continued)

Table 16.3 (Continued)

Step	Task	Remarks
13	Prepare two sets of contract documents	• Set 1 – For a contract between the owner and the designer/geotechnical consultant or the specialist monitoring contractor together with risk responsibility allocations; and • Set 2 – For a contract between the owner and the construction contractor should also refer the first contract and include risk responsibility allocations.

Table 16.4 Attributes of successful instrumented monitoring program (after Dunnicliff et al. 2012 and de Rubertis 2018)

Attribute	Description
Simplicity	Overall simplicity is a key factor. A complex system has avenues to supply misleading observations and could often face data validity issues when anomalous data is generated.
Robustness and vandal resistance	Instruments should demonstrate remote risk of failure and should be resistant to vandals during its design life.
Range, resolution, sensitivity, accuracy and precision	Instruments are manufactured to operate within a specified range, resolution, accuracy, precision and repeatability. Therefore, the selection of instruments shall match program needs.
Reliability and compatibility with construction methods	Reliability of instruments is indispensable where replacements are realistically impractical. The system of selected instruments shall not adversely impact the safe and proper construction.
Insensitive to environmental conditions	The system including all its parts shall be designed to operate properly in environmental conditions to which they will be exposed.
Easy to read, access, replace and maintain	These are key factors that influence the cost of operation and maintenance of the program. Safety regulations shall be considered and confirmed when installing instruments in areas where access is difficult and dangerous.
Low cost of procurement, installation, data acquisition and maintenance	When selecting instruments, low cost of procurement, operation and maintenance shall be considered. However, cost saving achieved at the expense of other vital considerations discussed above is unwise. Out of the above considerations, the cost may be the least important consideration in the design of an instrument system.

In addition, in situ stresses, temperature, velocity and acceleration are the other key parameters often monitored in numerous geotechnical engineering projects. Out of the above, the instruments commonly adopted for soft soils and ground improvement works are further discussed in the following sections.

16.4.5 Instruments for deformation monitoring

Instruments needed for deformation monitoring in soft clay and ground improvement projects include:

- Bench mark or datum – For vertical and horizontal movement measurements. Hence it should be established in stable ground (Figure 16.4);
- Settlement markers (monuments), plates and anchors – Used to monitor surface movements (Figure 16.5 and Figure 16.6);
- Borros and spiral foot anchors – Used when compressible layer is thin and shallow overlying a competent layer, eliminates the need for survey monuments (Figure 16.7);
- Extensometers – Used to measure vertical movement at different depths (usually within a borehole) (see Eberhardt and Stead 2011). They differ based on single or multi points or type of readout (e.g., dial gauge, fiber optics etc.) (see Durham 2004);

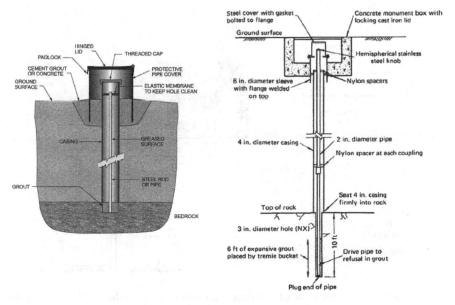

Figure 16.4 Typical arrangements of benchmarks installed in stable ground.

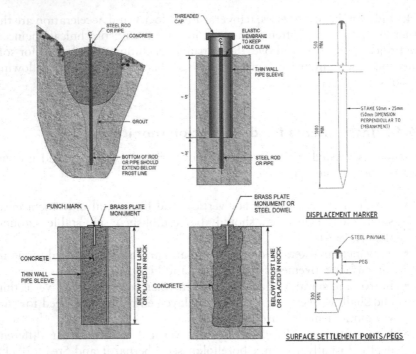

Figure 16.5 Examples of survey monuments used in construction (adapted from D'Appolonia 2010).

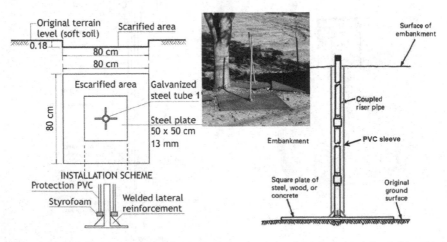

Figure 16.6 Schematic view of a typical settlement plate (adapted from Almeida and Marques 2013, and Dunnicliff 1993).

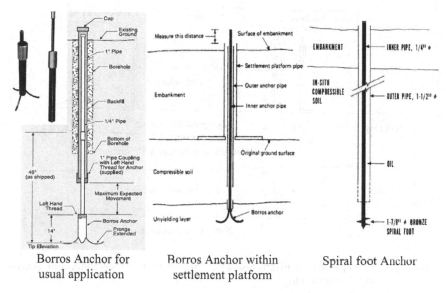

Borros Anchor for usual application

Borros Anchor within settlement platform

Spiral foot Anchor

Figure 16.7 Borros and spiral foot anchors or subsurface settlement points (adapted from Dunnicliff 1993, Durham 2004 and de Rubertis 2018).

- Settlement (or full) profile gauges – Provide a continuous, near surface settlement profile along a line. Appropriate where nonuniform vertical deformations are likely to occur, e.g., road embankment on soft clays. There are two main probe types currently available in the industry, viz. inclinometer and hydrostatic (or torpedo) probes.
- Inclinometers – Vertical inclinometers measure the lateral movement whereas horizontal inclinometers act similar to a profile gauge and measure the vertical movement profile. In-place inclinometers (IPI) are recent additions and allow automation.

An example of an instrumentation scheme for settlement monitoring of a typical embankment is shown in Figure 16.8. The advantage of using a settlement profiler is shown in Figure 16.9.

16.4.6 Instruments for water pressure measurements

Groundwater pressure is one of the key parameters necessary to understand the field performance of a number of geotechnical engineering structures in soft clay projects. Water pressure is monitored for several reasons (Sarsby 2013):

- To establish the location and fluctuation (especially in tidal affected areas) of the groundwater table;
- To assess settlement rate and degree of consolidation;

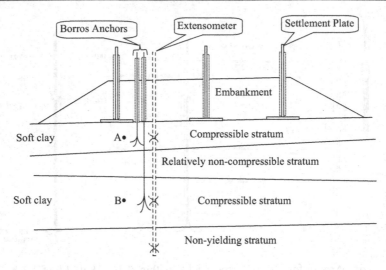

Figure 16.8 Example of an instrumentation scheme for an embankment settlement monitoring.

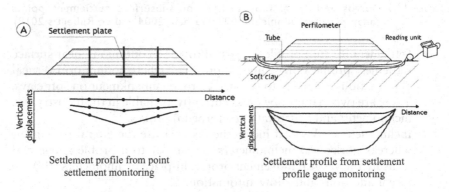

Figure 16.9 Settlement profiles from point and profile measurements (modified from Almeida and Marques 2013).

- To evaluate initial groundwater quality and to set baseline values prior to construction and to subsequently monitor any changes; and
- To assess the performance of groundwater control design measures, such as relief wells, blanket drains, construction dewatering efforts, and other drainage features.

Most commonly used types, such as groundwater wells, standpipes and vibrating wire piezometers, are briefly discussed here. Other types of instruments, such as liquid levels, hydraulic and pneumatic piezometers are described by de Rubertis (2018) and Dunnicliff (1993, 2012).

Observation wells and standpipes are very common features in soft soil projects and are shown in Figure 16.10. The major inherent drawback is their prolonged time lag (response time). These piezometers are, therefore,

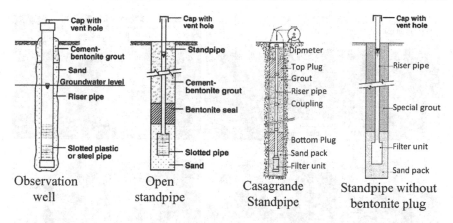

Figure 16.10 Schematic of observation well and standpipes installations (adapted from Dunnicliff 2012 and de Rubertis 2018).

best suited for the long-term groundwater monitoring works and in relatively permeable materials.

Due to relative merits of this system, vibrating wire (VW) piezometers are noted for long-term stability and reliability of measurements (de Rubertis 2018). Compared to standpipes, they have very short time lag (response time), and are thus well suited for monitoring rapid fluctuation of groundwater or excess pore water pressure. The data from VW piezometers can be either obtained manually with a portable data logger or could be stored in an Automated Data Acquisition System (ADAS). In the latter case, the instrumentation setup can be configured to acquire data in real-time and operate remotely via an appropriate telecommunication. Currently, VW piezometers are often the instrument of choice for monitoring groundwater pressure, especially in soft soils and ground improvement projects.

Pneumatic piezometers are also operating on a similar principle of membrane deflection. These instruments use gas, often nitrogen, to balance outside water pressures. Installation of this instrument is similar to that of VW piezometers and it produces a simple and reliable measurement that is free from zero drift.

16.4.7 Protection of instrumentation and maintenance

One of the key factors that should be considered in the design of an appropriate instrumentation system is the protection against damage during and after installation. The main contributors for instrumentation damage are:

- Construction activities – Components of instruments above ground are often subjected to damage by construction machinery;
- Vandalism;

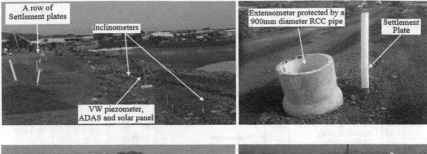

Figure 16.11 Above ground instrument protection measures on earthworks projects.

- Harsh environments and climates, water and humidity, shock or excessive vibrations, dust, lightening exposure, chemical precipitates and loss of access; and
- Attack by animals, e.g., rodents.

Importantly, the data collected from the field shall also be protected against loss from accidental misplacement or physical damage.

Figure 16.11 shows means of protecting above ground components of instruments. For further details, refer to Dunnicliff (1993) and de Rubertis (2018).

16.4.8 Future trends in instrumented monitoring

Latest advances in electronics and sensing have paved way to innovate various types of deformation monitoring instruments. These instruments are continuously evolving and are briefly discussed in this section.

16.4.8.1 Settlement systems

A settlement system generally consists of a liquid-filled settlement reservoir attached to a settlement plate. A nylon liquid line originating from the reservoir connects to a vibrating-wire sensor at the other end. A VW sensor measures the water head at the reservoir that changes as it settles together with the embankment (Figure 16.12).

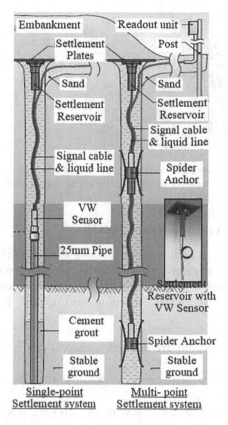

Figure 16.12 A settlement system configuration (adapted from Almeida and Marques 2013; Eberhardt and Stead 2011).

16.4.8.2 Shape accelerometer arrays (or ShapeAccelArray)

A shape accelerometer array (SAA) generally consists of a number of slender, rigid segments of typically 0.3 m to 0.5 m in length and connected end-to-end by special torque resistance flexible joints. Each segment of this SAA contains three micro electro mechanical system (MEMS) accelerometers to measure the axis tilt in 3D and vibrations. SAAs are installed in trenches and boreholes when used to monitor horizontal as well as vertical displacements, respectively. This system does not require any special casing other than a conduit to be installed in the ground (Figure 16.13). Further information on SAA systems can be found in de Rubertis (2018).

16.4.8.3 Fiber optics instruments

Fiber optics is an emerging geotechnical instrument technology. Currently, fiber optic-based instrumentation includes displacement gauges, piezometers, strain gauges and temperature gauges.

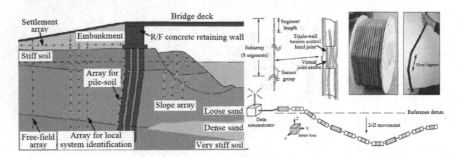

Figure 16.13 Applications of ShapeAccelArrays at a bridge abutment and its typical components (adapted from Eberhardt and Stead 2011 and de Rubertis 2018).

16.4.8.4 Wireless sensor network (WSN)

The latest trend in geotechnical monitoring instruments is to go wireless. This has become an attractive alternative to currently wired networking due to the possible elimination of surface cable installation within boundaries of the construction site. This emerging technology has the following advantages:

- Sensors in areas that are difficult to access can be connected without cables;
- Wide range data acquisition is possible that would eventually help swift decision making;
- Each node is self-sustainable with its own long-lasting internal battery, which can serve long durations without a need for battery change; and
- Advanced WSN contains mesh networks where nodes can serve as sensors as well as relays. This makes the installation, operation and maintenance of the system less time consuming and cost-effective.

The presence of high-voltage power lines or radio transmission towers could make the site too noisy for wireless links. Zhang et al. (2018) provide a detailed account of this emerging technology.

16.4.8.5 Automated data acquisition

Automated Data Acquisition Systems (ADAS) are becoming reliable and cost-effective for monitoring the performance of construction works and their surroundings. It is expected that with time, ADAS will continue to improve and become a common feature at work sites. Main features of a typical ADAS are that:

- They are programmed to retrieve data on a prescribed schedule without human intervention;
- They are designed to accommodate more than one transducer;

- Signal conditioning could be incorporated to amplify the output and present in terms of engineering dimensions; and
- Data can be recorded or transmitted elsewhere for recording.

ADAS allows real-time monitoring of critical parameters. Although ADAS cannot replace engineering judgement, the following distinct advantages can be gained:

- Increase the frequency of readings;
- Retrieve data from remote/inaccessible locations;
- Instantaneous transmittal of data over long distances, either by wired links or wireless communication network;
- Measurement of rapid changes or fluctuations in monitored parameters;
- Increase reading sensitivity;
- Reduce measurements and data acquisition errors;
- Increased flexibility in selecting, managing and storing data; and
- Increased safety due to less interference with construction activities during monitoring.

In addition, the control software is typically customized for each application to facilitate automatic processing of readings, alert checks, graphic displays and report generation. The software is usually programmed to highlight alert levels through scanning the incoming data and also to store raw readings in project databases.

There are some disadvantages of ADAS as well:

- More susceptible to lightening damage;
- Very dependable on the reliability of the communication networks;
- Threat of vandalism is greater – monitoring stations need to be hard fenced; and
- Loggers need to be in lockable enclosures to protect from possible vandalism and adverse environmental effects.

Detailed descriptions on various types of ADAS and their uses and limitations are presented by de Rubertis (2018).

16.5 INSTRUMENTATION MONITORING PLAN

The main objective of the monitoring plan is to provide adequate information to the designer on KPIs identified at the design stages to help establish the validity of the initial design and, thereby, to implement planned modifications or emergency or contingency plans where necessary.

The main elements of a well-developed monitoring plan are (Nicholson et al. 1999):

- Identification of observations that are needed, i.e., KPIs pertaining to the postulated failure modes, and also for 'general health monitoring' of the works;
- Defining the instrumentation system that measures the required KPIs and their range of expected performance, directly and/or indirectly via primary and/or secondary instrumentation systems, including methods of monitoring, data acquisition, transmission, processing and management;
- Establishing reporting methods and frequency of monitoring;
- Assigning duties and responsibilities within the project team, in particular, the actions to be taken and by whom in an event of an emergency, when alert levels being approached, or unexpected adverse trends recognized in the results;
- Deducing construction control procedures and to produce construction control instructions and guidelines, e.g., safe work method statements, manuals and charts for the use of site personnel; and
- Determining ways and means of data evaluation that assist to provide trend data for later comparison and refinement of the design, if required.

Importantly, planning a monitoring program using geotechnical instrumentation should begin with defining the objectives and end with planning how the measurement data will be used. A 13-stepped systematic approach, as shown in Table 16.3, has been proposed for successful planning and implementation of geotechnical monitoring works and it is expected that all these steps should, if possible, be completed prior to commencing geotechnical monitoring works in the field.

Dunnicliff (2012) further stresses that, although all of above steps are important, steps 3, 4, 7 and 8 (in Table 16.3) are critical to implement and deliver a successful instrumentation monitoring plan. He repeatedly highlights that singling out these four steps in no way implies that the other steps are unimportant.

16.5.1 Instrument numbering

Instrument numbering systems are not often discussed in geotechnical instrumentation literature. However, establishing appropriate instrument numbering systems for large-scale projects is very important for a successful monitoring plan. Figure 16.14 presents an example of an instrumentation system adapted for a road widening project consisted of ground improvement works.

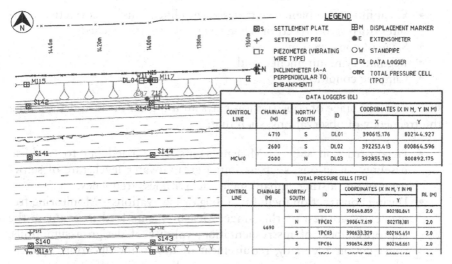

a. Plan view of an instrumented section for a road widening works

b. Sample showing designated instruments at primary instrumented sections

Figure 16.14 Example of an instrument numbering system adapted for a road widening.

Appropriate numbering system for a linear infrastructure project generally includes:

- Location of the primary instrumented section, e.g., chainage of road;
- Location of the instrument with respect to the centerline of the road, e.g., north or south (optional);
- Type of instrument, e.g., typical abbreviations: TPC for Total Pressure Cell; N or IN for inclinometers (see the legend in Figure 16.14b); and
- Instrument number with '– sub number' for multiple instruments of the same type found at the same borehole or location, e.g. VW piezometers, extensometer magnets, etc.

16.5.2 Frequency of monitoring

Type, quantity and the frequency of monitoring should be established at the time of the initial design stage depending on the instrumentation systems and locations. Typically, the monitoring frequency depends on the following:

- Degree of uncertainty in the design assumptions and modelling;
- Complexity of the ground and loading conditions;
- Potential risk of failure during construction;

- Feasibility of implementing design modifications or corrective measures during construction when trigger limits are reached;
- Stage of construction;
- Regulatory requirements; and
- Climatic conditions.

In general, the stage of construction plays an important role in determining the frequency of observations, such that:

- Preconstruction stage – Measurements are carried out to establish baseline conditions, such as initial groundwater level, rate of settlement of an existing embankment, etc. with less frequency.
- During construction including filling or excavations – Monitoring frequency is greatest at this stage. Measurements lead to evaluate the appropriateness of instrumentation, validity of design assumptions, construction adequacy, requirement of design modification, application of contingency measures and overall safety.
- During service operation – Instruments are measured less frequently. Monitoring of some of the instruments may even cease. Additional instrumentation/monitoring may be needed (e.g., in a road embankment, once the pavement is placed settlement marks on the pavement and/or kerb basically replace the requirement for settlement plate monitoring).

16.5.3 Data processing and presentation

The current trend in the industry is to collect instrumentation data digitally as much as possible. However, there could be some occasions where manual data collection is still warranted such as, records of visual observations, standpipe piezometers, extensometers, and, at times, settlement readings. Such raw data from manual recording should appropriately be stored in the instrumentation data management system, where they can be easily located, used in the interpretation of observations and archived to maintain historical records.

There are a number of objectives in the data processing and presentations:

- Provide rapid assessment of data in order to detect data anomalies and changes that require immediate action, such as construction control;
- Provide information to verify design assumptions confirming the predicted performance and/or guidance for a need of design change; and
- Fulfil obligations towards stakeholders.

Therefore, when assessing and evaluating data, the following ways of data presentation would be beneficial (de Rubertis 2018):

- Plots to assist with data screening;
- Graphing of observations versus time;

- Graphs comparing observed versus predicted behavior;
- Plots to compare observations with construction activities, e.g., settlement with filling and filling rates;
- Graphs to examine 'cause and effect' relationships, e.g., pore water pressure changes due to nearby dewatering works; and
- Summary graphs.

Data interpretation utilizing the data processing works should be carried out continuously and as frequently as possible until clear understanding of the behaviors of the geotechnical elements under surveillance are established. *'The legitimate uses of instrumentation are so many, and the questions that instruments and observation can answer so vital, that we should not risk discrediting their value by using them improperly or unnecessarily'* (Peck quoted in Dunnicliff et al. 2012: 1363).

16.6 TRIGGER LIMITS

In common engineering practice, the trigger criteria of KPIs, such as values of forces, stresses, displacements, water pressures, etc., are the limits for implementing emergency plans, i.e., discovery-recovery model. However, in the light of the OA, the trigger criteria are considered as the limit for implementing planned design or construction modifications, which is also known as *progressive design model*.

The numerical values of these KPIs can be assessed by calculations for the range of conditions considered in the initial design of the OA based on levels of uncertainties associated in quantifying their impact on design and construction. Understanding of the timing required to implement the recovery cycle is important when establishing trigger limits. This may include discovery time, review and decision-making time and the modification implementation time. The time available to implement the planned modification depends on the rate of deterioration (Nicholson et al. 1999).

The likelihood of failure is governed by the rate at which risk increases. If the risk is developing rapidly, where there is no reasonable time to identify the initiation and onset of failure, i.e., 'brittle failure,' the OA is not suitable. In conditions where the risk is developing at a rate allowing sufficient time between discovery and recovery, then the OA could be best suited, and the feasibility of its successful implementation should be assessed.

In addition, primary and secondary monitoring systems are often deployed to observe the field performance of KPIs. The primary monitoring system is monitored routinely and reviewed by site staff while comparing the results against the trigger criteria. The secondary system often measures the KPIs indirectly and provides additional information to the designer. It also acts as a check on the primary system and as a back-up system and supplement to the primary system (Nicholson et al. 1999).

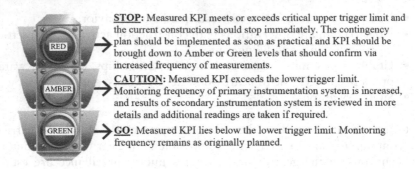

STOP: Measured KPI meets or exceeds critical upper trigger limit and the current construction should stop immediately. The contingency plan should be implemented as soon as practical and KPI should be brought down to Amber or Green levels that should confirm via increased frequency of measurements.

CAUTION: Measured KPI exceeds the lower trigger limit. Monitoring frequency of primary instrumentation system is increased, and results of secondary instrumentation system is reviewed in more details and additional readings are taken if required.

GO: Measured KPI lies below the lower trigger limit. Monitoring frequency remains as originally planned.

Figure 16.15 Typical traffic-light system applies to geotechnical monitoring systems.

The typical and convenient engineering practice is to consider the requirements of action via traffic-light analogy as shown in Figure 16.15. The measured KPI values less than the lower threshold level are considered to be GREEN, which signals to continue with the construction. If the KPI measurements exceed the lower trigger level, the action level becomes AMBER, which signals caution and continue. Measurements passing the upper trigger level are critical and considered to be RED, which signals stop work immediately.

The trigger limits for the OA traffic-light systems depend on the project type and the site location. For greenfield sites, the serviceability limit state of proposed geotechnical elements would be of main concern and thus, the limits of KPIs relevant to this requirement are established at the initial design stage. Whereas, for construction in built-up areas, the serviceability limit state requirements of the adjacent structures often dictate the trigger criteria that should be implemented in the OA traffic-light systems.

A good example of using the OA traffic-light system in multistage construction of an embankment over soft clay is shown in Figure 16.16 (Patel 2012). The SLS trigger values of a predicted/monitored parameter, i.e. KPIs, have been established adopting 'most probable' and 'moderately conservative' conditions. The ULS value of the KPI is established adopting the 'most unfavourable' set of parameters or conditions in the geotechnical modelling.

A couple of observation scenarios are shown in this figure. First, when the monitored KPI is consistently plotting in the green region, the planned improvements to the design could be carried out such that the observed KPI would move up towards the amber region in the next stages. Carrying out predetermined design modifications when the observed KPI poses low risk is called progressive modification in the OA.

However, if the KPI is plotted in the red zone consistently, then the construction should be altered via pre-planned contingency measures to bring the KPI back to amber region or better. This type of scenario, where

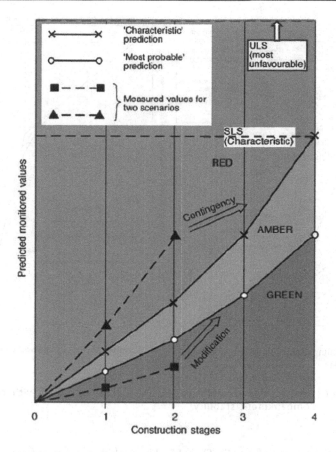

Figure 16.16 Multistage construction trigger limits (adapted from Patel 2012).

predetermined remedial measures are used to reduce high risk towards moderate or low, is known as discovery-recovery procedure in the OA.

Numerous case histories can be found on the application of the traffic-light system in construction control of geotechnical elements. In the case of settlement prediction, there are several methods as discussed in Chapter 5. These techniques can be used to back calculate the settlement parameters and thereby, the necessary modifications to the initial designs can be carried out progressively during construction to achieve optimum designs.

Figure 16.17a shows Ladd's (1991) attempt to demonstrate the possible relationship of deformation parameters to the stability of the embankment. He proposes that the incremental deformation ratio (IDR), i.e., ratio of incremental horizontal deformation at the toe against the incremental settlement, could be about 0.8 to 0.15 for embankments without vertical drains during filling and over the pre-loading period, respectively, while maintaining a

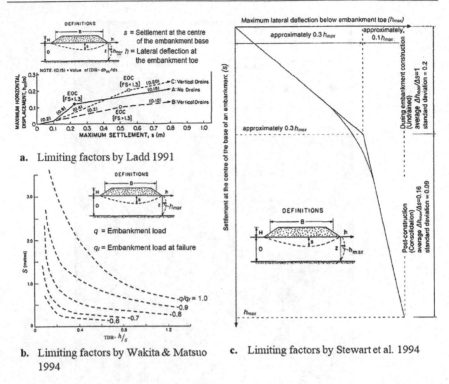

a. Limiting factors by Ladd 1991

b. Limiting factors by Wakita & Matsuo 1994

c. Limiting factors by Stewart et al. 1994

Figure 16.17 Predictions of deformation ratio (IDR or TDR) with respect to embankment stability.

stability Factor of Safety (*FOS*) at 1.3. He further states that for a case of an embankment over ground with vertical drains, IDR could be lower, i.e., 0.2 and 0.1 (*FOS*>1.3) and slightly higher, i.e., 0.4 and 0.2 (*FOS*<1.3) during filling and over the preloading period, respectively.

Stewart et al. (1994) reported IDR to be 1 (with a standard deviation of 0.2) during construction and 0.16 (with a standard deviation of 0.09) post construction for embankments on soft clays (see Figure 16.17c) and admitted that these values could be varied with the stability conditions of embankments.

Wakita and Matsuo (1994) referred to total deformation ratio (TDR) rather than IDR and presented the threshold values of TDR in relation to total settlement of an embankment and ratios of embankment loading (q) with respect to the failure load (q_t) of an embankment (q/q_f) as presented in Figure 16.17b. They suggested that the construction would be in the green region if TDR values and the total settlements plots lie below the 0.6 of q/q_f zone. When TDR vs. s value plots within 0.7 to 0.9 of q/q_f curves then the embankment conditions could be considered as amber. When TDR vs. s value plots beyond 0.8, the condition of the embankment is considered to be in the critical red zone.

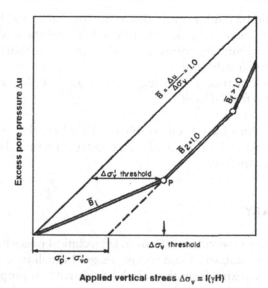

Figure 16.18 Limiting pore pressure ratios for embankment stability (adapted from Eberhardt and Stead 2011 and de Rubertis 2018).

Buggy (2013) claims that IDR offers a reliable method for controlling the stability of the multi-staged embankment construction when used in conjunction with pore water pressure measurements. He suggested to keep IDR under 0.6 for satisfactory performance with acceptable stability.

Hunter and Fell (2003) assessed the deformation behavior and excess pore pressure responses of well-monitored embankments on soft ground and presented graphically as shown in Figure 16.18. They suggest that if the pore pressure ratio, i.e., $\bar{B} = \Delta u/\Delta \sigma_1$ (where Δu is the excess pore pressure generated in the foundation and $\Delta \sigma_1$ is the change in principal stress) is less than 0.75, the embankment construction works is in the green zone. The embankment construction is considered to be in the amber zone if the measured $\bar{B}$ is between 0.75 and 1. If the tendency of observed $\bar{B}$ is reaching towards 1 and increasing beyond, then the status of embankment is considered to be in the red zone, which requires contingency measures to be implemented immediately to prevent impending failure.

The authors' experience on establishing trigger levels for embankments during construction over soft ground areas are:

- Incremental deformation ratio, i.e., IDR = $(\Delta y_m/\Delta h_m)$ will be up to 0.4 for green limit, 0.4 to 0.6 for amber and greater than 0.6 for red, for sites with high risk of failure and where the details of geotechnical investigation are not sufficient to establish geotechnical models with adequate confidence, e.g., sparsely spaced investigations on linear infrastructure projects, such as roads;

- IDR would be 0.6 for green, 0.6 to 0.8 for amber and beyond 0.8 for red limits where the risk of failure is low and the information from geotechnical investigations are sufficient to establish geotechnical models confidently; and
- Pore pressure ratio, $\bar{B} = \Delta u/\Delta \sigma_1$, is 0.7 for green, 0.7–1.0 for amber and beyond 1.0 for red limits.

These typical values have been established based on the thorough assessments of numerous risks associated with construction of embankments and, thus, are project specific.

16.7 SUMMARY

Conventionally, engineering solutions to geotechnical issues have been based on a single, fully established and robust design. In such instances, the inherent uncertainties arising from the variability of material properties, the lack of understanding and knowledge of the behavior of geotechnical design elements due to natural variability of ground and external factors are dealt with by the introduction of often overly excessive factors of safety into the designs. It has been recognized that geotechnical design on the basis of excessive safety factors and unfavorable assumptions are inevitably uneconomical.

In contrast, it has been demonstrated around the world that the Observational Approach (OA) can successfully be adopted and implemented to overcome difficulties or unknowns in construction. OA techniques developed to deal with uncertainties in the design process related to applied soil mechanics are widely known to geotechnical engineers. Depending on the circumstances, OA could often lead to project savings in terms of time and cost without compromising the performance and safety. If the OA is assessed to be suitable, then the implementation plans including trigger levels, construction control, instrumentation and monitoring, design modifications and contingency plans should be developed with an agreement of all stakeholders in the project organization. The emphasis on prediction, observation, feedback, evaluation, and teamwork are the key factors for a successful implementation of the OA which also creates a strong opportunity for learning. This chapter discusses the current definition of the OA related to geotechnical engineering in detail, outlining the frameworks associated with its successful implementation.

Appropriate instrumentation and monitoring of KPIs are an inherent requisite for the successful implementation of OA design. Such instrumentation and monitoring schemes not only provide the needed assurance concerning safety but also feedback for the reviewers and designers to implement necessary contingency plans via design modifications as construction progresses. Therefore, this chapter discusses the current state and practices in geotechnical instrumentation and monitoring works in the industry. Information on a number of frequently deployed geotechnical instruments is also presented.

In the light of OA, trigger criteria are considered as the limits for implementing planned design or construction modifications. This chapter discusses how trigger values could be established for critical KPIs and what actions should be taken in addition to typical trigger values proposed in the past as well as the experience of the authors.

Special attention is given to the timing required to implement recovery measures. If the risk is developing rapidly, where there is no reasonable time to identify the initiation and onset of failure, i.e., 'brittle failure,' it is proposed that OA is not suitable. Conditions where the risk is developing at a rate allowing sufficient time between discovery and recovery, then the OA could be best suited, and the feasibility of its successful implementation should be assessed.

REFERENCES

Almeida, M. de S.S. and Marques, M.E.S. (2013). 'Design and performance of embankments on very soft soils.' *Monitoring Embankments on Soft Soils*, CRS Press, London, 159–174.

Baynes, F.J. (2010). 'Sources of geotechnical risks.' *Quarterly Journal of Engineering Geology and Hydrology, Geological Society of London*, 43, 321–331.

British Standards Institution (BSI) EN 1997-1 (2004). '*Eurocode 7: Geotechnical design – Part 1: General rules*.' BSI Standards Publication.

Buggy, F. (2013). '*Deformation performance and stability control of multi-staged embankment in Ireland*.' *Proceedings of the 18th International Society of Soil Mechanics and Geotechnical Engineering Conference, Paris*, 1241–1244.

Christian, J.T. (2004). 'Geotechnical engineering reliability: How well do we know what we are doing?' Thirty-Ninth Terzaghi Lecture in 2003, *ASCE Journal of Geotechnical and Geoenvironmental Engineering*, 130(10), 985–1003.

Christian, J.T., Ladd, C.C., and Beacher, G.B. (1994). 'Reliability applied to slope stability analyses.' *ASCE Journal of Geotechnical Engineering*, 120(12), 2180–2207.

Chung, S.G., Kweon, H.J. and Jang, W.Y. (2014). 'Observational method for field performance of prefabricated vertical drains'. *International Journal of Geotextiles and Geomembranes, Elsevier*, 42, 405–416.

de Rubertis, K. (2018). 'Monitoring dam performance: Instrumentation and measurements.' *ASCE Manuals and Reports on Engineering Practice No. 135*. ASCE, American Society of Civil Engineers Publication, Virginia, USA.

D'Appolonia Engineering (2010). 'Instrumentation and performance monitoring.' *Chapter 13 and Appendix 13A: Engineering And Design Manual: Coal Refuse Disposal Facilities, Second Edition*. Mine Safety and Health Administration (MSHA), US Department of Labor. https://arlweb.msha.gov/Impoundments/ DesignManual/Chapter-13,_Appendix-13A.pdf.

DIN 1054 (2010). 'Subsoil – Verification of the safety of earthworks and foundations – supplementary rules to DIN EN 1997-1.' German Standards, German Institute for Standardisation (Deutsches Institut für Normung), Germany.

Dunnicliff, J. (1982). 'Synthesis of Highway Practice 89'. *National Corporative Highway Research Program (NCHRP)*, Transport Research Board, National Researchg Council, Washington, DC.

Dunnicliff, J. (1988). *Geotechnical Instrumentation for Monitoring Field Performance.* John Wiley and Sons, New York.

Dunnicliff, J. (1993). *Geotechnical Instrumentation for Monitoring Field Performance.* John Wiley and Sons, New York.

Dunnicliff, J. (2012). 'Types of geotechnical instrumentation and their usage'. *Chapter 95: ICE Manual for Geotechnical Engineering,* Institute of Civil Engineers, London, 1379–1403.

Dunnicliff, J., Marr, W.A. and Standing, J. (2012). 'Principles of geotechnical monitoring.' *Chapter 94: ICE Manual for Geotechnical Engineering,* Institute of Civil Engineers, London, 1363–1377.

Dunnicliff, J. and Powderham, A.J. (2001). 'Recommendation for procurement of geotechnical instrumentation and field instrumentation services'. *Geotech News,* BiTech Publishers, 19(3), 30–35 www.geotechnicalnews.com/instrumentation_news.php or www.geotechnicalnews.com/pdf/GeoTechNews/2001/GIN_Sept2001.pdf.

Durham Geo Slope Indicator (2004). 'Guide to geotechnical instrumentation. 2004.' *Durham Geo Slope Indicator* (https://durhamgeo.com/)

Eberhardt, E. and Stead, D. (2011). 'Geotechnical instrumentation.' *Chapter 8.5: SME Mining Engineering Handbook,* 3rd edition, Peter Darling (ed.), Society for Mining, Metallurgy, and Exploration (SME), 551–571.

Finno, R.J. and Calvello, M. (2005). 'Supported excavations: Observational method and inverse modelling.' *ASCE Journal of Geotechnical and Geoenvironmental Engineering,* 131(7), 826–836.

Fuentes, R., Pillai, A. and Ferreira, P. (2018). 'Lessons learnt from a deep excavation for future application of the observational method.' *International Journal of Rock Mechanics and Geotechnical Engineering, Elsevier,* 10, 468–485.

Hunter, G. and Fell, R. (2003), 'Prediction of impending failure of embankments on soft ground.' *Canadian Geotechnical Journal,* 40, 209–220.

Ladd, C. (1991). 'Stability evaluation during staged construction'. *ASCE Journal of Geotechnical Engineering,* 117(4), 540–615.

Marr, W.A. (2007). 'Why monitoring performance?' *ASCE Seventh International Symposium on Field Measurements in Geomechanics (FMGM 2007),* DiMaggio, J.A.& Osborn, P. (Eds), Geo-Institute of the American Society of Civil Engineers Publication Reston, VA, 1–27.

Nicholson, D.P. (2011). 'Observational methods – use of 'review' and 'back analysis' to implement the best way out approach', *Presentation at Joint meeting BGA and CFMS,* Paris.

Nicholson, D.P. and Low, A. (1994). 'Performance of Foyle Bridge east abutment.' *Géotechnique,* 44(4), 757–769.

Nicholson, D., Tse, C. and Penny, C. (1999). 'The observational methods in ground engineering: Principles and applications.' *Construction Industry Research and Information Association (CIRIA) Report 185.*

Palmström, A. and Stille, H. (eds) (2015). *Rock Engineering (2nd Edition),* ICE Publication.

Patel, D. (2012). 'Observational methods.' *Chapter 100: ICE Manual for Geotechnical Engineering,* Institution of Civil Engineers, London, 1489–1501.

Patel, D., Nicholson, D., Huybrechts, N. and Maertens, J. (2007). 'The observational method in geotechnics.' *Proceedings of the 14th European Conference on Soil Mechanics and Geotechnical Engineering*, 2, 365–370.

Peck, R.B. (1969). 'Advantages and limitations of the observational methods in applied soil mechanics.' *Ninth Rankine Lecture, Geotechnique*, 19(2), 171–181.

Peck, R.B. (2001). 'The observational methods can be simple.' *Proceedings of Institute of Civil Engineers, London, Geotechnical Engineering*, 149(2), 71–74.

Powderham, A.J. (2002). 'The observational method – learning from projects.' *Proceedings of the ICE – Geotechnical Engineering*, 155(1), 59–69.

Powderham, A.J. and Nicholson, D.P. (1996). 'The way forward.' *The Observational Methods in Geotechnical Engineering*, Institution of Civil Engineers (ICE), Thomas Telford, London, 195–204.

Prästings, A., Müller, R. and Larsson, S. (2014). 'The observational method applied to a high embankment founded on sulphide clay,' *International Journal of Engineering Geology*, 181, 112–123.

Sarsby, R.W. (2013). 'Instrumentation and monitoring.' *Chapter 11: Environmental Geotechnics*, 2nd edition, 211–229.

Spross, J. (2014). 'A critical review of the observational methods.' *Licentiate Thesis*, Department of Civil and Architectural Engineering, KTH Royal Institute of Technology, Stockholm.

Spross, J. (2016). 'Toward a reliability framework for the observational method.' Doctoral thesis, KTH Royal Institute of Technology, Sweden.

Spross, J. and Johansson, F. (2017). 'When is the observational method in geotechnical engineering favourable?' *International Journal of Structural Safety*, 66, 17–26.

Stateler, J.N., Kelsic, R.H., Deway, R.L., Dewayne, A.C., Knight, K. and Luebke, T.A. (2014). 'Instrumentation and monitoring.' *Chapter 11: Design Standards No.13 – Embankment Dams*, Final reclamation design standards, Technical Service Centre, Bureau of Reclamation, U.S. Department of the Interior, www.usbr.gov/tsc/techreferences/designstandards-datacollectionguides/designstandards.html.

Stewart, D.P., Jewell, D.J. and Randolph, M.F. (1994). 'Design of piled bridge abutments on soft clay for loading from lateral soil movements.' *Geotechnique*, 44(2), 277–296.

Szavits-Nossan, A. (2006). 'Observations on the observational methods.' *Proceeding of XIII Danube-European Conference on Geotechnical Engineering*, Logar, J., Gaberc, A. and Majes, B. (eds.), Active Geotechnical Design in Infrastructure Development, Ljubljana, Slovenian Geotechnical Society, 1, 171–178.

Terzaghi, K. (1943). *Theoretical Soil Mechanics*. John Wiley and Sons Inc.

Terzaghi, K. and Peck, R. B. (1967). *Soil Mechanics in Engineering Practice*, 2nd Edition, John Wiley, New York.

van Baars, S. and Vrijling, J.K. (2005). 'Geotechnical applications and conditions of the observational method.' *HERON Electronic Journal*, 50(3), 155–172, http://heronjournal.nl/backissues.html.

Wakita, E. and Matsuo, M. (1994). 'Observational design method for earth structures constructed on soft ground.' *Géotechnique* 44(4), 747–755.

Werner, R. (2013). 'I-20 Mississippi River Bridge – Vicksburg geotechnical solutions.' *Observational Methods in Geotechnical Engineering*, Louisiana Transportation Conference (presentation), www.ltrc.lsu.edu/ltc_13/pdf/presentations/S17_I-20%20Mississippi%20River%20Bridge%20at%20Vicksburg%20Geotechnical%20Solutions_LTC2013.pdf.

Whitman, R.V. (1984). 'Evaluating calculated risk in geotechnical engineering.' *ASCE Journal of Geotechnical Engineering*, Seventeenth Terzaghi Lecture 1981, 110(2), 143–188.

Zhang, C.C., Zhu, H.H., Liu, S.P., Shi, B. and Zhang, D. (2018). 'A kinematic method for calculating shear displacements of landslides using distributed fiber optic strain measurements.' *International Journal of Engineering Geology, Elsevier*, 234(21), 83–96.

Chapter 17

Geotechnical risk management

C. Bridges

17.1 INTRODUCTION

When the word 'risk' is mentioned in a project, one invariably thinks of workplace health and safety. This is a subject applicable to any project and not limited to civil projects only and, in fact, is related to any workplace. Workplace health and safety is covered in many publications, especially government publications, and therefore not covered in this chapter, which is limited to geotechnical risk.

The fact that ground-related problems are a significant contributor to construction project delays and cost overruns (Chapman 2008; Clayton 2001b; Clayton 2001a; Plimmer 2019) indicates that geotechnical risks are still not fully appreciated and, as such, cannot be easily mitigated.

Geological models and geotechnical hazards are often discussed in the literature and in these ground-related risks are identified (Baynes, 2010). However, there are a number of other risks around the people and processes that geotechnical engineers use that are often inadequately addressed or ignored in the geotechnical literature (e.g., experience, leadership, interface issues, separation of roles, design methodology, interpretation of parameters, software) (Bridges 2019).

The Nicoll Highway collapse in Singapore is an extreme example where the various professionals involved have not perceived risk appropriately or communicated risk effectively, and this has resulted in catastrophic outcomes. In this case, a number of workers were killed due to errors made by design and construction professionals, including in the design process. Even when significant distress was observed, the risks to the project were not fully appreciated (Committee of Inquiry 2005a; Committee of Inquiry 2005b).

The Heathrow tunnel collapse in London in 1994 is another example where there was an operational failure in that risks were not managed nor recognized (Health and Safety Executive 2000). The geotechnical consultant on that project was responsible for design, monitoring and interpretation of the monitoring and was found guilty of, among other things, 'failure to warn' (Health and Safety Executive 2000).

More recently, failures to fully appreciate ground risks on the HS2, high speed rail project in the UK, have led to a massive blowout in project costs by over £20bn (Plimmer 2019).

17.2 RISK

The origin of risk is said to date back to the mid-seventeenth century from the French *risque* (noun) and *risquer* (verb), which were derived from Italian *risco* 'danger' and *rischiare* 'run into danger' (Oxford Dictionaries 2018). The concept of risk was ably described in 1662 by Arnaud (1850) as 'the fear of an evil ought to be proportionate, not only to its magnitude, but also to its probability.'

A more recent definition is that risk is the '*effect of uncertainty on objectives*' and it is often expressed similarly in engineering as the '*combination of the consequence of an event and the associated likelihood of occurrence*' (Standards Australia / Standards New Zealand 2009). The consequence can be in terms of loss of life or financial impact. In numerical terms, annual risk, R, can be defined as:

$$R = P_{(O)} \times P_{(P)} \times V \tag{17.1}$$

where:
$P_{(O)}$ = Probability of an occurrence (i.e. a failure)
$P_{(P)}$ = Probability of the presence of a target (building/person etc.)
V = Probability of a consequence (i.e., loss of life, loss of value)

For a construction project the likely impact of an event (hazard) could have consequences to:

* Health and safety;
* The design, design program and associated costs;
* The construction, construction program and associated costs;
* The environment; and,
* Adjacent parties (including structures), the public, and local government/utilities.

In determining the risk of an event, risk matrices are often used (Table 17.1). For example, if the likelihood is considered 'Almost Certain' and the consequence is assessed to be 'Major,' they result in Extreme Risk as per the table.

Table 17.1 Example risk matrix

Probability (likelihood)	Consequence (impact)				
	1 Insignificant	2 Minor	3 Significant	4 Major	5 Extreme
5 – Almost certain	Medium 11	High 16	High 20	Extreme 23	Extreme 25
4 – Likely	Medium 7	Medium 12	High 17	High 21	Extreme 24
3 – Possible	Low 4	Medium 8	Medium 13	High 18	High 22
2 – Unlikely	Low 2	Low 5	Medium 9	Medium 14	High 19
1 – Rare	Low 1	Low 3	Low 6	Medium 10	Medium 15

Impact Band	Impact (e.g. Health & Safety)
5 - Extreme	Fatality
4 - Major	Permanent Injury
3 - Significant	Lost Time Injury
2 - Minor	Medical Treatment
1 - Insignificant	First Aid Treatment

Probability Band	Description
5 - Almost Certain	The threat can be expected to occur 75% - 99%
4 - Likely	The threat will quite commonly occur 50% - 75%
3 - Possible	The threat may occasionally occur 25% - 50%
2 - Unlikely	The threat could infrequently occur 10% - 25%
1 - Rare	The threat may occur in exceptional circumstances 0% - 10%

17.2.1 Geotechnical risk

The major part of the college training of civil engineers consists in the absorption of the laws and rules which apply to relatively simple and well-defined materials, such as steel or concrete. This type of education breeds the illusion that everything connected with engineering should and can be computed on the basis of a priori assumptions.

> Unfortunately, soils are made by nature and not by man, and the products of nature are always complex.

> (Terzaghi, 1936)

As Terzaghi stated in his Presidential Address over 80 years ago, many civil engineers, including geotechnical specialists, do not fully comprehend the uncertainty in what they do (Christian 2004). Geotechnical engineers must deal with heterogeneous ground conditions that can vary in thickness, strength and stiffness below every square meter of the earth's surface.

Unlike other branches of engineering, geotechnical engineers can be faced with two identical structures having different risks as their individual locations will have different resulting consequences. Cut slopes, for example, have different risk profiles depending on what is near, or at, the crest and toe of the slope.

Similarly, settlement of soft ground beneath a road embankment may not be critical if the whole embankment settles uniformly. However, settlement of an embankment adjacent to a structure, such as a bridge or culvert, can cause concern as the structure will tend to settle much less or not at all. An important element of geotechnical risk is, therefore, that it will be different for every location, with the type of risk being as variable as the ground itself and the individuals working with it.

17.2.2 Ground risk

To manage the geotechnical risks appropriately, a site investigation of appropriate scale and scope will need to be designed. Failure to do so can lead to significant cost overruns or delays (Clayton 2002; Clayton 2001b; Clayton 2001a; Mott MacDonald and Soil Mechanics Ltd 1994; National Audit Office 1992; Plimmer 2019). The typical ground related hazards are shown on Figure 17.1.

In terms of soft ground engineering, there are several obvious geotechnical risks as discussed elsewhere in this book. Examples include:

- Ground model inaccuracies;
- Groundwater and seepage;
- Earthworks – global stability and ongoing settlements, classification of materials for reuse/disposal;

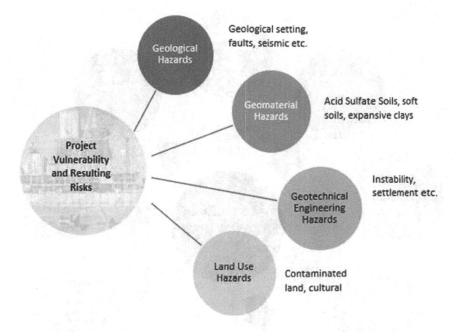

Figure 17.1 Main groups of ground hazard (Clayton and Smith 2013).

- Pavements – inadequate subgrade, removal and replace of unsuitable materials;
- Piled foundations – lack of lateral support and soil movements causing lateral pile displacements;
- Retaining walls, shallow foundations – stability and ongoing settlement; and
- Adjacent infrastructure.

17.2.3 Team risks

The role of the geotechnical design lead or manager on a major project is incredibly important and should not be left to chance. As can be seen on Figure 17.2, the geotechnical manager can co-ordinate with many internal and external stakeholders. The communication risk is, therefore, high as information can get distorted if not clear. So, the geotechnical manager must also be politically savvy in dealing with these various parties, for example building alliances within the design team.

In addition, the geotechnical manager needs to have visibility over the work of the team, ensuring all deliverables are met, providing guidance where necessary and ensuring that sensible solutions are produced. However,

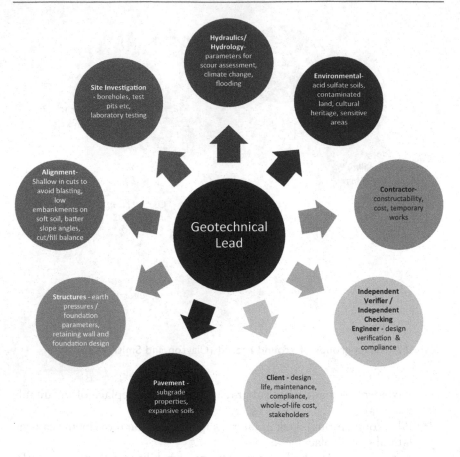

Figure 17.2 Typical interfaces with the geotechnical lead on a major project (Bridges 2019).

the geotechnical manager should not have their own deliverables, as this will distract from successfully managing the team.

Inexperienced leaders can sometimes feel isolated and struggle to get their opinions heard or respected without adequate support from the wider design team leadership. This can result in the geotechnical risks on the project not being fully appreciated by the project team. Constant pressure on the team that is typical in a tender or detailed design environment can also negatively impact the team's performance if not well led (Bakker et al. 2006; Chong et al. 2012).

When managing project risks it is important to have the right team but, companies appoint individuals they have available rather than who is most suitable. However, having a range of experience in terms of skill and years of experience is critical. Senior team members, who may be part-time on the project, need to be engaged early to ensure that project risks are understood at the outset. Junior team members need to be adequately mentored and developed, and peer and senior level reviews should occur frequently.

Design documentation needs to fully explain the design process and assumptions, and present a clear and accurate representation of the works (Tilley and McFallan 2000). It is often clear that design reports are not read by all parties, so drawings need to be clear with assumptions stated. When early contractor involvement is available, the design should incorporate information and input from the contractor. Where site information needs to be checked during construction (such as foundation conditions) then relevant *HOLD POINTS* need to be on the drawings. Similarly, where instrumentation and monitoring are critical to the design (as in the observational approach (Peck 1969; Peck 2001)), then this and the inherent risks also need to be explained.

17.2.4 Data collection

Clients rarely see the value of data collection (Leo 2018). However, good quality data is imperative for geotechnical design. Unfortunately, site investigation is a price sensitive field with as little money as possible spent by clients and the result is the lowest priced site investigation contractor is awarded the work and poor-quality data results. An example is shown on Figure 17.3, which shows the logs of two adjacent boreholes – one showing sand and the other sandstone. The one logged as sandstone had a geologist on site logging as the borehole was drilled, whereas the other was logged after receiving samples from site from a driller.

Geotechnical investigations are of value to the designer, but usually limited, due to the variability and complexity of the ground with relatively disturbed soil samples and rock core taken at wide spacing across a site.

On top of this the quality of laboratory testing can be a significant issue, with obviously incorrect test certificates leaving laboratories without adequate checking. There are also inconsistent results between different laboratories (Anon 2018).

17.2.5 Design risks

Following on from an investigation, geotechnical parameters need to be determined for design. Bond and Harris (2008) asked around 100 engineers to determine the characteristic values of the soils based on given test data. An example is given in Figure 17.4 which shows undrained shear strength with depth and the range of interpretations, which was large and inconsistent.

When it comes to design, further variability exists as geotechnical engineers use a wide variety of standards, papers, textbooks and software to derive solutions. An example of this variability can be seen in the use of simple bearing capacity equations in Europe. For a 1 m square foundation with a 0.6 m

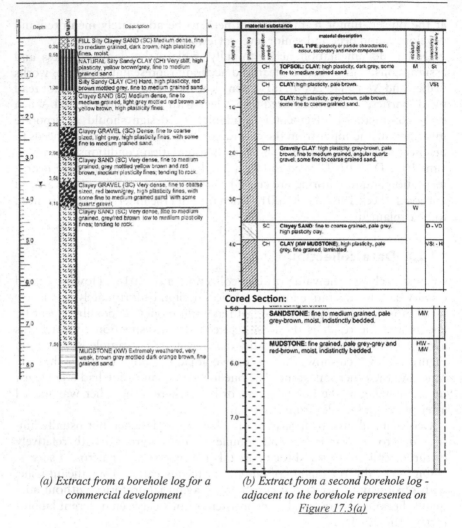

(a) Extract from a borehole log for a commercial development

(b) Extract from a second borehole log - adjacent to the borehole represented on Figure 17.3(a)

Figure 17.3 **Extract from a borehole logs for a commercial development (Bridges 2019).**

embedment on a soil with $c' = 10$ kPa, $\phi' = 35°$ and $\gamma = 19$ kN/m^3, Sieffert and Bay-Gress (2000) identified marked differences in calculated bearing capacity depending on which country you were in or which method you used (Figure 17.5).

Prediction competitions are also useful to see how different design approaches compare. Various prediction symposia have been published in the literature (Morgenstern 2000; Massachusetts Institute of Technology 1975; Kelly et al. 2018; Anon 1999; Doherty et al. 2018) for various typical geotechnical problems. The results of two of these symposia are shown on

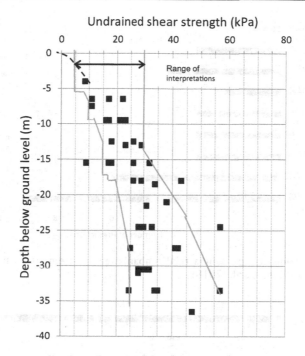

Figure 17.4 Range of interpretations from a given set of data (Bond and Harris 2008).

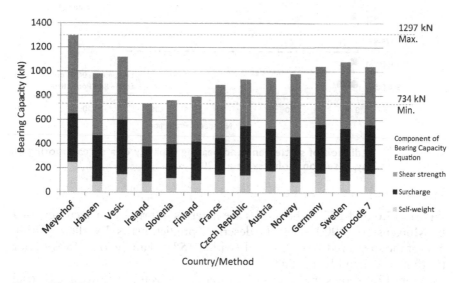

Figure 17.5 Comparison of calculated bearing capacity by different methods (Sieffert and Bay-Gress 2000).

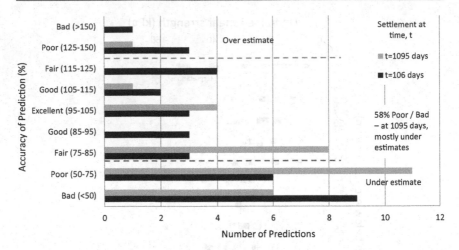

Figure 17.6 Prediction of Ballina trial embankment settlement with time, based on Kelly et al. (2018).

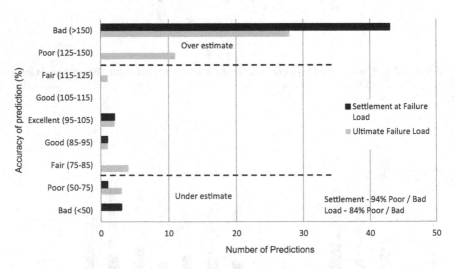

Figure 17.7 Prediction of settlement and bearing capacity of a spread footing on Clay, based on Doherty et al. (2018).

Figures 17.6 and 17.7 – the results are in terms of prediction accuracy used by Morgenstern (2000) where he described predictions as: Excellent (within 5% of the measured result); Good (within 15%); Fair (within 25%); Poor (within 50%); and Bad (>50%).

Figure 17.6 shows the results of a recent symposium that involved a trial embankment on soft clay at Ballina in New South Wales, Australia. Predictions of settlement were one of the compared outcomes and the results are given. Nearly 60% of the predictions were poor or bad, i.e. over- or under-predicted settlement by 50%. An underestimation of settlement, particularly over the longer term, is noticeable.

Figure 17.7 also shows the results of a recent prediction exercise that involved a 1.8 m square shallow foundation at a depth of 1.5 m, on soft clay ($c_u \approx 10$ kPa) also at Ballina in New South Wales, Australia. In this case, predictions of settlement and load were compared outcomes and the results are given. Over 90% of the settlement predictions, and over 80% of the predictions of ultimate failure load, were poor or bad.

The results of this and the other symposia show that geotechnical prediction is difficult. However, the results are worse than they appear. In prediction symposia, the ground is usually investigated and tested to a far greater extent and to a higher (research level) quality than any real project. In addition, the predictors are often able to use state-of-the-art software and theories that have not always permeated down to the average geotechnical engineer. Practicing geotechnical engineers also have significant time and pressure disadvantages. Therefore, the quality of predictions by practicing geotechnical engineers in routine projects can be a lot poorer.

17.3 KNOWLEDGE TRANSFER AND RISK COMMUNICATION

It is common practice that the geotechnical investigation, laboratory testing and the associated reporting is usually carried out by one company and the design carried out by another. Even when undertaken by the same company, the investigation and design components may be separated. This brings with it knowledge transfer risks. For example, a test pit may have been relocated on site due to the presence of obstructions on the surface. The relocated test pit will be clear of the obstructions and unless the report carries site photos and the log states the reason for the relocation, this information may be lost to the designer.

Similarly, during construction the geotechnical site staff as well as the contractor may not have been involved with the design. Interpreting the drawings and design intent by the site team can be difficult and lead to further risk of misinterpretation. It is imperative, therefore, that the design drawings and reports are clear as to the parameters assumed during design, the purpose of the design and the risks assumed. It is impossible for the site team to understand if certain actions can be undertaken on site if they do not appreciate the design intent.

It is, therefore, implicit on the geotechnical lead to communicate the geotechnical risks effectively to the wider project team but also ensure that the information required to determine those risks is obtained from that wider team (Figure 17.2). Drawings containing all the relevant information and risks are key to the safe delivery of a project.

17.4 RISK MANAGEMENT SYSTEMS

Risk management systems are designed to manage project risk. They can be used through both the design and construction and should be constantly monitored and updated.

The risk management process as developed in ISO 31000 (Standards Australia/Standards New Zealand 2009) follows the approach of identify – analyze/evaluate – treat (Figure 17.8). It is clear that this is not a solitary task and that engineers and constructors from different disciplines need to be involved so that all project risks can be identified, responsibility allocated, and appropriate action taken.

This process is expanded in Figure 17.9, which details the stages of the geotechnical risk management process and how they fit into the design and construction program. It is often considered that more than one stage of investigation should be undertaken – a preliminary investigation at concept stage, followed by a more detailed investigation when the design and understanding of the project risks are better developed. In addition, access issues, site events or monitoring data indicating greater geological variability than originally assumed, can lead to additional investigations during construction.

Risks should be identified and evaluated before detailed design is undertaken. At detailed design the risk is to be fully addressed through mitigation, but elimination of risks through actions such as alignment or location changes will need to be resolved earlier. The proposed mitigation may itself

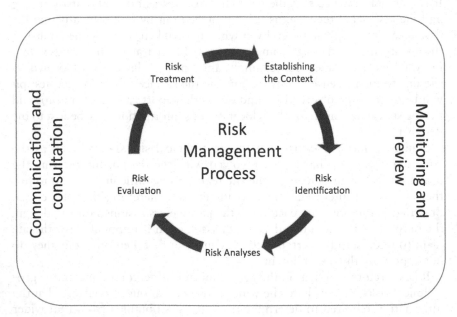

Figure 17.8 Risk management process (Standards Australia/Standards New Zealand, 2009)

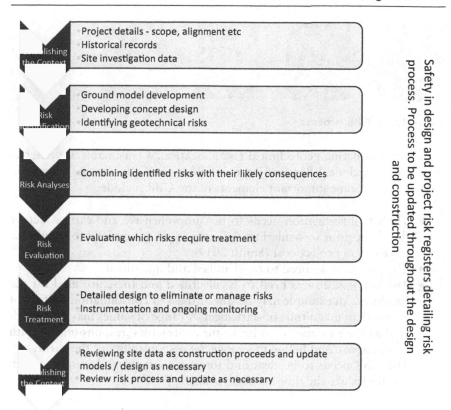

Figure 17.9 Geotechnical risk management process.

have its own risks and these will also need to be addressed. It should be emphasized that even insignificant risks should be addressed if easily achieved.

During construction, on-site evaluation of ground conditions as they are exposed, as well as ongoing monitoring, should feed back into the risk process as risks are addressed or eliminated, or new ones identified.

17.5 RISK ALLOCATION AND THE GEOTECHNICAL BASELINE REPORT

To obtain project financing, many clients push as many project risks, including geotechnical, to the contractor. This approach can give the client a fixed project cost, but it will ultimately result in a higher total cost as contractors price in the additional risks, which may be unquantifiable at tender.

Figure 17.10 GBR process.

When considering geotechnical risk allocation, a reasonable approach is to use a Geotechnical Baseline Report (GBR). The GBR process is given on Figure 17.10. Some important elements of the GBR include:

- The site investigation needs to be comprehensive and carried out by the client prior to tendering. A failure to do so can lead to significant increases in project cost (Smith 2019).
- The project risks need to be identified and quantified.
- The key risk drivers need to be identified and these are used for the baselines. An example may be soil stiffness or consolidation coefficient rather than magnitude of settlement. A range of values may be considered as the baseline – and hence the contractors' responsibility – with values above and below this range the responsibility of the client.
- The GBR needs to be clear and forms a contractual document which should reduce the time spent on claims.

17.6 SUMMARY

It is clear that there are many risks in ground engineering, from the variability in the geological conditions to that of the structures we build. These conditions make many of the problems faced by geotechnical engineers and geologists unique. But before we can communicate them to various stakeholders we deal with, we need to understand them ourselves.

Addressing the ground conditions consists not only of understanding the geological conditions, but also requires a team with the right experience and skills. Limitations of our design methods and software need to be appreciated and processes put in place to deal with the resulting ambiguity and unforeseen conditions as they come to light.

REFERENCES

Anon (1999). 'Uncertainty principle.' *Ground Engineering*, 32, 32–34.
Anon (2018). 'Laboratory testing: Revealing consistency concerns.' *Ground Engineering*.

Arnaud, A. (1850). *Logic, or the Art of Thinking being the Port-Royal Logic* (Translated from the French by Baynes). Edinburgh, UK: Sutherland and Knox.

Bakker, A.B., Van Emmerik, H. and Euwema, M.C. (2006). 'Crossover of burnout and engagement in work teams.' *Work and Occupations*, 33, 464–489.

Baynes, F.J. (2010). 'Sources of geotechnical risk.' *Quarterly Journal of Engineering Geology and Hydrogeology*, 43, 321–331.

Bond, A. and Harris, A. (2008). *Decoding Eurocode 7*. Abingdon, UK, Taylor & Francis.

Bridges, C. (2019). 'Geotechnical risk: It's not only the ground.' *Australian Geomechanics*, 54, 27–38.

Chapman, T.J.P. (2008). The relevance of developer costs in geotechnical risk management. In: Brown, M.J., Bransby, M.F., Brennan, A.J. and Knappett, J.A. (eds.) *Foundations: Proceedings of the Second BGA International Conference on Foundations, ICOF 2008*. IHS BRE Press.

Chong, D.S.F., Van Eerde, W., Rutte, C.G. and Chai, K.H. (2012). 'Bringing employees closer: The effect of proximity on communication when teams function under time pressure.' *Journal of Product Innovation Management*, 29, 205–215.

Christian, J.T. (2004). 'Geotechnical engineering reliability: How well do we know what we are doing?' *Journal of Geotechnical and Geoenvironmental Engineering*, 130, 985–1003.

Clayton, C.R.I. (2001a). *Managing Geotechnical Risk: Improving Productivity in UK Building and Construction*. London, UK: The Institution of Civil Engineers.

Clayton, C.R.I. (2001b). 'Managing geotechnical risk: Time for change?' *Proceedings of the Institution of Civil Engineers – Geotechnical Engineering*, 149, 3–11.

Clayton, C.R.I. (2002). 'Managing geotechnical risk: Time for change? – Discussion.' *Proceedings of the Institution of Civil Engineers – Geotechnical Engineering*, 155, 79–80.

Clayton, C.R.I. and Smith, D.M. (2013). *Effective Site Investigation*. London, UK: Site Investigation Steering Group.

Committee of Inquiry (2005a). Report of the Committee of Inquiry into the Incident at the MRT Circle Line worksite that led to the collapse of the Nicoll Highway on 20 April 2004 (Volume 1 (Part I)).

Committee of Inquiry (2005b). Report of the Committee of Inquiry into the Incident at the MRT Circle Line worksite that led to the collapse of the Nicoll Highway on 20 April 2004 (Volume 1 (Part II)).

Doherty, J.P., Gourvenec, S. and Gaone, F.M. (2018). 'Insights from a shallow foundation load-settlement prediction exercise.' *Computers and Geotechnics*, 93, 269–279.

Health and Safety Executive (2000). The collapse of NATM tunnels at Heathrow Airport.

Kelly, R.B., Sloan, S.B., Pineda, J.A., Kouretzis, G. and Huang, J. (2018). 'Outcomes of the Newcastle symposium for the prediction of embankment behaviour on soft soil.' *Computers and Geotechnics*, 93, 9–41.

Leo, G. (2018). *Bridge that collapsed six hours after opening was built without geotech investigation of riverbed: Reeve [Online]*. Canada: CBC. Available: https://bit.ly/2OScsA4 [Accessed September 27 2018].

Massachusetts Institute of Technology (1975). *Proceedings of the Foundation Deformation Predictions Symposium*. Washington, DC: Federal Highway Administration.

Morgenstern, N.R. (2000). *Performance in Geotechnical Practice*. Hong Kong: The Inaugural Lumb Lecture.

Mott Macdonald and Soil Mechanics Ltd (1994). *Study of the Efficiency of Site Investigation Practices*. Crowthorne, Berkshire, UK: Transport Research Laboratory.

National Audit Office (1992). *Department of Transport: Contracting for Roads*. London, UK: National Audit Office.

Oxford Dictionaries (2018). Oxford Living Dictionary [Online]. Oxford University Press. Available: https://en.oxforddictionaries.com/definition/risk [Accessed September 29 2018].

Peck, R.B. (1969). 'Advantages and limitations of the observational method in applied soil mechanics.' *Géotechnique*, 19, 171–187.

Peck, R.B. (2001). 'The observational method can be simple.' *Proceedings of the Institution of Civil Engineers – Geotechnical Engineering*, 149, 71–74.

Plimmer, G. (2019). 'Why HS2 rail line is way over budget and badly delayed.' *Financial Times*, 19 September 2019.

Sieffert, J.-G. and Bay-Gress, C. (2000). 'Comparison of European bearing capacity calculation methods for shallow foundations.' *Proceedings of the Institution of Civil Engineers – Geotechnical Engineering*, 143, 65–74.

Smith, C. (2019). HS2 plans to reduce ground risk cost described as 'carnage' [Online]. UK: Ground Engineering. Available: www.geplus.co.uk/news/hs2-plans-to-reduce-ground-risk-cost-described-as-carnage-28-01-2019/ [Accessed May 14 2019].

Standards Australia/Standards New Zealand (2009). Risk Management – Principles and guidelines AS/NZS ISO 31000:2009. Sydney, Australia: Standards Australia / Standards New Zealand.

Terzaghi, K. (1936). Presidential Address: Relation between soil mechanics and foundation engineering. International Conference on Soil Mechanics and Foundation Engineering. Harvard University, USA.

Tilley, P.A. and McFallan, S.L. (2000). *Design and Documentation Quality Survey*. Highett, Victoria, Australia: CSIRO Division of Building, Construction and Engineering.

Index

Printed in the United States
by Baker & Taylor Publisher Services